Green Energy and Technology

Climate change, environmental impact and the limited natural resources urge scientific research and novel technical solutions. The monograph series Green Energy and Technology serves as a publishing platform for scientific and technological approaches to "green"—i.e. environmentally friendly and sustainable—technologies. While a focus lies on energy and power supply, it also covers "green" solutions in industrial engineering and engineering design. Green Energy and Technology addresses researchers, advanced students, technical consultants as well as decision makers in industries and politics. Hence, the level of presentation spans from instructional to highly technical.

Indexed in Scopus.

Indexed in Ei Compendex.

Weiqi Hua · Xiao-Ping Zhang · David C. H. Wallom
Editors

Digitalisation of Local Energy Systems

Springer

Editors
Weiqi Hua
School of Engineering
University of Birmingham
Birmingham, UK

Xiao-Ping Zhang
School of Engineering
University of Birmingham
Birmingham, UK

David C. H. Wallom
Department of Engineering Science
University of Oxford
Oxford, UK

ISSN 1865-3529 ISSN 1865-3537 (electronic)
Green Energy and Technology
ISBN 978-3-031-77832-2 ISBN 978-3-031-77833-9 (eBook)
https://doi.org/10.1007/978-3-031-77833-9

This Springer imprint is published by the registered company Springer Nature Switzerland AG
The registered company address is: Gewerbestrasse 11, 6330 Cham, Switzerland

Preface

The global energy systems are undergoing a profound transformation, driven by the urgent need to mitigate climate change and transition towards more sustainable energy systems. At the heart of this transition is the increasing digitalisation of energy infrastructure, which offers innovative ways to enhance efficiency, integrate renewable energy sources and empower local communities in managing their energy needs. This book provides a comprehensive overview of the technologies, strategies and policies that are shaping the future of energy systems at the local level.

In recent years, digital technologies such as the Internet of Things (IoT), Artificial Intelligence (AI), blockchain and 5G networks have emerged as powerful tools in the energy sector. These technologies enable more precise control over energy consumption, enhance the integration of distributed renewable energy sources and facilitate new business models such as Virtual Power Plants (VPPs). Local energy systems, which were once largely passive, are evolving into dynamic ecosystems where prosumers, i.e. individuals and communities generating their own energy, can actively participate in energy markets. The digitalisation of these systems is critical for achieving global decarbonisation goals and addressing the challenges posed by climate change.

This book is the result of collaborative efforts from leading researchers and experts across the globe, who have contributed their insights into various aspects of local energy systems and their digitalisation. The chapters cover a wide range of topics, from the integration of renewable energy and energy storage solutions to demand-side management, electric vehicle adoption and the role of law and policy in shaping the future of energy systems. Each chapter is grounded in the latest research and includes real-world examples of how digitalisation is being applied in practice.

Our aim with this book is to provide both a broad and detailed understanding of how digitalisation is revolutionising local energy systems. It is designed for a wide audience, including researchers, industry professionals, policymakers and students who are interested in the intersection of digital technologies and energy. We hope that this book will serve as a valuable resource for those seeking to understand the key trends, challenges and opportunities in this rapidly evolving field.

We would like to express our gratitude to the contributing authors, whose expertise and dedication have made this book possible. We also thank our institutions and the wider research community for their support in this endeavour. It is our hope that this book will inspire further innovation and collaboration in the digitalisation of energy systems, ultimately contributing to a more sustainable and resilient energy future.

Birmingham, UK
September 2024

Dr. Weiqi Hua

Contents

Introduction to Digitalization and Local Energy Systems

Weiqi Hua, Elnaz Azizi, Xiao-Ping Zhang, and David C. H. Wallom

Abstract This chapter explores the significant shift in the local energy landscape as power systems transition from traditional centralized models to modern, distributed frameworks. Historically, electricity flowed unidirectionally from large power plants to passive consumers, but today's energy systems are increasingly dynamic and decentralized. The rise of distributed energy resources, such as solar panels, wind turbines, and energy storage, has transformed consumers into active participants, or "prosumers," who both produce and consume energy. This transition introduces challenges, particularly in adapting existing infrastructure to handle bidirectional power flows and integrating variable renewable energy sources. Enhanced visibility and control in low-voltage distribution networks are essential to managing these complexities. The chapter also addresses the need for resilient energy systems in the face of global challenges like aging infrastructure, market volatility, and climate-related disruptions. Innovative approaches, including digitalization and decentralized energy solutions, are crucial for building resilient, equitable energy systems. The discussion includes the UK's Future Energy Scenarios (FES) and Distribution Future Energy Scenarios (DFES), which guide planning and investment in the evolving energy landscape. Ultimately, digitalization plays a vital role in enabling real-time monitoring, predictive analytics, and the seamless integration of distributed resources, paving the way for a sustainable and reliable energy future.

W. Hua (✉) · X.-P. Zhang
School of Engineering, University of Birmingham, Birmingham, UK
e-mail: w.hua@bham.ac.uk

X.-P. Zhang
e-mail: x.p.zhang@bham.ac.uk

E. Azizi · D. C. H. Wallom
Department of Engineering Science, University of Oxford, Oxford, UK
e-mail: elnaz.azizi@eng.ox.ac.uk

D. C. H. Wallom
e-mail: david.wallom@oerc.ox.ac.uk

W. Hua et al. (eds.), *Digitalisation of Local Energy Systems*, Green Energy and Technology, https://doi.org/10.1007/978-3-031-77833-9_1

1 The Evolving Local Energy Landscape

This section introduces the significant shift from traditional, centralized power systems to modern, distributed energy frameworks. It highlights how the integration of distributed energy resources has transformed the power flow from a unidirectional system into a more dynamic and interactive model, where consumers actively participate in energy production and consumption.

1.1 Transition to a Modern Power System

Historically, the architecture of power systems has been rooted in a centralized framework, characterized by a unidirectional flow of electricity. In this conventional system, large-scale power plants harness primary energy sources to generate electrical power, which is then transmitted over long distances through high-voltage transmission lines. The journey of power concludes as it is stepped down via substations to lower voltages suitable for distribution networks, ultimately reaching passive consumers who utilize the electricity for various needs. This traditional model is typified by its central generation, passive consumers, and notably low visibility at the distribution networks [1]. The latter is due to historically less monitoring and control over low-voltage distribution networks as opposed to their high-voltage transmission counterparts.

In contrast to the conventional model, the current global and Great Britain's energy landscape is undergoing a profound transformation toward a more dynamic and distributed system. This evolution is driven by the integration of an increasing number of distributed generation resources, such as wind turbines and solar panels, energy storage devices, and smart meters at the consumer end of the power system. In this modern paradigm, power flow is bidirectional. Consumers are no longer mere passive participants but take an active role as "prosumers [2]"—both consuming and producing energy. The surplus energy generated by prosumers can be fed back into the utility grids, supporting a more resilient and sustainable energy infrastructure. The distributed generation model elevates the importance of visibility in the operation of distribution networks. Understanding the dynamic changes in consumer behavior and the integration impacts of low-carbon technologies is vital. Unlocking visibility at the "last mile"—the final leg of the electricity journey to the consumers—is paramount to capturing the microscale perspective of energy systems.

Embracing this transition is not without its challenges. The existing power network was not originally designed to accommodate reverse flows of electricity [3] or the variability introduced by renewable energy sources. Both the physical hardware, such as wires and transformers, and the software infrastructure, including information and communication technologies, must be adapted or redesigned to support the new functionalities required by a modern, interactive power system. Moreover, the modern energy system necessitates sophisticated control systems capable of managing distributed energy resources and complex power flows. These control systems, often referred to as smart grids [4], incorporate real-time monitoring and rapid response capabilities to maintain grid stability and efficiency.

The shift from a conventional energy system to a modern, distributed power system represents a pivotal step toward a sustainable and resilient energy future. This transition, already underway in Great Britain and across the globe, marks a move toward an electrical grid that is more reflective of today's technological advancements and environmental considerations. As we navigate the complexities of this transformation, the goal remains to create an energy system that is flexible, reliable, and capable of meeting the diverse needs of a changing world.

1.2 Energy System Resilience in Facing Global Challenges

As illustrated in Fig. 1, globally, our electrical power systems are at a critical juncture, especially in under-developed and developing regions. These systems are grappling with multifaceted challenges that undermine the reliability, affordability, and sustainability of electricity supply. A primary concern is the outdated electrical infrastructure that is currently in place, which necessitates expensive reinforcement to meet today's energy demands [5]. Coupled with the inadequate grid connectivity [6], particularly in rural areas, the situation becomes dire, leaving many communities without reliable access to energy services.

Moreover, the volatility of gas and electricity markets has resulted in unaffordable energy costs, disproportionately affecting the underprivileged and amplifying the divide between the energy-secure and the energy-poor [7, 8]. Compounding these issues are the increasing frequency and intensity of extreme weather events [9], e.g., floods, wildfires, heatwaves, that not only disrupt energy systems but also disproportionately impact the economically disadvantaged, fuel-poor, and health-vulnerable groups. The cumulative effect of these challenges signals an urgent need for a power system that can withstand and adapt to such disruptions.

Addressing these pressing issues requires innovative solutions that can transform the current state of energy systems. Embracing distributed renewable energy resources is a key strategy. This involves decentralizing energy production to allow for local generation and consumption, thereby reducing dependency on the central grid and enhancing the resilience of the system against failures and natural disasters. Place-based energy systems are another piece of the puzzle. Tailoring energy solutions to the specific geographic, social, and economic contexts of a location can yield systems that are more attuned to the needs and capacities of local communities, further strengthening the grid's resilience and ensuring energy justice.

Digitalization of energy systems emerges as a powerful enabler in this transformation. Digital technologies can optimize energy distribution, improve demand response, and facilitate real-time management of energy flows. These technologies not only support the integration of renewable energy sources but also enable

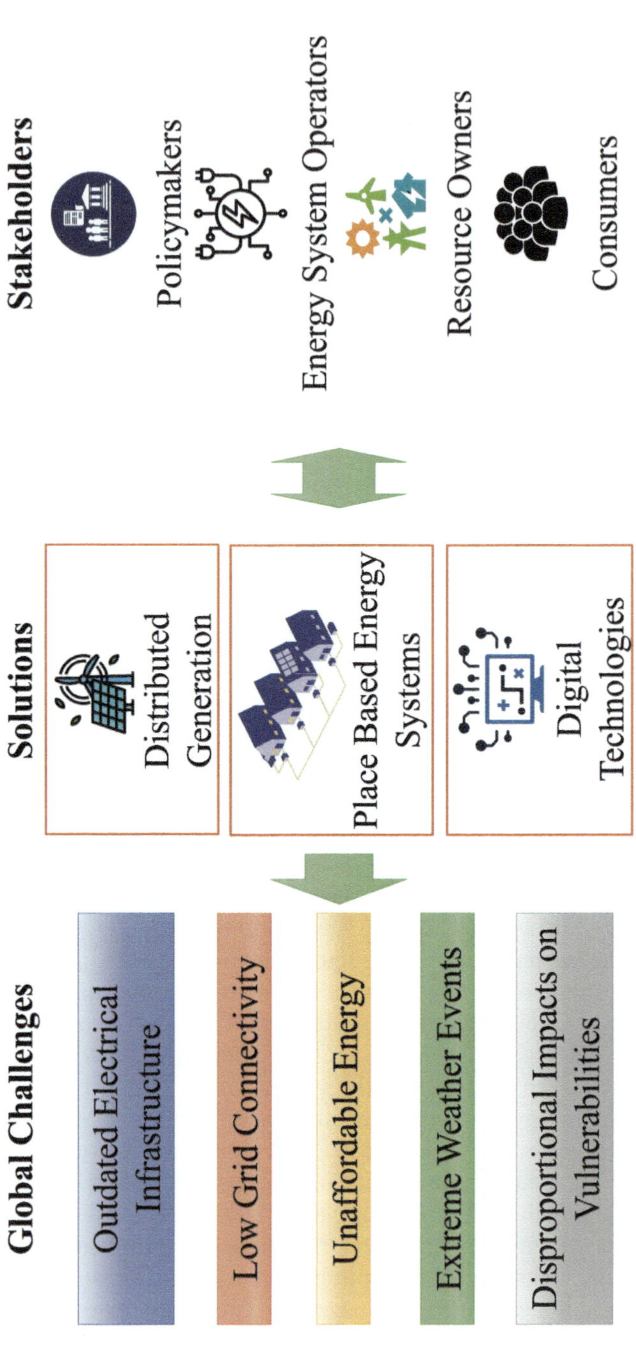

Fig. 1 Schematically illustration of challenges, solutions, and relevant stakeholders for building resilient energy systems in the face of global challenges

energy systems to become more adaptive and intelligent, capable of anticipating and responding to varying conditions and demands.

The transition to a resilient and equitable energy system is not the sole responsibility of a single entity but a collaborative effort involving multiple stakeholders:

- Policymakers have a pivotal role in setting ambitious net-zero targets and creating the regulatory environment that incentivizes investment in renewable energy and energy efficiency. They can also drive initiatives that ensure the energy transition is inclusive and equitable.
- Energy system operators are at the forefront of managing the increasingly complex grid, balancing supply with fluctuating demand, and incorporating new sources of distributed generation without incurring prohibitive costs for grid reinforcement.
- For resource owners, distributed generation opens new avenues for profit generation and faster payback on investments. It enables a more active participation in the energy market through mechanisms such as feed-in tariffs and net metering.
- Consumers stand to benefit significantly from the shift toward distributed generation and smart energy systems. Through improved energy efficiency, the use of smart appliances, and the potential to generate their own power, consumers can expect to see lower energy bills and greater control over their energy consumption.

The global challenges facing our energy systems are daunting, yet they present an opportunity for a radical reimagining of how we generate, distribute, and consume energy. By leveraging distributed generation, place-based systems, and digital technologies, we can construct energy infrastructures that are not only more resilient to external shocks but also more equitable, sustainable, and aligned with the needs of the twenty-first century. As we March toward this energy future, the collaboration of all stakeholders—policymakers, operators, resource owners, and consumers—will be crucial in steering the global community toward energy security and environmental stewardship.

1.3 Future Energy Scenarios

The United Kingdom's energy sector is undergoing a significant transition, aiming to balance the trilemma of providing sustainable, affordable, and secure energy. This transition is guided by comprehensive planning and forecasting efforts, notably through the UK Future Energy Scenarios (FES) and the Distribution Future Energy Scenarios (DFES) [10], which offer critical insights and data to stakeholders across the energy landscape.

The UK FESs is a document developed by the National Grid Electricity System Operator (ESO), which serves as a cornerstone for understanding the possible trajectories of the UK energy system. The FES outlines a series of potential futures, each based on a complex interplay of factors including, but not limited to, policy decisions, technological advancements, consumer behavior, and economic trends. These scenarios aim to inform the industry, investors, and policymakers about potential outcomes, enabling them to prepare for and influence the future energy mix. The FES document highlights pathways to achieve various goals, such as meeting emission targets, ensuring energy security, and maintaining affordability. The FES delineates four distinct pathways that the UK might follow, ranging from a scenario where actions toward decarbonization are taken late to one where green measures are adopted rapidly in a coordinated effort across all sectors.

Complementing the national perspective provided by the FES, the DFES offers a more granular view of the UK's energy future, focusing on the distribution networks. The DFES considers local energy demands, distributed energy resources, and the interactions between national and local energy systems. It provides detailed insights into how the distribution of energy may change, including the impact of increased adoption of electric vehicles, heat pumps, and localized renewable energy generation. The connection between the UK FES and DFES is foundational to the comprehensive understanding of the UK's energy trajectory. The FES provides the overarching narratives that guide the national energy strategy, while the DFES drills down into how these narratives will affect local energy systems. The scenarios presented in the FES inform the development of the DFES, ensuring that local distribution operators can align their planning with the national vision. This alignment is vital for ensuring coherence in the transition to a low-carbon, flexible, and reliable energy system that meets both current and future demands. The interplay between the FES and DFES is crucial for integrating the energy needs of different regions, accounting for local variations in energy resources, consumption patterns, and infrastructure development. It ensures that the unique characteristics and requirements of local energy systems are factored into the national energy planning processes.

1.4 Regulatory Barriers for Demand Side Engagement

The modernization of energy systems and the rising involvement of consumers in energy generation—often referred to as "prosumers"—are significantly influenced by policy frameworks. This shift toward the electrification of heat and transportation sectors has illuminated various barriers impeding demand-side engagement. To comprehensively address these obstacles and harness the potential of prosumers, we must consider a multifaceted approach that reconciles policy support with market dynamics and technological innovation.

First, the energy landscape is experiencing a pivotal transition from large, centralized power producers to a more distributed and decentralized model, where individual consumers also act as energy producers. This shift brings to light the discrepancy in market power between large energy suppliers and individual prosumers, with the latter often lacking the influence to significantly impact market prices or policies.

Second, access to energy markets is traditionally designed for established generation companies and suppliers that can navigate complex licensing and regulatory environments. In contrast, prosumers face a fragmented landscape. The creation of standardized, consumer-centric licenses could streamline market entry for prosumers, enhancing their participation in the energy economy. Standardization also promotes replicability, allowing for the rapid scaling of prosumer integration across the market.

Third, incentivizing prosumers to engage in local energy systems trading is essential [11]. There are several approaches: utilizing market-based energy prices to influence consumer behavior, offering additional monetary incentives, or employing regulatory mechanisms that facilitate participation. Subsidy schemes and pricing strategies can effectively motivate prosumers to invest in renewable technologies and use digital technologies.

Fourth, the allocation of responsibilities for prosumers participating in local energy markets raises several considerations:

- *Carbon Accounting*: How can we quantify and credit the carbon reduction efforts of prosumers [12], such as those installing air-source heat pumps or rooftop solar panels?
- *Grid Use*: Should prosumers contribute to the costs associated with the use of distribution networks [13] when they exchange energy?
- *Energy Delivery and Security of Supply*: Ensuring that prosumers provide accurate predictions of their energy generation and maintain supply security is crucial [14]. Determining who is liable in cases of under-delivery is also essential for a reliable power system.

Fifth, traditional economic models often fail to capture the nuanced decision-making processes of households and individual consumers [15]. A promising solution is to employ data-driven approaches like digital twins or machine learning to predict and understand prosumer behavior based on historical and real-time data. Such models must be consumer-centric, reflecting the reality that household investments are not always based on purely economic rationales.

Sixth, a comprehensive accounting framework must consider both physical assets, such as generation equipment owned by prosumers, and non-physical assets, like the data and transaction records stemming from energy trading [16]. This accounting is vital for recognizing the value contributions of prosumers and ensuring transparency in the energy marketplace.

To foster a resilient and responsive energy system, policy support must align with technological advancements and market designs that accommodate the growing presence of prosumers. Overcoming the identified barriers necessitates a collaborative approach that engages stakeholders across the spectrum, from policymakers to energy suppliers, and from system operators to consumers themselves. By addressing the dual aspects of market power and access, incentivizing engagement, delineating responsibilities, understanding economic behavior, and accounting for assets comprehensively, we pave the way for a more inclusive and sustainable energy future. The integration of prosumers into the energy market not only democratizes energy production but also contributes to the broader goals of carbon reduction and energy resilience.

2 Digitalization of Local Energy Systems

This section introduces the application of digitalization in energy systems, emphasizing its role in enhancing the efficiency and reliability of power networks. It will explore how digital technologies, including digital twins, can be strategically deployed across different voltage levels of the power grid, offering tailored solutions that address the specific challenges of each segment, from high-voltage transmission to low-voltage distribution networks.

2.1 *Application of Digitalization in Energy Systems*

As we delve into the topic of digitalization, including the concept of a digital twin in the power system, it is important to recognize that while digitalization is a potent tool, it is not a universal solution applicable to every component of the power system indiscriminately. Identifying the precise areas within the power system infrastructure where digital technologies can yield the greatest benefit is crucial. This section will discuss the different voltage levels in a typical GB power grid distinguishing between the high-, medium-, and low-voltage networks, as shown in Fig. 2, each with its unique monitoring and control requirements.

At the highest level, we have transmission networks characterized by high voltage (>33 kV), primarily linking large-scale fossil fuel-based power plants and substantial industrial consumers. These segments of the grid are already subject to intensive monitoring and control, leveraging advanced systems like Supervisory Control and Data Acquisition (SCADA), smart sensors, and power quality monitors. Such systems are pivotal for real-time tracking of various parameters like voltage, current, phase angle, fault status, and temperature of the power infrastructure.

Descending to the medium voltage distribution networks (1–33 kV), we find a network that connects industrial or sizeable commercial consumers and integrates the majority of distributed energy sources. Like their high-voltage counterparts, these networks benefit from significant monitoring and control, which supports the seamless integration of distributed resources and maintenance of grid stability.

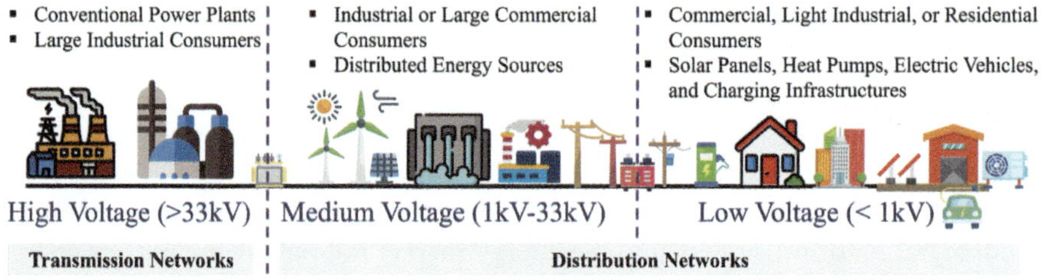

Fig. 2 Schematic illustration of different voltage levels in a typical GB power grid

The low-voltage distribution networks (<1 kV), however, present a different narrative. They are the direct link to commercial, light industrial, and residential consumers and the nexus at which solar panels, heat pumps, EVs, and charging infrastructures will proliferate. Despite the tens of thousands of cables and transformers operating at this level, these networks are traditionally the least monitored and controlled parts of the power system.

One of the primary challenges with low-voltage networks is the presence of legacy assets, which often have uncertain specifications and are massively heterogeneous in design. This inconsistency and lack of transparency pose significant hurdles for the efficient and effective operation of the grid, especially as we push toward more decentralized and renewable-focused energy systems. Given these challenges, low-voltage networks are ripe for the application of digital technologies. Digitalization can provide the needed visibility and control through accurate prediction, efficient operation, and systematic coordination [17]. Implementing digital twins, i.e., virtual replicas of physical assets [18], can allow for sophisticated simulations and analyzes, enabling operators to anticipate and respond to potential issues before they manifest in the physical network.

Digital tools can facilitate better asset management, predictive maintenance, and the integration of distributed energy resources, such as rooftop solar and home energy storage systems. They can also support advanced demand response strategies and enhance the coordination of EV charging to avoid overloading local transformers and cables. The path to a fully digitalized power grid encompasses upgrading existing infrastructure, embedding smart sensors, and implementing advanced control systems. This shift promises an array of improvements, including:

- *Enhanced Grid Reliability*: Through real-time monitoring and automated response systems, grid operators can quickly isolate and address faults, reducing downtime and service interruptions.
- *Optimized Asset Utilization*: Digital tools allow for better management of the grid's physical assets, ensuring they are used efficiently and within their operational limits.
- *Integrated Renewable Energy*: With the help of digital technologies, the grid can smoothly integrate variable renewable energy sources, maintaining stability despite the fluctuating nature of solar and wind power.
- Empowered Consumers: Digitalization enables consumers to become active participants in the energy market through smart meters and home energy management systems, allowing for more informed energy use and potentially lower costs.

Next, let us estimate the cost of monitoring infrastructure for low-voltage distribution networks across Great Britain (GB) and the time required to collect sufficient data to build a digital twin model. It is noted that due to the variability in factors such as the size of the Distribution Network Operator (DNO) license areas, the current state of their infrastructure, and the level of detail required for the digital twin model.

- *Step 1 Define a Sample DNO License Area*: For our example, we will use a hypothetical DNO license area in the UK, similar to areas managed by companies like UK Power Networks or Northern Power Grid. Let us assume this area covers about 1 million households and businesses, which is within the range of what a large DNO might manage.
- *Step 2 Estimating Monitoring Infrastructure Cost per Household/Business*: Based on existing projects and studies, the cost of installing monitoring equipment can vary widely. For estimate, let us assume: *(a)* Low voltage smart meters or sensors cost around £100–£150 per installation. This includes the cost of the device and the installation but does not cover advanced analytics or integration into wider network management systems. *(b)* Additional infrastructure and integration costs might double this figure, so let us estimate £300 per point of monitoring.
- *Step 3 Calculating Cost for One DNO License Area*: For 1 million points of monitoring (assuming one per household/business), the cost would be £300,000,000
- *Step 4 Scaling to All GB*: Assuming there are around 28 million households and businesses in GB, we scale the cost to be £8,400,000,000.

Therefore, it is approximately £8.4 billion to monitor all low-voltage points in GB.

2.2 *Cost-Effectiveness for Local Energy System Digitalization*

As nations vigorously pursue net-zero emissions to combat climate change, the energy sector stands at the forefront of this global endeavor. Digitalization, particularly within low-voltage distribution networks, emerges as a crucial enabler in this transition. It is not merely a technological upgrade but a paradigm shift toward a more flexible, efficient, and consumer-responsive energy system.

The drive toward a digitalized energy future is underpinned by substantial economic and societal benefits. In the UK, the Department for Energy Security and Net-Zero (DESNZ), which has evolved from the Department for Business, Energy & Industrial Strategy (BEIS), has projected that by the mid-21st century, the augmentation of system flexibility through digitalization can yield annual savings of up to £10 billion in UK energy system costs. This transformation is also expected to be a catalyst for job creation, potentially adding up to 24,000 new roles across the sector [19, 20]. Digitalization lays the groundwork for the decentralization of energy systems. This transition sees small-scale generation assets installed closer to where energy is consumed, which not only shortens the distance energy must travel, thereby reducing losses, but also empowers local generation and consumption dynamics. The interconnection of electricity, heating, and transportation sectors, facilitated by digital platforms, propels efficiency and offers consumers increased control and choice over their energy sources and usage.

According to the RIIO-ED2 regulatory framework from Scottish & Southern Electricity Networks (SSEN), the challenge of overloading in distribution networks is significant, especially during high-demand periods such as winter [21]. Two of the Distribution Future Energy Scenarios (DFESs)—"Consumer Transformation" and "Leading the Way"—highlight this issue, with concerns extending across all scenarios during spring and autumn. To mitigate these challenges, the adoption of digital-based flexible services presents a compelling proposition. These services, which rely on advanced data analytics and real-time grid management tools, offer a cost-effective alternative to conventional infrastructure upgrades. By leveraging digital technologies, grid operators can dynamically manage loads, integrate distributed energy resources, and optimize asset utilization.

A detailed economic analysis underlines the superiority of digital interventions. Employing digital-based flexible services is demonstrated to offer superior net present value (NPV) and require less investment when juxtaposed with traditional approaches such as replacing or adding new circuits and transformers. For instance, by 2050, flexible digital services are projected to achieve an NPV of £k1464 and necessitate an investment of £k3074. In contrast, traditional reinforcement methods would yield a lower NPV of £k2572 and a higher investment cost of £k4093. The most expensive traditional option, involving the addition of new infrastructure, would incur an NPV of £k4253 and an investment of £k4457 [21]. Digitalization within the energy sector, especially at the local energy system level, is not merely an option but a necessity to meet future demands and sustainability goals. By harnessing the power of digital technologies, the sector can realize significant economic savings, reduce environmental impacts, and contribute to job growth. The strategic integration of digital-based solutions ensures that the energy systems of the future are not only robust and capable of handling increased and variable loads but are also aligned with the overarching goals of energy security, affordability, and carbon neutrality.

2.3 Functions of Digitalization in Fulfilling Requirements of Local Energy Systems

Modern energy systems are gravitating toward greater complexity and decentralization, which necessitates an evolution in the management of distribution networks. Particularly for low voltage distribution networks, which are integral in interfacing with the end consumers, maintaining a state of informed operations is critical. The utility of digitalization in this domain is profound, and it is encapsulated in five key functions necessary for the informed operation of distribution networks as presented in Fig. 3:

- *Planning*: Investment in network infrastructure and generation capacity must be meticulously orchestrated to meet present and future demands.
- *Operation*: Continuous monitoring of system conditions to maintain system constraints and ensure reliable service.
- *Market*: A platform where energy and ancillary services are efficiently traded.
- *Asset Accountability*: Ensuring both physical assets, such as distributed generators, and non-physical assets, such as data, are traceable and optimally utilized.

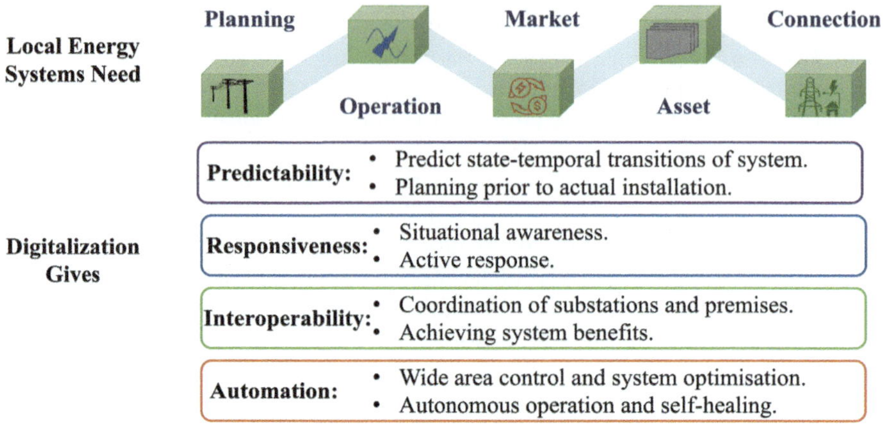

Fig. 3 Requirements of local energy systems fulfilled by digitalization

- *Connection Management*: Facilitating new connections for distributed generation in a seamless manner.

Digitalization provides multifaceted support to these functions through four key dimensions [22]:

- *Predictability*: Utilizing predictive analytics, digital twins can model and forecast state-temporal transitions within the network. By simulating load pattern changes and other dynamic events, they inform planning decisions well ahead of actual infrastructure installations.
- *Responsiveness*: Digital twins enhance operational agility by providing real-time situational awareness and enabling rapid, informed responses to system changes. This responsiveness is crucial for maintaining system reliability in the face of fluctuating demand and supply.
- *Interoperability*: As a nexus for various stakeholders, including consumers, prosumers, distributed network operators, and policymakers, digital twins facilitate coordination and communication. This ensures that individual actions are harmoniously aligned with broader system benefits and objectives.
- *Automation*: Beyond mere coordination, digitalization brings the power of automation to distribution networks. This encompasses wide area control and system optimisation, leading to autonomous operations and self-healing capabilities that reduce the need for human intervention and enhance system efficiency.

Digital twins serve as a bridge between physical operations and digital potential. In planning, digital twins can test different infrastructure investments in virtual environments, foreseeing their impacts on the network. During operation, these virtual models can detect anomalies, prescribe corrective actions, and even predict their outcomes before implementation. In the market sphere, digital twins can simulate trading scenarios, optimize grid services in response to supply and demand, and inform asset management strategies. Finally, in managing new connections, they can streamline the integration of distributed resources, ensuring that the grid can adapt to new inputs without compromising performance.

The role of digitalization in local energy systems is multifaceted and indispensable. Digital twins, by enabling predictive analytics, real-time monitoring, and seamless interoperability, transform the network into a responsive and adaptable system capable of meeting modern energy challenges. As we push toward the net-zero emissions goal, the attributes of predictability, responsiveness, interoperability, and automation that digitalization provides will be instrumental in ensuring that the distribution networks remain robust, efficient, and aligned with the evolving landscape of energy production and consumption.

2.4 Benefits of Digitalization

Digitalization in energy distribution networks is not a one-time event but a continuous process with benefits that span various timeframes—from real-time operations to long-term strategic development. By leveraging advanced digital tools and analytics, distribution network operators can reap significant benefits across all facets of their operations.

In the real-time operational context, digitalization enables:

- *Accurate Predictions*: By employing AI and machine learning techniques, distribution network operators can more accurately predict the uncertainties associated with renewable energy sources and fluctuating loads [23]. This facilitates better integration of renewables, ensuring grid stability despite their intermittent nature.
- *Ancillary Services and Flexibility*: Digitalized networks can provide ancillary services more efficiently. Through real-time data analysis, networks can offer flexibility services such as demand-side response, which adjusts consumers' energy usage patterns to balance the grid [24].
- *Dynamic Power Flow Analysis*: With digital tools, distribution network operators can conduct dynamic power flow analysis to instantly detect and rectify operational constraints. This real-time adjustment helps maintain an uninterrupted power supply and prevents outages.

Looking ahead, in the mid-term, digitalization aids in:

- *Enhanced Network Understanding*: For distribution networks that have traditionally been under-monitored, digitalization can illuminate network behavior. By generating and analyzing data, distribution network operators can create more accurate and representative network structures.
- *Strategic Planning and Investment*: Digital tools facilitate strategic system planning and help inform investment decisions. Predictive modeling can identify future network needs and direct capital to where it will be most effective, enhancing the overall value and efficiency of network upgrades [25].

Over the long term, digitalization can lead to:

- *Integration with Smart Transmission Systems*: Digitalization enables low-voltage networks to augment and synergize with smart transmission systems [26]. This results in a more cohesive and integrated grid capable of handling the complexities of future energy demands.
- *Cross-Sectoral Co-benefits*: The benefits of a digitalized network are not confined to the energy sector alone. Stakeholders across various sectors–including transportation [27], heating [28], public services [29], and housing [30]—can leverage these advancements for broader societal benefits, such as reduced emissions, enhanced energy access, and economic growth.

Digitalization presents an array of benefits that cater to the immediate and evolving needs of energy distribution networks. It equips distribution network operators with the tools to make informed decisions, adapt to changes swiftly, and plan for a future that is sustainable, resilient, and aligned with broader socio-economic goals. The digitalization of energy networks is a key driver in the transition toward a smarter, cleaner, and more customer-centric energy future.

Acknowledgements This work was supported by the EPSRC Innovation Launchpad Network + Researchers in Residence Scheme through the project of "Digitally-Enabled Flexibility Assessment of Multi-Energy Systems Toward Net-Zero Transition" (RIR35231118-1), and The British Academy through the project of "Assembling Digital Solutions for Inclusive and Net-Zero Transition of Community Energy Systems (ADvaNCE)" (PPHE2400207).

References

1. Pukhrem S, Basu M, Conlon MF, Sunderland K (2016) Enhanced network voltage management techniques under the proliferation of rooftop solar PV installation in low-voltage distribution network. IEEE J Emerg Sel Top Power Electron 5(2):681–694
2. Parag Y, Sovacool BK (2016) Electricity market design for the prosumer era. Nat Energy 1(4):1–6
3. Jayaraman V, Patterson RA, Rolland E (2003) The design of reverse distribution networks: models and solution procedures. Eur J Oper Res 150(1):128–149
4. Borlase S (2017) Smart grids: infrastructure, technology, and solutions. CRC Press
5. Strbac G (2008) Demand side management: benefits and challenges. Energy Policy 36(12):4419–4426
6. Khan SM, Cottrell RL, Kalim U, Ali A (2008) Quantifying the digital divide: a scientific overview of network connectivity and grid infrastructure in South Asian countries. J Phys: Conf Ser 119(5), Article no. 052022. IOP Publishing
7. Bloomer J, Boateng D (2024) Prepaid into poverty? Exploring energy poverty and access to alternative electricity sources among urban poor households in Ghana. Energy Res Soc Sci 112:103510
8. Besada H, Stevens Y, Olender M (2015) Addressing the economic costs of sustainable energy in the global south. Background Research Paper, submitted to the United Nations High Level on the Post-2015 Development Agenda, vol 5

9. Panteli M, Mancarella P (2015) Influence of extreme weather and climate change on the resilience of power systems: impacts and possible mitigation strategies. Electr Power Syst Res 127:259–270
10. National Grid Electricity System Operator (2023) Future energy scenarios. https://www.nationalgrideso.com/future-energy/future-energy-scenarios-fes, April 2024
11. Shandurkova I, Bremdal BA, Bacher R, Ottesen S, Nilsen A (2012) A prosumer oriented energy market. IMPROSUME: NCE Smart Energy Markets
12. Hua W, Sun H (2019) A blockchain-based peer-to-peer trading scheme coupling energy and carbon markets. In: Proceedings of the 2019 international conference on smart energy systems and technologies (SEST). IEEE, pp 1–6
13. Paudel A, Sampath L, Yang J, Gooi HB (2020) Peer-to-peer energy trading in smart grid considering power losses and network fees. IEEE Trans Smart Grid 11(6):4727–4737
14. Zhou Y, Wu J, Long C, Ming W (2020) State-of-the-art analysis and perspectives for peer-to-peer energy trading. Engineering 6(7):739–753
15. Mullainathan S, Shafir E (2009) Savings policy and decision-making in low-income households. In: Insufficient funds: savings, assets, credit, and banking among low-income households, vol 121. Russell Sage Foundation, New York, pp 140–142
16. Zhou Y, Manea AN, Hua W, Wu J, Zhou W, Yu J, Rahman S (2022) Application of distributed ledger technology in distribution networks. Proc IEEE 110(12):1963–1975
17. Baidya S, Potdar V, Ray PP, Nandi C (2021) Reviewing the opportunities, challenges, and future directions for the digitalization of energy. Energy Res Soc Sci 81:102243
18. Boschert S, Rosen R (2016) Digital twin–the simulation aspect. In: Mechatronic futures: challenges and solutions for mechatronic systems and their designers. Springer, pp 59–74
19. Department for Energy Security and Net Zero (2050) The future of smart energy in the UK
20. Department for Energy Security and Net Zero (2050) The economic impact of digitalisation in energy
21. Scottish & Southern Electricity Networks (2050) RIIO-ED2: our commitment to a sustainable future. DFES Analysis Report
22. Hua W, Stephen B, Wallom DC (2023) Digital twin based reinforcement learning for extracting network structures and load patterns in planning and operation of distribution systems. Appl Energy 342:121128
23. Hua W, Jiang J, Sun H, Tonello AM, Qadrdan M, Wu J (2022) Data-driven prosumer-centric energy scheduling using convolutional neural networks. Appl Energy 308:118361
24. Di Silvestre ML, Favuzza S, Sanseverino ER, Zizzo G (2018) How decarbonization, digitalization and decentralization are changing key power infrastructures. Renew Sustain Energy Rev 93:483–498
25. Rosetto N, Reif V (2021) Digitalization of the electricity infrastructure: a key enabler for the decarbonization and decentralization of the power sector. In: A modern guide to the digitalization of infrastructure. Edward Elgar Publishing, pp 217–265
26. Kjølle G, Sand K, Gramme E (2021) Scenarios for the future electricity distribution grid. IET
27. Qiu D, Wang Y, Hua W, Strbac G (2023) Reinforcement learning for electric vehicle applications in power systems: a critical review. Renew Sustain Energy Rev 173:113052
28. Hua W, Zhou Y, Qadrdan M, Wu J, Jenkins N (2022) Blockchain enabled decentralized local electricity markets with flexibility from heating sources. IEEE Trans Smart Grid 14(2):1607–1620
29. Cambini C, Congiu R, Jamasb T, Llorca M, Soroush G (2020) Energy systems integration: implications for public policy. Energy Policy 143:111609
30. Manfren M, Sibilla M, Tronchin L (2021) Energy modelling and analytics in the built environment—a review of their role for energy transitions in the construction sector. Energies 14(3):679

Local Energy Systems

Weiqi Hua, Bruce Stephen, Xiao-Ping Zhang, and David C. H. Wallom

Abstract This chapter discusses the role of, and advancements in, local energy systems, emphasizing the integration of distributed generation, energy storage, and digital technologies. The key focus is on how these local systems contribute to the whole or wider energy system through enhancing efficiency, reliability, resilience and sustainability. The chapter outlines the benefits of decentralized energy production, such as reduced transmission losses and increased local resilience, and highlights the importance of the role of energy storage in balancing supply and demand. Overarching all of this is the significance of digitalization in managing, controlling and optimizing energy systems, paving the way for a more flexible and robust energy infrastructure. The chapter also touches on the challenges and technological innovations necessary to support this transition towards a modern and sustainable energy future.

1 Introduction

Local energy systems represent a transformative approach to energy production, distribution, and consumption. They are characterized by the close geographic proximity of energy generation to energy consumption points and by the integration of various

W. Hua (✉) · X.-P. Zhang
School of Engineering, University of Birmingham, Birmingham, UK
e-mail: w.hua@bham.ac.uk

X.-P. Zhang
e-mail: x.p.zhang@bham.ac.uk

B. Stephen
Department of Electronic and Electrical Engineering, Institute for Energy and Environment, University of Strathclyde, Glasgow, UK
e-mail: bruce.stephen@strath.ac.uk

D. C. H. Wallom
Department of Engineering Science, University of Oxford, Oxford, UK
e-mail: elnaz.azizi@eng.ox.ac.uk; david.wallom@oerc.ox.ac.uk

W. Hua et al. (eds.), *Digitalisation of Local Energy Systems*, Green Energy and Technology, https://doi.org/10.1007/978-3-031-77833-9_2

energy resources, including renewables, to meet local demand. These systems are gaining traction as they enhance energy security, reduce transmission losses, and support the global transition to a low-carbon economy.

The ascendancy of local energy systems is inextricably linked to digitalization, which serves as the backbone of these systems. Digital technologies facilitate the management of distributed energy resources, demand response programs, and the integration of various energy vectors such as electricity, heat, and mobility services. By leveraging data analytics, IoT devices, and smart controls, local energy systems can dynamically balance supply and demand, optimize asset utilization, and improve system resilience.

2 Distributed Generation

Distributed generation plays a pivotal role in transforming local energy systems by decentralizing the production of electricity and increasing the resilience of energy supply. These systems, which include solar panels, wind turbines, and small-scale hydroelectric stations, allow electricity to be generated closer to where it is consumed, thereby reducing transmission losses and dependency on large-scale power plants. This proximity enhances energy efficiency and, importantly, supports the integration of renewable energy sources into the grid. As local energy solutions, distributed generation systems contribute to reducing greenhouse gas emissions and fostering energy independence. They also provide critical flexibility in managing energy loads and responding to peak demands, thus stabilizing the local grid. By empowering communities to produce their own energy, distributed generation supports sustainable development and can provide reliable power even in remote or underserved regions.

2.1 Advantages of Adopting Distributed Generation

Distributed generation offers a suite of benefits that address the pressing need for sustainable and resilient energy systems. This section delves into the key advantages of adopting distributed generation and the overarching impact on local energy systems as presented in Fig. 1.

- *Increased Efficiency and Reduced Transmission Losses*: Distributed generation's hallmark is its ability to enhance the efficiency of the entire energy system. By generating energy closer to the point of use, it minimizes losses that typically occur during long-distance transmission and distribution [1]. With less strain on transmission networks, the existing infrastructure can be used more effectively, postponing or negating the need for costly upgrades.
- *Enhanced Reliability and Grid Stability*: The resilience of the power grid is significantly bolstered by distributed generation. By diversifying the sources of energy,

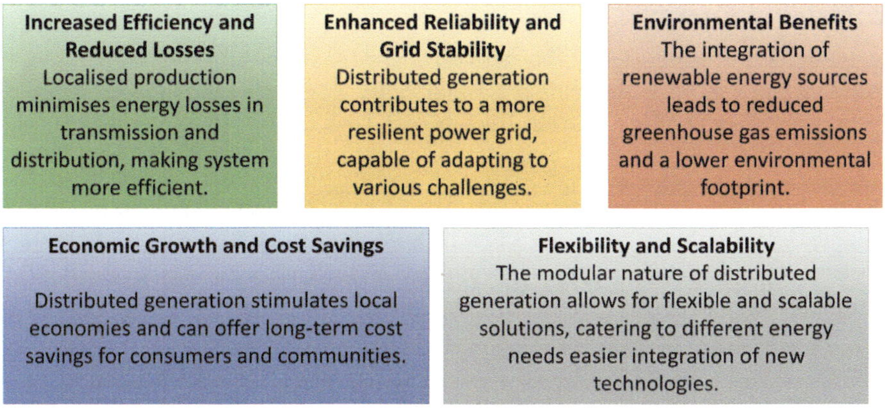

Fig. 1 Summary for advantages of adopting distributed generation technologies

the grid becomes less susceptible to single points of failure. Localized sources can isolate and contain disruptions, enabling faster restoration of service. Distributed systems can quickly adjust to changes in demand, maintaining stability even under variable conditions [2].

- *Environmental Benefits and Net Zero Transition*: As nations strive for net-zero carbon emissions, integrating renewable distributed energy sources becomes critical. Renewable distributed generation curtails reliance on fossil fuels, directly reducing emissions. Local generation projects often employ sustainable practices, contributing to broader environmental stewardship.
- *Economic Growth and Consumer Cost Savings*: Distributed generation is not only a technical solution but also an economic catalyst. It has a direct impact on local economies and consumer energy expenditures. Investment in local energy projects generates jobs and keeps capital within the community [3]. Consumers benefit from lower energy costs due to increased competition and reduced need for energy transportation.
- *Flexibility, Scalability, and Technological Integration*: The modular nature of distributed energy resources is a gateway to a more adaptable and scalable energy future [4]. Local generation can offer grid services such as frequency regulation and voltage support, enhancing grid flexibility. Distributed systems can more readily adopt emerging technologies, from energy storage solutions to smart grid innovations.

The adoption of distributed generation ushers in a new era of energy production that is efficient, reliable, environmentally friendly, and economically beneficial. By prioritizing local energy solutions, communities can shape their energy futures to be more sustainable, resilient, and in harmony with local needs and aspirations. The future of energy lies in empowering localities to harness their resources, paving the way for a greener, more prosperous energy landscape.

2.2 *Wind Energy*

Wind turbines stand as sentinels of sustainability, harnessing the kinetic energy of the wind and converting it into electrical power. Understanding the components and operation of a wind turbine is essential for appreciating its role in the local energy system and the broader push towards renewable energy.

2.2.1 Components of a Wind Turbine

As presented in Fig. 2, the main components of a wind turbine are designed to capture wind energy efficiently:

- *Rotor and Blades*: The rotor, consisting of two or three aerodynamically designed blades, captures wind energy. The blades, often rotating at tip speeds of 50–70 m/s [5], are the primary energy converters.
- *Transmission System*: The gearbox, a critical component of the transmission system, is connected to the rotor through a low-speed shaft. It steps up the rotational speed from about 20–60 RPM to a higher speed suitable for electricity generation, typically in the range of 1200–1500 RPM [6].
- *Generator*: Positioned within the nacelle, the generator converts the mechanical energy from the rotor into electrical power.
- *Braking System*: For normal or emergency shutdowns, the braking system can halt the rotor, bringing the turbine to a standstill.
- *Yaw System*: The yaw system adjusts the nacelle's position to ensure the rotor faces the wind directly, maximizing efficiency.

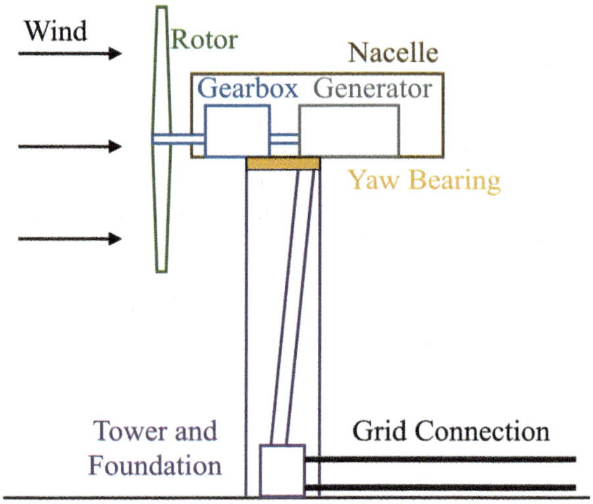

Fig. 2 Schematic illustration of different voltage levels in a typical GB power grid

- *Tower*: The tower, a tubular structure made of concrete or steel, not only supports the turbine but also houses access stairs to the nacelle.
- *Control System*: An electronic controller orchestrates the turbine's operation, managing startup, blade pitch, shutdown, and safety functions.

The operation of a wind turbine involves four key processes:

- *Aerodynamic Force*: Wind exerts aerodynamic forces on the blades, creating torque that initiates blade rotation.
- *Mechanical Energy Transfer*: The rotor's rotation, facilitated by the shaft and gearbox, drives the generator.
- *Energy Conversion*: The generator translates the mechanical energy into electricity.

The journey from wind energy to electricity is a testament to human ingenuity. Initially, the rotor's motion activates the generator through the transmission system. The generator then converts the kinetic energy into electrical energy, which is conditioned and controlled to meet utility standards.

2.2.2 Power Output

The calculation of power output from a wind turbine is integral to the design and efficiency of wind energy systems. This part explores the factors determining power generation from wind turbines and the mathematical principles governing this process.

Wind turbines convert the kinetic energy of wind into mechanical energy, and subsequently, electrical energy. The power P generated by a wind turbine is derived from the kinetic energy formula:

$$P = \frac{1}{2} m V^2, \tag{1}$$

where m is the mass of the air, and V is the velocity of the wind. The mass m of the wind passing through the turbine can be expressed as:

$$m = \rho A V, \tag{2}$$

where ρ represents the air density (typically about 1.225 kg/m^3 at sea level), A is the swept area of the turbine's blades, and V is the wind velocity. Substituting the expression for m into the power formula and incorporating the swept area A, which is calculated as $\pi \times r^2$ (where r is the radius of the rotor blades), the power equation becomes:

$$P_{\text{total}} = \frac{1}{2} \rho A V^3. \tag{3}$$

This formula illustrates that the power output from a wind turbine is proportional to the cube of the wind speed, indicating that even small changes in wind speed can cause significant variations in power output.

Not all the kinetic energy in the wind can be converted to mechanical power. The Power Coefficient (C_p) represents the fraction of the wind's kinetic energy that can be converted by a turbine. The theoretical maximum efficiency a wind turbine can achieve, known as Betz's Limit, is 59.3%. This limit is derived from fluid dynamics principles, stating that:

$$P_{\text{available}} = P_{\text{total}} C_p = \frac{1}{2}\rho A V^3 C_p. \quad (4)$$

Betz's limit indicates that no wind turbine can capture more than 59.3% of the kinetic energy in wind under ideal conditions.

Wind speed varies with height above the ground, affecting the turbine's power output. The Power Law, an empirical relationship, helps estimate wind speed at different heights, crucial for accurate turbine design and site selection. The law states:

$$V_2 = V_1 \times \left(\frac{H_2}{H_1}\right)^{\alpha}. \quad (5)$$

Here, V_1 is the wind speed at the original height H_1, and V_2 is the wind speed at the turbine hub height H_2. The exponent α, typically around 1/7th, describes how wind speed changes with altitude.

Understanding these fundamental principles is crucial for engineers and designers in optimizing turbine size, blade design, and placement relative to typical wind speeds and site characteristics. The ability to accurately calculate and predict power output ensures that wind turbines are designed and located to maximize efficiency and energy generation.

2.2.3 Variable Versus Fixed-Speed Wind Turbines

Among the diverse range of wind turbine technologies, two distinct categories emerge: fixed-speed wind turbines and variable-speed wind turbines. Each of these categories has its own set of characteristics, advantages, and challenges as summarised in Table 1, which are pivotal for stakeholders in the energy sector to understand. The comparison of simplified diagrams between fixed-speed wind turbine and variable-speed wind turbine is presented in Fig. 3.

Fixed-speed wind turbines are the traditional workhorses of the wind energy industry. With their design focused on simplicity and durability, they operate at a constant rotor speed regardless of changing wind conditions. This design choice stems from their primary method of synchronization with the electrical grid–matching the grid frequency is crucial for the seamless export of electricity. The name 'fixed-speed' is derived from the turbine's operational characteristic of maintaining a constant

Table 1 Comparison of fixed-speed and variable-speed wind turbines

Characteristic	Fixed-speed wind turbine	Variable-speed wind turbine
Speed control	Constant speed regardless of wind conditions	Can adjust rotor speed to match wind conditions
	Usually synchronized with the grid frequency	Optimizes energy capture
Design	Simpler and older, often using a gearbox to match the rotor speed to the generator speed	More complex, using power electronics and sometimes gearless (direct drive) designs
Cost	Less expensive to manufacture and maintain due to simplicity	Higher initial cost due to more complex components and control systems
Efficiency	Less efficient at converting wind energy	More efficient across a wider range of wind speeds
	Captures energy best at a specific designed speed	Adapts to changing wind conditions
Grid interaction	More stress on the grid, leading to power quality issues	Better for grid stability, adjusts output to balance supply and demand, reduces fluctuations
Noise	More noise due to the gearbox	Generally quieter, especially in designs without a gearbox

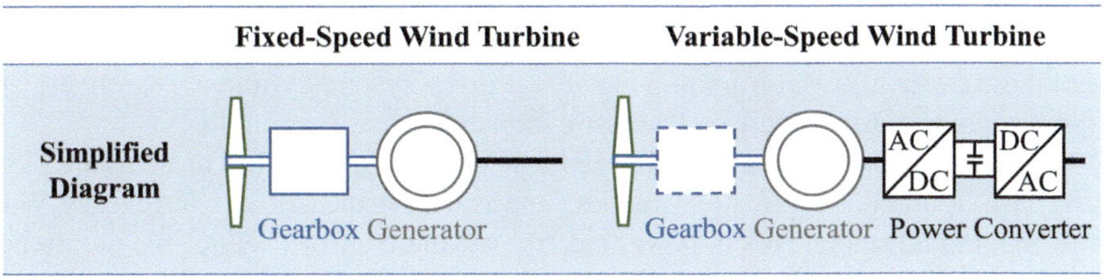

Fig. 3 Comparison of simplified diagrams between fixed-speed wind turbine and variable-speed wind turbine

speed, which aligns with the frequency of the power grid. This feature implies that the turbine is optimized to capture energy efficiently at a specific wind speed for which it has been designed. Inherently simpler in design, fixed-speed turbines typically employ a gearbox that mechanically adjusts the rotor speed to match that of the generator. The maintenance of such turbines is more straightforward, leading to reduced manufacturing and maintenance costs.

Variable-speed wind turbines represent a more advanced and adaptive approach to wind energy generation. Unlike their fixed-speed counterparts, these turbines can alter their rotor speed to harmonize with fluctuating wind conditions, thereby enhancing energy capture over a broader range of wind speeds. A key differentiator for variable-speed turbines is the inclusion of sophisticated power electronics. These systems enable fine-tuned control over the turbine's operation. The power converters, which include rectifiers and inverters, play a pivotal role in this process. To address the primary question of why variable-speed turbines require power converters to transition the generated AC power to DC and then back to AC, we must consider the nature of the electricity produced. Without conversion, the frequency of the produced electricity would be variable, fluctuating with the wind speed and hence incompatible with the grid. The conversion system ensures that the frequency and voltage of the exported power are regulated and synchronized with the grid's requirements. The process begins with the rectification of variable AC generated by the turbine into DC, allowing for easier manipulation of voltage and current. Subsequently, the DC is inverted back into AC, matching the grid's precise frequency and voltage. This bidirectional conversion facilitates a seamless integration with the power grid, safeguarding its stability and ensuring the quality of power supply. While variable-speed turbines boast improved efficiency and adaptability, they incur higher initial costs due to the complexity of their components and control systems. The long-term benefits of optimized energy production, however, can offset these initial investments. Efficiency in energy capture is where variable-speed turbines shine. By adjusting to the dynamic wind environment, these turbines maintain optimal operation across a spectrum of wind conditions, unlike fixed-speed turbines, which are bound to a 'sweet spot' of wind speed. Furthermore, the grid interaction of variable-speed turbines is characterized by their ability to alleviate power quality issues through the regulation of voltage and current, contributing positively to grid stability. Lastly, the operational noise is a non-trivial aspect of wind turbine technology. Fixed-speed turbines, with their dependency on gearboxes, are generally noisier. Variable-speed turbines, especially those employing direct-drive systems without gearboxes, are quieter, contributing to reduced acoustic impact on their surroundings.

The distinctions between fixed-speed and variable-speed wind turbines are clear and have significant implications for their application in the field of wind energy. With the ongoing advancements in power electronics and control systems, the preference for variable-speed technology is likely to grow. Nevertheless, the choice between the two technologies must be made judiciously, considering factors such as local wind conditions, economic constraints, and grid requirements.

2.3 Solar Photovoltaic

Solar photovoltaic (PV) technology harnesses sunlight and converts it directly into electrical energy. At the heart of this technology lie semiconductors–materials with electrical conductivity properties that fall between conductors and insulators. Semiconductors have revolutionized the way we capture and utilize solar energy.

2.3.1 Semiconductor Physics and Bandgap Energy

The key to understanding PV operation lies in semiconductor physics, particularly the concept of bandgap energy. The bandgap is a fundamental property of semiconductors, defining a range of forbidden energies that neither electrons (negatively charged) nor holes (positively charged, due to the absence of an electron) can possess. The structure of atoms in a semiconductor and their interaction forms this bandgap, dictating the material's electrical characteristics.

Adjacent to the bandgap are two critical energy bands as shown in Fig. 4:

- *Conduction Band*: This is the higher energy level where free electrons reside. When electrons are in the conduction band, they can move freely, and thus, contribute to electrical conduction.
- *Valence Band*: The lower energy band is filled with electrons in their stable state. These electrons are not free to move under normal conditions.

In PV devices, energy from photons (light particles) is absorbed by electrons in the valence band. If the photon energy is greater than or equal to the bandgap energy, it can excite an electron to jump from the valence band to the conduction band, leaving behind a hole and creating an electron-hole pair–this is the fundamental process of PV action.

The core of a PV cell is the p-n junction. In silicon, the most commonly used semiconductor, an n-type region is created by doping silicon with phosphorous, which has more valence electrons than silicon. This excess of electrons results in a region with a negative character (n-type) near the valence band. Conversely, a p-type region is created by doping with boron, which has fewer valence electrons, creating a deficit or 'holes' near the conduction band. When these two types of doped silicon are put together, they form a p-n junction. The junction facilitates an electric field as electrons move from the n-type region to fill holes in the p-type region, creating a diode that allows current to flow in one direction. When the cell is exposed to light, the electron-hole pairs are separated by the electric field at the junction, driving the electrons towards the conduction band, and thus, creating an electrical current. While the bandgap energy describes the essential threshold for electron excitation,

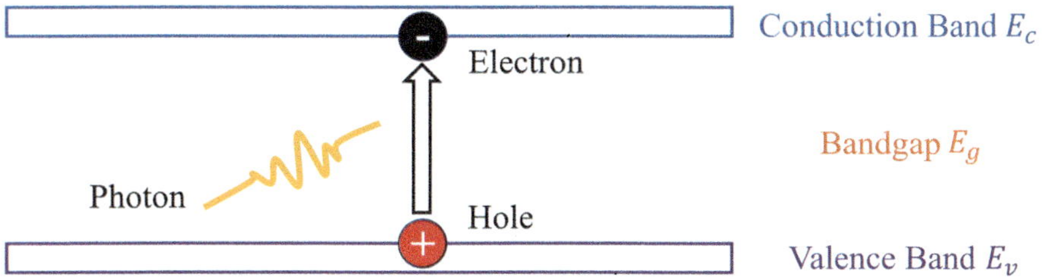

Fig. 4 Schematic illustration of conduction band, valence band, bandgap energy, and mechnimism for semiconductors

itis important to note that not all photons will have sufficient energy to cross this threshold. The efficiency of a PV cell depends significantly on the semiconductor's ability to absorb photon energy and convert it into electrical energy.

2.3.2 The Photovoltaic Hierarchy

A deeper understanding of the PV hierarchy–from individual cells to expansive arrays as shown in Fig. 5—provides insight into how solar power is transformed from a minute electrical contribution to a substantial power source.

At the foundation of the PV hierarchy lies the silicon PV diode. With an open circuit voltage of approximately 0.7 V [5], a single diode provides a minimal amount of electricity, sufficient for low-power applications such as electronic calculators. However, the true potential of PV technology is realized when these cells are combined into more complex structures. A PV cell's voltage may be modest, but when multiple cells are connected in series within a module, the voltages add up, increasing the total output. This series connection is fundamental to the design of PV modules, ensuring that they produce a higher voltage necessary for more significant applications. Modules are the intermediary step in the PV hierarchy. They consist of a number of individual PV cells wired together. The design is not only about achieving an additive voltage effect but also about configuring a practical and installable unit that can be efficiently manufactured, transported, and deployed.

For applications that require more power than a single module can provide, PV modules are further connected in parallel to form an array. This parallel configuration boosts the current and, consequently, the power output, without increasing the voltage beyond the operational specifications. Arrays can be tailored to meet specific energy needs and are scalable to accommodate everything from residential rooftops to vast solar farms. The practical assembly of PV cells into modules and arrays is dictated by both electrical requirements and real-world constraints. Factors such as available space, shading, local climate conditions, and intended use cases all influence how PV

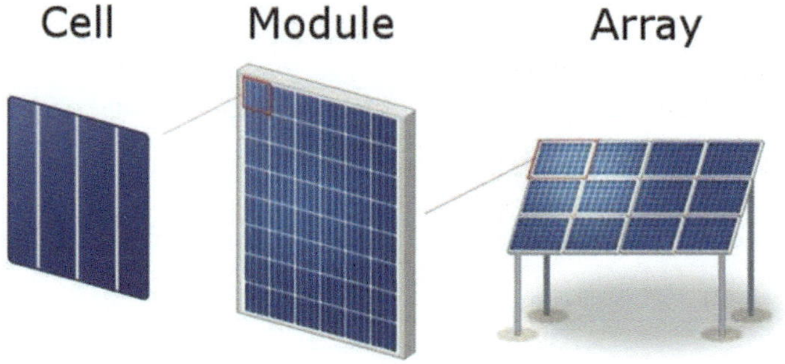

Fig. 5 PV hierarchyfrom individual cells to expansive arrays [7]

systems are designed and implemented. Moreover, the efficiency of each PV cell and the overall system plays a critical role in the viability of solar power. Technological advancements in cell materials, manufacturing processes, and system design continue to enhance the performance and reduce the costs of PV systems, further solidifying their role in the global energy portfolio.

2.3.3 Structure of A Photovoltaic Module

This part provides a detailed look into the construction and functionality of a standard PV module, illustrated through a cutaway diagram as shown in Fig. 6 that reveals the internal components and their assembly.

- *PV Cells*: The core components of a PV module are the photovoltaic cells themselves. These cells are responsible for converting sunlight into electrical energy. Typically made from silicon, these cells are arranged in a series to form a string within the module.
- *Transparent Window*: The top layer of a PV module is a transparent window, usually made of tempered glass. This window allows sunlight to reach the photovoltaic cells while providing protection from environmental elements. Its transparency is crucial for maximizing light penetration and thus the efficiency of energy conversion.
- *Reflective Backing*: Beneath the array of PV cells lies a reflective backing. This layer enhances the module's efficiency by reflecting any light that has passed through the PV cells back towards them, providing a second chance for absorption.
- *Encapsulation*: Encapsulation materials, typically made from ethylene-vinyl acetate (EVA), are used to bond the glass top to the rest of the module and to secure the PV cells in place. This encapsulation ensures the durability of the cells and protects them from moisture and mechanical damage.

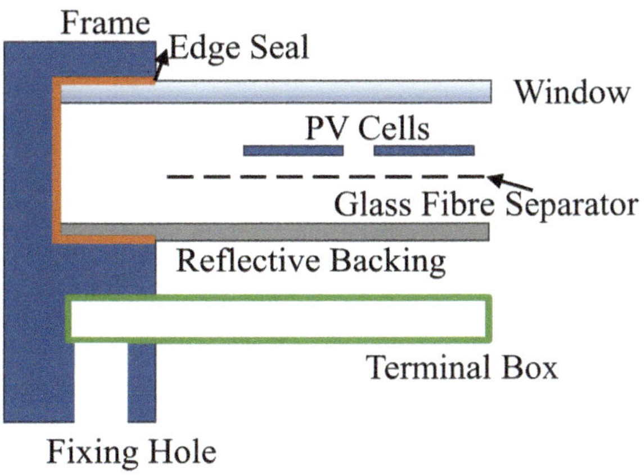

Fig. 6 Cutaway diagram of a solar photovoltaic module

- *Metal Frame*: Surrounding the edges of the PV module is a robust metal frame, commonly aluminum. This frame provides structural stability, enabling the module to withstand various environmental stresses such as wind, rain, and snow. The metal frame also facilitates the mounting of the module onto supporting structures through pre-drilled fixing holes.
- *Sealing for Moisture Protection*: To prevent moisture ingress, which can significantly degrade the performance and lifespan of the PV cells, the edges of the module where the frame meets the glass and backing are sealed. Effective sealing is critical in maintaining the module's operational integrity over its lifespan.
- *Terminal Box*: The electrical connections of the PV cells are routed to a terminal box located typically at the back of the module. This box houses the bypass diodes that prevent the cells from overheating and serves as the connection point for wiring the module to a load, to other modules, or to the grid.
- *Fixing Holes*: At the bottom of the module, fixing holes are strategically placed within the frame to secure the module to a supporting structure. These are critical for ensuring the module remains stationary and correctly aligned for optimal sun exposure.

The design and configuration of PV modules are pivotal in determining their suitability for different installation environments and energy requirements. By connecting multiple modules in series or parallel, installers can meet specific voltage and current requirements for residential, commercial, or utility-scale solar projects.

2.3.4 Balance of System Components in Photovoltaic Systems

The PV system's efficacy is not solely dependent on the solar panels themselves but also on a crucially supportive infrastructure known as the Balance of System (BOS). The BOS encompasses all components of a solar system other than the photovoltaic panels. This includes wiring, mounting systems, inverters, the battery storage system, and additional components that are essential for operating a fully functional PV system. Details of a BOS system is given as follows:

- *Wiring, Cables, and Fuses*: The wiring and cables transport the electrical current generated by the PV panels to the inverter and from the inverter to the load or grid. The selection of appropriate gauge, insulation, and material is crucial for safety, efficiency, and durability. Fuses are critical for protecting the PV system from overcurrent that can lead to overheating and potential fires. They act as a safety device that breaks the circuit if the current flowing through a system exceeds safe levels.
- *Tracking Systems*: Tracking systems are used to maximize power collection by adjusting the orientation of solar panels to follow the sun's path across the sky. This can significantly increase the energy output of the PV panels by keeping them optimally aligned with the sun, especially in installations with high direct sunlight exposure.

- *Battery System*: Batteries store the electricity generated by PV panels for later use, addressing the issue of solar energy's intermittency. They are particularly crucial for off-grid systems or for providing power during peak demand times in grid-tied systems. Charge regulators, also known as charge controllers, ensure that the right amount of current flows to and from the batteries to prevent overcharging and excessive discharge, thus extending battery life.
- *Inverters*: Inverters play a pivotal role by converting the DC electricity generated by solar panels into AC electricity, which is the standard electrical current used by most household appliances and for grid distribution.
- *Maximum Power Point Tracking (MPPT)*: Modern inverters often include MPPT technology which continuously adjusts the electronic load to maximize the power output from the PV panels. MPPT is essential for optimizing the efficiency of solar panels under varying sunlight and temperature conditions.
- *PV Controller*: The PV controller manages the flow of electricity from the PV panels to the battery and the load. It ensures that the energy produced by the solar panels is either consumed immediately or stored for future use. It also decides when to draw energy from the storage batteries to power the load, optimizing the system's overall performance.

The BOS components are integral to the safe, efficient, and effective operation of PV systems. They not only facilitate the basic function of converting solar energy into usable power but also enhance the system's performance through smart energy management and protective measures. The cost of BOS can account for a significant portion of the total system costs, which makes choosing high-quality components and efficient designs to be important.

2.3.5 Photovoltaic Connection

In solar PV systems, the connection of PV modules significantly impacts the overall system performance, including voltage levels, current output, and power generation capacity. Understanding the difference between parallel and series connections is essential for designing efficient and reliable solar power systems.

In the parallel connection as shown in Fig. 7, multiple PV modules are connected such that the voltage across each module remains constant, while the currents are additive. This arrangement is preferred when a specific voltage needs to be maintained across the system, and a higher current output is required. According to Kirchhoff's Current Law, the total current (I_T) output from the parallel connection of PV modules is the sum of the currents through each module ($I_1 + I_2 = I_T$). This is particularly useful when scaling the current capacity in large installations without altering the voltage across each module.

As shown in Fig. 8, When identical PV modules are connected in parallel, the short-circuit current doubles (if two modules are used), but the open-circuit voltage remains the same. This is visualized in the I-V curves where the curve extends vertically, increasing current capacity while maintaining the same voltage. If one of the

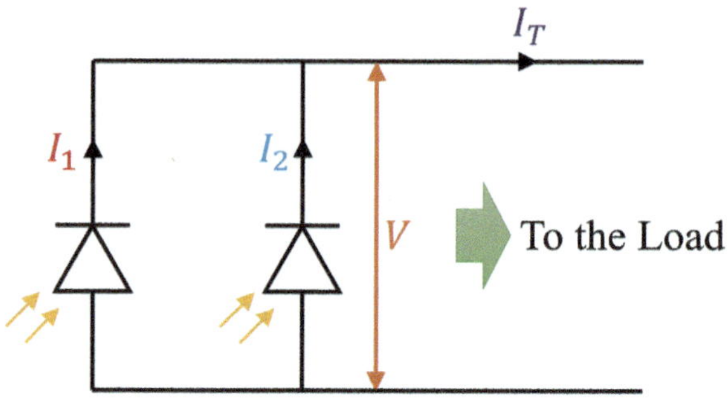

Fig. 7 Diagram of a two PV modules connected in parallel

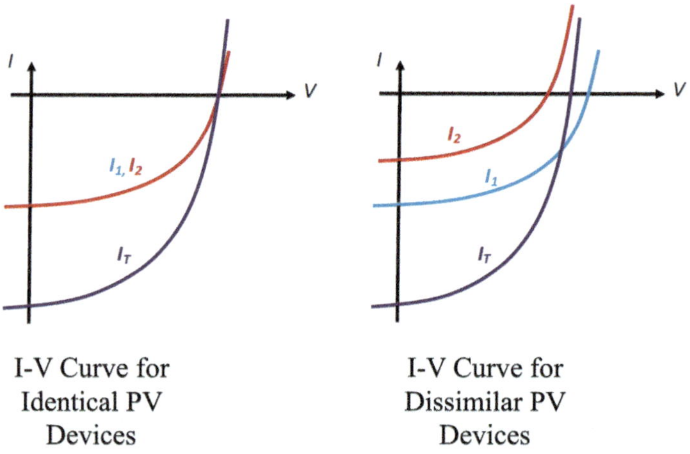

Fig. 8 Comparison of I-V curves between identical and dissimilar PV Devices connected in parallel

PV modules is damaged or underperforming (e.g., due to shading), the total current capacity is still the sum of individual currents, but the effective open-circuit voltage is capped by the least performing module. In this scenario, the overall performance is degraded, and the system operates less efficiently.

In the series connection as shown in Fig. 9, PV modules are connected end-to-end, resulting in additive voltages and a constant current flow through the string. This type of connection is used to increase the voltage to appropriate levels for inverters or grid interconnection without altering the current. Using Kirchhoff's Voltage Law, the total voltage (V_T) across a series string of PV modules is the sum of the voltages across each module ($V_1 + V_2 = V_T$). This is advantageous for applications requiring higher operating voltages.

As shown in Fig. 10, in a series connection of identical PV devices, the voltage doubles while the current remains constant, effectively shifting the I-V curve horizontally to higher voltages. The presence of a weaker or damaged module in a series connection limits the overall current to that of the least capable module. This

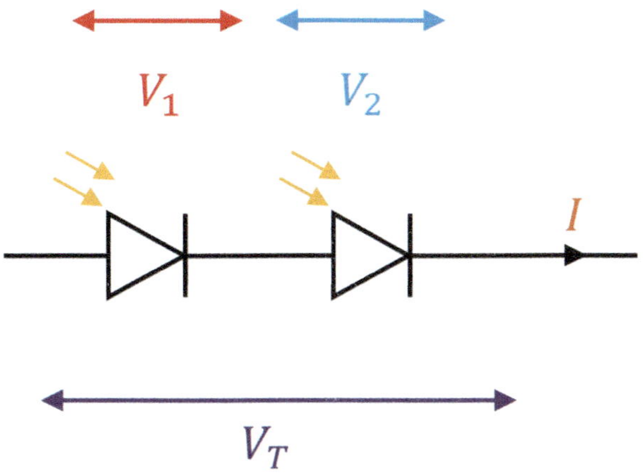

Fig. 9 Diagram of a two PV modules connected in series

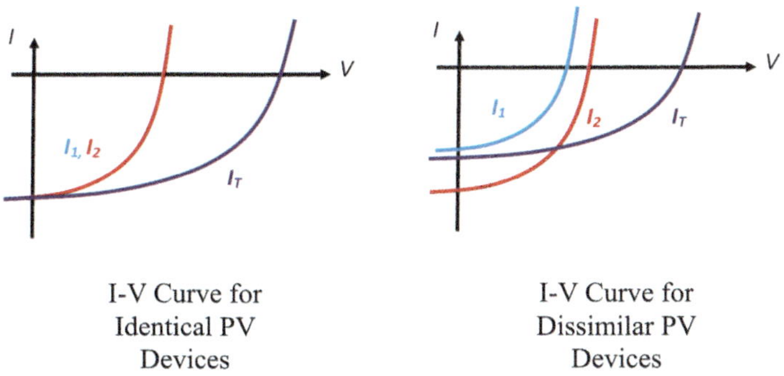

Fig. 10 Comparison of I-V curves between identical and dissimilar PV Devices connected in series

phenomenon results in a bottleneck effect, where the 'weakest link' determines the overall performance, potentially causing energy losses and localized heating within the underperforming module.

The above two cases is for a single PV device. Now, think about for the connection of entire module. It will follow the same principle. In parallel connections as shown in Fig. 11, PV modules are configured so that the voltage across each module remains constant, and the currents are additive. This setup ensures that the total output current of the array is the sum of the currents from each module. For instance, if each module produces 3 amps at 15 V, three modules connected in parallel will deliver a total of 9 amps at the same voltage. Blocking diodes are crucial in parallel configurations to prevent reverse current flow. These diodes are typically placed at the positive terminals of the PV modules. Their primary function is to suppress reverse leakage currents that can occur, particularly when some modules are shaded or underperforming. Without blocking diodes, shaded or faulty modules could reverse-bias and dissipate power, potentially damaging the module and reducing the overall system

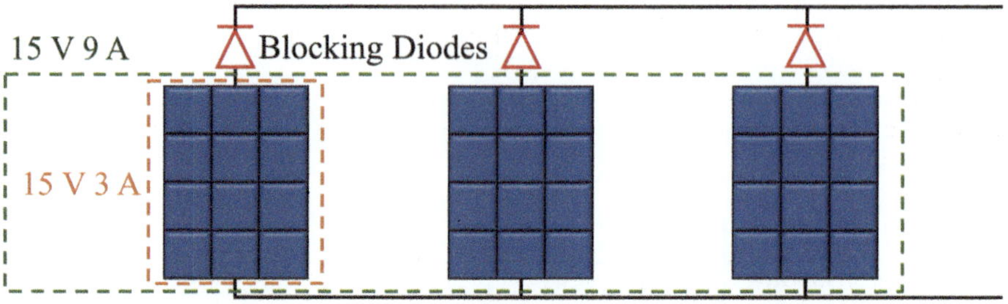

Fig. 11 Schematic illustration of the parallel connection of PV modules and application of blocking diodes

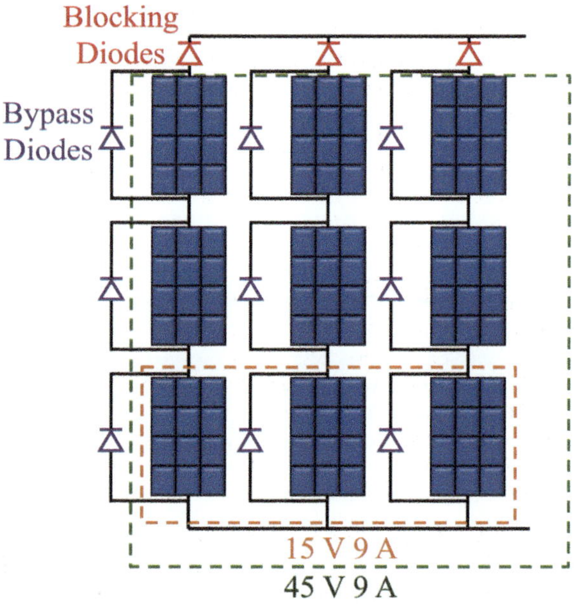

Fig. 12 Schematic illustration of the series connection of PV modules and application of blocking diodes and bypass diodes

efficiency. Consider a scenario where one of the three modules in a parallel configuration is shaded or fails. The blocking diodes ensure that the healthy modules continue to operate efficiently, maintaining the system output at 6 amps and 15 V, rather than allowing the entire system's performance to degrade.

In a series configuration as shown in Fig. 12, the voltage outputs of the modules are additive, while the current remains constant through the string. This arrangement is beneficial for applications requiring higher voltages, such as inverter input specifications or grid-tied systems. Bypass diodes play a critical role in series connections. They are installed across each PV module to provide an alternative path for the current if a module becomes faulty or is significantly shaded. This prevents the entire string from being disabled due to the failure of a single module.

In a typical series setup of three modules each producing 15 V and 9 amps, the total output would be 45 V at 9 amps. However, if one module fails or is shaded, bypass diodes activate to allow current to pass around the faulty module, ensuring the rest of the string continues to operate. Without bypass diodes, the failure of one module could result in zero output from the entire string.

2.3.6 Array Mountings

PV systems require mountings that not only support the weight of the panels but also position them optimally towards the sun to maximize energy production. The choice of mounting type, whether fixed or tracking, is a significant decision in system design due to its impact on the cost, efficiency, and maintenance of the installation.

Fixed mountings are commonly chosen for their simplicity and cost-effectiveness. They typically involve less complexity in both design and deployment, which translates into lower upfront costs and reduced maintenance requirements. While traditionally static, some fixed mounting systems offer the flexibility to adjust the tilt of the panels. This adjustment is usually made semi-annually to account for the seasonal variation in the sun's position. During summer, the tilt is lower to capture the high-standing sun, and in winter, the angle is increased to optimize the capture of solar radiation when the sun is lower in the sky.

Tracking mountings offer a dynamic approach to maximizing solar energy collection by following the sun's path across the sky. These systems can increase the energy output of PV installations significantly, depending on the geographic location and specific environmental conditions. There are two types of tracking systems:

- *Single-Axis Trackers*: These trackers rotate on one axis, generally aligned from north to south. This allows the panels to follow the sun's path from east to west throughout the day, which is particularly effective in regions with high direct sunlight.
- *Dual-Axis Trackers*: More sophisticated than single-axis, these systems can pivot both east-west and north-south. This dual movement capability allows the panels to maintain optimal perpendicularity to the sun's rays, enhancing energy capture during times when the sun's angle changes rapidly, such as during the mornings and evenings, or seasonally.

By continuously orienting the PV panels directly towards the sun, tracking systems can improve the energy yield of a PV system. Despite their higher efficiency, tracking systems are more expensive to install and maintain than fixed systems. They involve complex mechanisms and controls that can increase the initial investment and ongoing maintenance costs. Additionally, their moving parts are more susceptible to wear and tear, potentially reducing their lifespan compared to fixed mounts.

2.4 *Bio-energy*

Biomass energy systems harness the power of organic materials, turning forestry products, energy crops such as Miscanthus, and agricultural byproducts like straw into energy. This transformation process not only produces valuable energy but does so in a manner that is nearly carbon neutral, offering a sustainable alternative to fossil fuels.

2.4.1 Types of Biomass and Processing

Three typical types of biomass products are as follows:

- *Forestry Products*: These include various forms of wood residues, such as branches and leaves, collected during forestry operations.
- *Energy Crops*: Crops specifically grown for energy production, such as Miscanthus, switchgrass, and fast-growing woods like willow and poplar.
- *Agricultural Byproducts*: Organic materials leftover from agricultural processes, including straw, husks, and animal manure.

The process of converting these materials into energy involves three steps:

- *Biomass Intake*: Initial collection and preparation of biomass materials for processing.
- *Shredding and Cutting*: Reducing the size of biomass materials to ensure uniformity and improve combustion or decomposition rates.
- *Drying and Storage*: Removing moisture from biomass to increase the energy content per unit weight and storing it safely for future use.

2.4.2 Biomass Combustion Process

As shown in Fig. 13, the process of biomass combustion in a power plant setting involves several key components and stages:

- *Storage for Biomass*: Biomass is stored in a way that maintains its quality and facilitates easy feeding into the combustion system.
- *Hot Water Boiler*: Biomass fuel is combusted in a boiler, generating heat that turns water into steam.
- *Turbine and Generator*: The high-pressure steam produced in the boiler is then used to drive a turbine, which in turn rotates a generator to produce electricity.
- *Fuel Gas Cleaning*: Before the combustion gases are released through the chimney, they pass through cleaning processes to reduce pollutants, ensuring the process remains environmentally friendly.

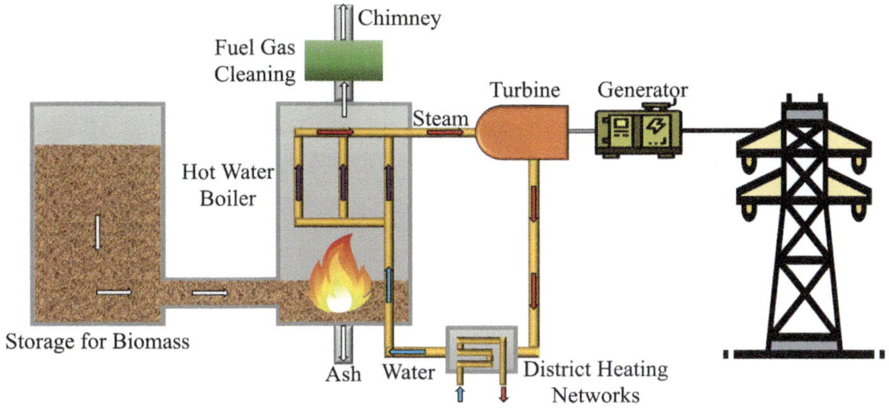

Fig. 13 Schematic illustration of biomass combustion

- *District Heating Networks*: Post-electricity generation, the residual heat from the steam can be directed to nearby homes and businesses as part of a district heating scheme, significantly improving the efficiency of the system.
- *Ash Collection*: Ash, a byproduct of biomass combustion, is collected for use in various secondary applications, such as in the creation of fertilizers or as a component in building materials.

The integrated use of biomass for both heat and power generation, known as Combined Heat and Power (CHP), is particularly efficient. By using the waste heat from electricity generation for heating, CHP systems can achieve efficiencies that significantly exceed those of separate heat and power generation. The dual use not only maximizes energy output from the biomass but also reduces environmental impact, aligning with global efforts towards sustainability.

2.4.3 Net-Zero Cycle

Biomass plays a pivotal role in the global strategy to reduce carbon emissions through its inherently renewable nature and contribution to the carbon cycle. Unlike fossil fuels, which release carbon that has been sequestered for millions of years, biomass utilizes carbon that is part of the current carbon cycle, making it a net-zero carbon energy source when managed sustainably.

As shown in Fig. 14, during their growth, biomass sources such as forestry products, energy crops, and agricultural byproducts absorb carbon dioxide from the atmosphere. This process of photosynthesis captures atmospheric carbon and incorporates it into the organic matter of the plants. When biomass is used for energy production, the carbon dioxide absorbed during the growth phase is released back into the atmosphere. This occurs through the combustion process in biomass power plants, where

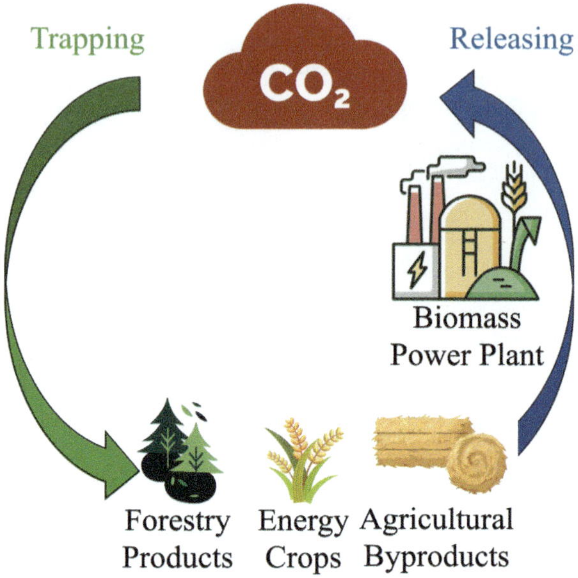

Fig. 14 Net-zero carbon cycle for biomass combustion

organic material is burned to generate heat and electricity. The carbon released is the same carbon that was recently captured; thus, it does not contribute additional net carbon to the atmosphere like fossil fuels do. By comparison, fossil fuels, such as coal and oil, are formed from organic material that was buried and transformed under high pressure over millions of years. When burned, fossil fuels release this "ancient" carbon into the atmosphere, which was otherwise locked away, thus increasing the overall carbon dioxide levels and contributing to global warming. Biomass, by contrast, is part of a much shorter and renewable carbon cycle. The key to maintaining its status as a net-zero carbon source lies in the rate of its utilization compared to its growth. Sustainable management ensures that biomass is replenished at the same rate or faster than it is consumed.

Ensuring that biomass harvesting does not exceed the rate of regeneration is crucial. This prevents the depletion of biomass stocks and avoids deforestation, which can lead to a significant increase in carbon emissions and loss of biodiversity. Proper balance in the use of biomass for energy is essential. If biomass is consumed more quickly than it can regrow, it can lead to a net increase in atmospheric CO_2 levels. Conversely, if managed properly, biomass can actually contribute to carbon sequestration, as growing plants continue to absorb more CO_2.

Without careful management, increased demand for biomass can lead to deforestation, particularly in regions where regulations may be inadequate or not enforced.

2.4.4 Bio-energy with Carbon Capture and Storage

Bio-energy with carbon capture and storage (BECCS) represents a pivotal technology in the shift towards net negative carbon emissions. This innovative approach harnesses the natural carbon-absorbing power of biomass while enhancing the sustainability of energy production by capturing and sequestering carbon dioxide produced during biomass combustion.

As shown in Fig. 15, the process begins with the growth of biomass, such as forestry products, energy crops like Miscanthus, and agricultural byproducts such as straw. These materials inherently capture atmospheric CO_2 as they grow, embedding the carbon within their structures. When these biomaterials are processed and combusted in biomass power plants, they release CO_2. However, unlike the combustion of fossil fuels—which reintroduces long-sequestered carbon into the atmosphere—the CO_2 emitted from biomass is part of a shorter carbon cycle, having been recently captured from the atmosphere.

The critical component of BECCS is the capture of CO_2 during the biomass combustion process. Before this CO_2 can be released back into the atmosphere, it is captured at the combustion site. Following capture, the CO_2 is transported to secure storage sites, where it is often injected into geological formations for long-term sequestration. This process effectively removes CO_2 from the atmosphere, making the overall impact of BECCS net negative in terms of carbon emissions.

The BECCS has the following advantages:

- *Negative Carbon Emissions*: BECCS is one of the few technologies capable of achieving negative carbon emissions, actively reducing the amount of CO_2 in the atmosphere and contributing to climate change mitigation goals.

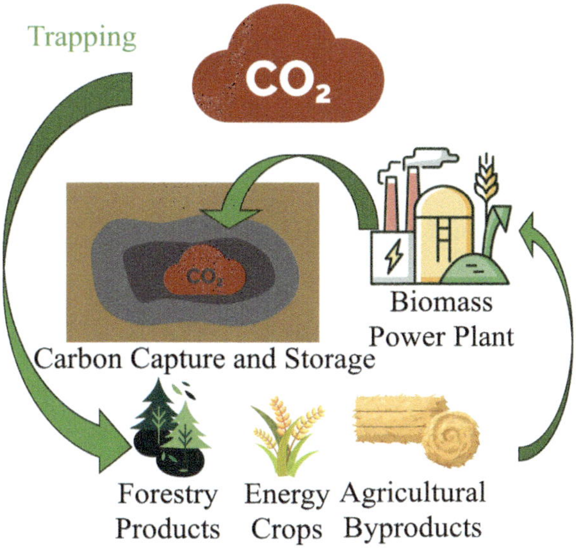

Fig. 15 Net-negative carbon cycle for bio-energy with carbon capture and storage

- *Renewable Energy Source*: As a renewable energy source, biomass is continuously replenished and plays a critical role in the natural carbon cycle. Its use in BECCS further enhances its sustainability credentials.
- *Utilization of Existing Infrastructure*: BECCS can often be integrated into existing power generation infrastructure with minimal modifications. This adaptability makes it a cost-effective solution compared to building new power plants from scratch.
- *Energy Security*: Utilizing locally sourced biomass reduces dependence on imported fossil fuels and enhances national energy security.
- *Economic Development*: BECCS promotes economic development, particularly in rural areas where biomass resources are abundant. The construction and operation of CCS facilities can create significant job opportunities.
- *Waste Management*: By utilizing organic waste materials for energy production, BECCS provides a practical solution to waste management challenges.
- *Co-benefits with Agriculture and Forestry*: Integrating BECCS with agricultural and forestry practices can provide additional revenue streams for farmers and foresters, promoting sustainable land management practices.
- *Flexibility in Scale and Application*: BECCS can be implemented on various scales and is adaptable to different sectors, including electricity generation, heating, and industrial processes.

While BECCS offers many advantages, it also faces several implementation challenges. The technology requires significant up-front investment in carbon capture facilities and infrastructure for transporting and storing CO_2. Moreover, careful management of biomass resources is essential to prevent adverse environmental impacts such as deforestation and biodiversity loss. Ensuring the sustainability of biomass supply chains and the integrity of carbon storage sites is crucial for the long-term viability of BECCS.

As we advance towards a carbon-neutral future, the role of technologies like BECCS becomes increasingly critical. By turning biomass energy production into a net negative emitter through sophisticated carbon capture and storage strategies, BECCS not only contributes to reducing atmospheric CO_2 levels but also enhances the sustainability of renewable energy sources.

2.4.5 Combined Heating and Power

CHP, or cogeneration, significantly enhances the efficiency of energy systems by simultaneously generating electricity and capturing byproduct heat for use in heating and industrial processes. This system can operate on a diverse range of fuels including biomass, natural gas, and coal, among others. The advantages of CHP systems include:

- *Higher Efficiency*: By capturing and utilizing waste heat, CHP systems can achieve efficiencies of 50–70%, which is considerably higher than traditional power plants that average 40–50% efficiency.

- *Reduced Emissions*: The high efficiency translates to lower per unit energy carbon emissions.
- *Cost Savings and Energy Security*: Localized generation reduces transmission losses and energy costs, while also enhancing energy reliability for critical facilities.
- *Flexibility*: CHP systems can be adapted to different scales and operational needs, making them suitable for a variety of industrial and residential applications.

Expanding on the concept of CHP, combined cooling, heat, and power (CCHP) systems, also known as trigeneration, include the production of chilled water for cooling alongside heat and electricity. This process utilizes an absorption chiller that converts waste heat into cooling, providing an efficient air conditioning solution. The primary benefits of CCHP include:

- *Improved Overall Efficiency*: CCHP systems can achieve total energy utilization efficiencies of 80–90%.
- *Versatility in Energy Production*: By producing electricity, heat, and cooling from a single fuel source, CCHP systems offer a comprehensive energy solution that can significantly reduce operating costs and environmental impact.

2.5 Fuel Cell

Fuel cells, as a transformative technology in the energy sector, offer several compelling advantages and specific characteristics that distinguish them from conventional energy conversion devices like batteries. This section aims to provide insights into the operation, types, and strategic significance of fuel cells in achieving a sustainable energy future.

2.5.1 Operating Mechanism

Fuel cells convert chemical energy directly into electrical energy through an electrochemical reaction involving oxygen and a fuel, typically hydrogen. As shown in Fig. 16, the key components of a typical fuel cell include:

- *Anode*: The anode is the negative post of the fuel cell where the fuel (commonly hydrogen) is oxidized. The catalyst at the anode helps split hydrogen molecules into protons and electrons.
- *Cathode*: At the cathode, the positive post, oxygen from the air and protons arriving through the electrolyte combine with electrons coming through the external circuit to produce water and release heat.
- *Electrolyte (Proton Exchange Membrane)*: The electrolyte's critical function is to allow only the protons to pass through it from the anode to the cathode while

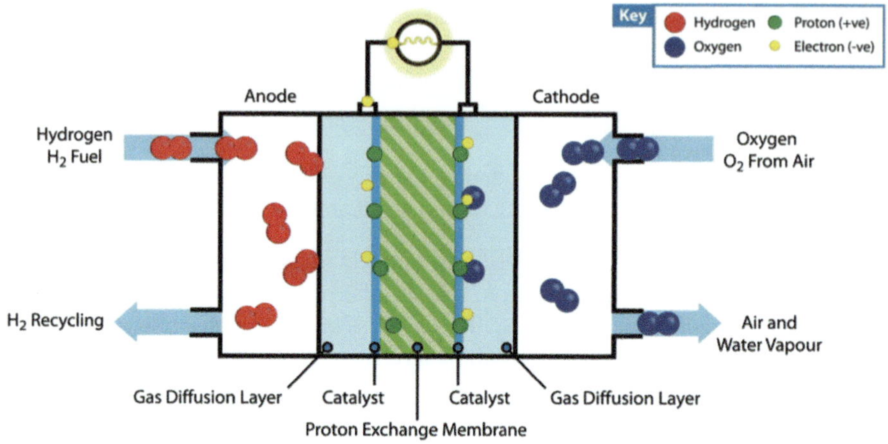

Fig. 16 Schematical illustration of chemical reactions of a fuel cell [8]

blocking electrons. This separation forces electrons to travel through an external circuit, generating an electric current.

- *Catalyst*: Typically made from platinum, the catalyst speeds up the reactions at both the anode and cathode.

This system does not involve combustion and operates at high efficiency levels because it directly converts chemical energy into electrical energy, minimizing energy loss that usually occurs during conversion.

2.5.2 Types and Applications of Fuel Cells

Fuel cells are designed in various forms to cater to different applications, each characterized by its electrolyte type and operating temperature:

- *Proton Exchange Membrane Fuel Cells (PEMFCs)*: Known for their low operating temperatures and fast start-up, PEMFCs are ideal for vehicles and other mobile applications as they can quickly adjust to changes in power demand.
- *Solid Oxide Fuel Cells (SOFCs)*: These cells operate at very high temperatures, which allows them to use a wider variety of fuels and makes them suitable for large-scale stationary applications where heat output can also be utilized.
- *Molten Carbonate Fuel Cells (MCFCs)*: and Phosphoric Acid Fuel Cells (PAFCs): Both types are used primarily in stationary applications, providing electricity and heat for commercial buildings or industrial plants. MCFCs can even utilize CO_2 as part of their input, potentially helping to reduce greenhouse gas emissions.

Table 2 Comparison of different types of fuel cells [9]

Type	Fuel	Electrolyte	Charge carrier	Operating temp. (°C)	Efficiency (%)
Proton Exchange membrane fuel cell	H_2	Solid polymer (Nafion)	H^+	80–100	35–50
Solid oxide fuel cell	H_2, CO, CH_4	Solid oxide (yttria-stabilized zirconia)	O^{2-}	800–1000	50–60
Molten carbonate fuel cell	H_2, CO, CH_4	Lithium and potassium carbonate	CO_3^{2-}	650	45–55
Phosphoric acid fuel cell	H_2	Phosphoric acid (H_3PO_4 solution)	H^+	175–200	35–45
Alkaline fuel cell	H_2	Potassium hydroxide (KOH)	OH^-	80–120	35–55

- *Alkaline Fuel Cells (AFCs) and Direct Methanol Fuel Cells (DMFCs)*: AFCs, commonly used in space applications due to their high efficiency with pure reactants, and DMFCs, suitable for portable power due to their ability to use methanol directly, highlight the versatility of fuel cell technologies.

Fuel cells are thus a crucial part of the narrative on renewable energy, representing a bridge towards low-carbon energy systems. Their ability to deliver reliable, clean energy in a range of applications from automotive to stationary power makes them a promising solution for the future energy mix, particularly in scenarios demanding high efficiency and environmental sustainability. The comparison of main characters for each type of fuel cell is presented in Table 2.

Fuel cells offer several distinct advantages:

- *High Efficiency*: They can achieve electrical efficiencies between 35 and 60%, significantly higher than traditional combustion engines.
- *Low Emissions*: When running on hydrogen, fuel cells emit only water vapor and heat, making them an environmentally friendly option.
- *Flexibility and Scalability*: Fuel cells can be modular, allowing for systems to be scaled up or down to meet specific power needs without extensive reconfiguration.

Despite their benefits, fuel cells face certain challenges:

- *High Initial Costs*: The production of fuel cells involves advanced materials and technologies, which currently come at a high cost.

- *Durability and Maintenance*: High operating temperatures and the corrosive nature of the electrolytes used can lead to faster wear and tear, necessitating frequent maintenance.
- *Fuel Purity Requirements*: Fuel cells are sensitive to impurities in the hydrogen fuel, which can poison the catalysts used in the reaction.

2.6 Internal Combustion Engines

Internal combustion engines (ICEs) are a pivotal component in local energy systems, particularly due to their flexibility and ability to provide reliable power. An internal combustion engine operates by combusting a fuel (such as fossil fuels including petrol, diesel, and natural gas) with an oxidizer (air) within a combustion chamber. This process generates the mechanical energy necessary to drive various applications.

2.6.1 Key Components of Internal Combustion Engines

The key components of ICEs are detailed as follows:

- *Piston*: The piston moves up and down within the cylinder, converting chemical energy from fuel combustion into mechanical motion. This motion drives the crankshaft, creating rotational energy.
- *Cylinders*: These house the pistons and provide the space where combustion occurs. The number of cylinders can vary, affecting the engine's power output and efficiency.
- *Crankshaft*: The crankshaft transforms the pistons' linear motion into rotational motion, which can then be used to power machinery or generate electricity.
- *Fuel System*: This system delivers fuel to the combustion chamber. It includes components like the fuel tank, pump, and injectors, ensuring a consistent and controlled fuel supply.
- *Ignition System*: In spark ignition (SI) engines, the ignition system ignites the air-fuel mixture, initiating the combustion process. This system typically involves spark plugs and an ignition coil.
- *Exhaust System*: The exhaust system expels the combustion gases from the cylinder after the combustion process. This system includes components like the exhaust manifold, catalytic converter, and muffler to manage emissions and reduce noise.

2.6.2 Four-Stroke Cycle of Internal Combustion Engines

As shown in Fig. 17, ICEs commonly operate on a four-stroke cycle, which consists of two revolutions of the crankshaft and one complete thermodynamic cycle:

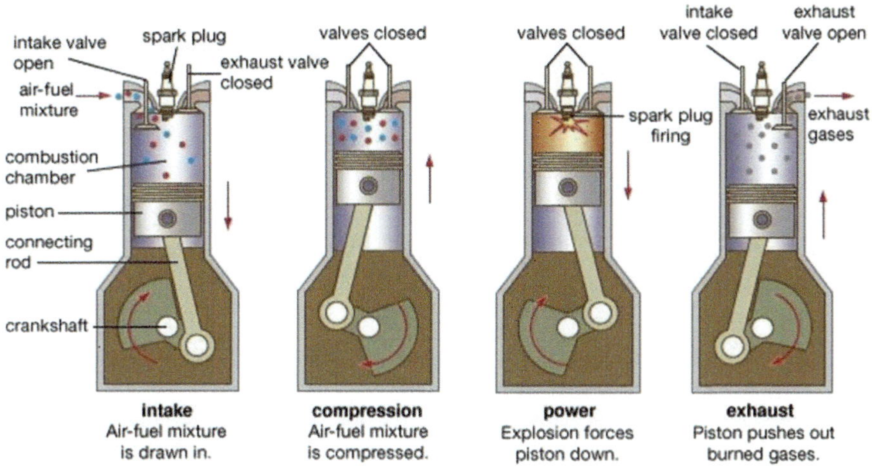

Fig. 17 Four-stroke cycle of internal combustion engine [10]

- *Intake Stroke*: The piston moves down, the intake valve opens, and the air-fuel mixture is drawn into the cylinder.
- *Compression Stroke*: The piston moves up, compressing the air-fuel mixture, increasing pressure and temperature.
- *Power Stroke*: The spark plug ignites the compressed air-fuel mixture, causing an explosion that forces the piston down, driving the crankshaft.
- *Exhaust Stroke*: The piston moves up again, the exhaust valve opens, and the spent gases are expelled from the cylinder.

2.6.3 Types and Features of Internal Combustion Engines

There are two main types of internal ICEs:

- *Spark Ignition (SI) Engines*: These engines use a spark to ignite a pre-mixed air-fuel mixture, typically found in gasoline engines.
- *Compression Ignition (CI) Engines*: These engines compress air to a high pressure, increasing its temperature, and then inject fuel into the hot air, commonly used in diesel engines.

ICEs, often referred to as gensets, are extensively used in distributed generation due to their reliability and flexibility. They serve as backup power supplies and can support the grid by providing base-load generation or peak-shaving. Their ability to start up quickly and adjust to changing loads makes them ideal for integration into local energy systems, ensuring stability and continuous power supply.

The key features of ICEs include:

- *Capacity*: Internal combustion engines have a wide range of capacities, from 0.5 kW to 10 MW, making them suitable for various scales of operation.
- *Efficiency*: ICEs typically achieve efficiencies between 25.
- *Start-up Time*: These engines can start up relatively quickly, within 0.5–15 min, and tolerate frequent start and stop cycles.

2.7 Micro Turbines

Incorporating micro turbines into energy generation offers a compact and efficient way to produce both electricity and heat. These turbines operate similarly to large gas turbines but on a smaller scale, making them suitable for decentralized energy production. The cut-view of a micro turbine is shown in Fig. 18. Micro turbines typically utilize an air compressor to draw in and compress air from the environment. This compressed air is then mixed with fuel in the combustion chamber where it ignites, producing high-temperature, high-pressure gas. This gas expands through the turbine, spinning the blades to generate mechanical energy, which is converted into electrical energy by the generator. An integral component of micro turbines is the recuperator, a type of heat exchanger that enhances efficiency by preheating incoming

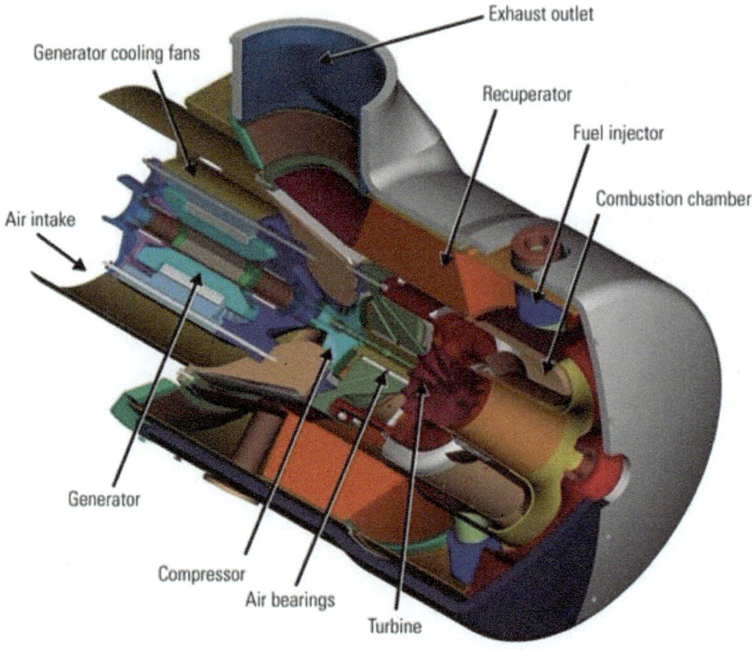

Fig. 18 Cut-view of a micro turbine [11]

air with exhaust heat, thus saving fuel and reducing operational costs. The flexibility in fuel choice–from natural gas to biogas–further enhances the applicability of micro turbines in various settings.

Micro turbines are known for their relatively high efficiency, especially when equipped with a recuperator, allowing energy efficiencies up to 30%. They can also be integrated into CHP systems to achieve even greater efficiencies, potentially up to 85%. This capability to recover and utilize waste heat makes micro turbines a viable option for boosting energy efficiency in industrial processes and reducing overall environmental impact. Moreover, micro turbines offer benefits like low emissions and noise levels due to their clean-burning and quiet operation. However, they also face challenges such as high initial costs and the need for precise maintenance and fuel purity. Their operational efficiency is influenced by rotor speed, which affects the frequency of the electricity generated, typically around 2 kHz. To integrate with conventional power grids, which operate 50 Hz, micro turbines require AC/DC/AC converters, adding complexity to their installation.

2.8 Stirling Engine

The Stirling engine operates on a closed cycle that consists of heating and cooling a gas (such as air or helium), which causes it to expand and contract. This expansion and contraction drive a piston, converting thermal energy into mechanical work. The engine's efficiency and quiet operation make it unique among heat engines. As shown in Fig. 19, key components and operation of the Stirling Engine are detailed as follows:

- *Displacer Piston*: This component moves the working fluid (gas) between hot and cold areas within the engine's cylinder. It's crucial for managing the heat flow without directly compressing or expanding the gas.
- *Power Piston*: This piston does the mechanical work. It converts the energy from the expanding gas into motion, typically linked to a flywheel or other mechanical load.
- *Regenerator*: Positioned between the hot and cold spaces, the regenerator is a temporary heat store that significantly boosts the engine's efficiency. It captures heat from the gas as it passes from the hot side to the cold side and then returns this heat to the gas as it moves back to the hot side.

Four typical cycles of a Stirling engine includes:

- *Heating and Expansion (Stroke 1–2)*: The working fluid is heated in the hot space, causing it to expand at constant temperature. This expansion drives the power piston and does work.
- *Displacement and Cooling (Stroke 2–3)*: The displacer piston moves the expanded gas to the cold space, where it is cooled at constant volume, causing its pressure to drop.

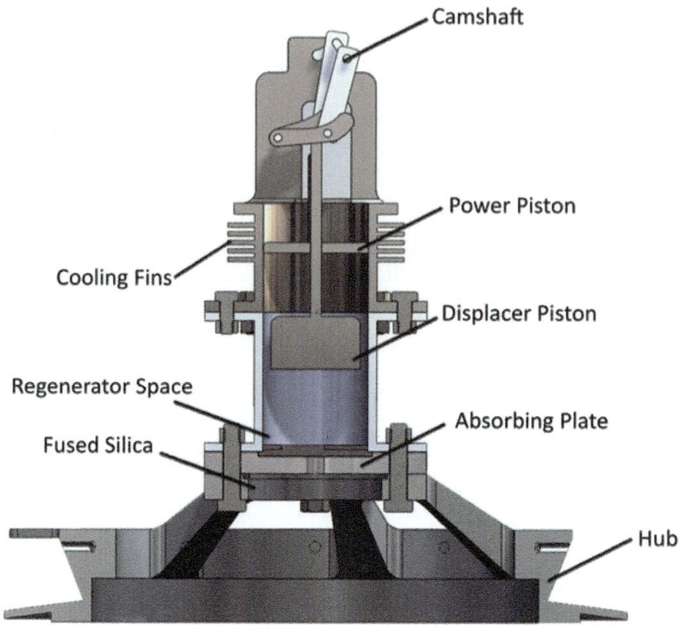

Fig. 19 Components of a Stirling engine [12]

- *Compression (Stroke 3–4)*: In the cold space, the gas is further cooled, contracting it and pulling the power piston inward, which prepares the gas for reheating.
- *Heat Reabsorption (Stroke 4–1)*: The cooled gas is moved back to the hot space by the displacer piston, passing through the regenerator which reheats the gas by transferring back the heat stored from the previous cycle.

While theoretical efficiencies of Stirling engines can approach 40%, practical efficiencies are lower, typically between 20 and 30%. This efficiency can be enhanced with the use of a regenerator. The applications of Stirling engines range from small generators and pumps to potential uses in solar power and waste heat recovery systems due to their ability to operate on a variety of heat sources.

2.9 *Distributed Generation Synchronisation and Connection*

In the context of DG, synchronization with the electrical grid is a critical requirement for ensuring stable and reliable power delivery. This process involves matching the phase, frequency, and voltage of the DG system with those of the grid before connection. Proper synchronization prevents transient disturbances, enhances power quality, and protects both the DG units and the grid infrastructure.

2.9.1 Synchronization Requirements

To achieve successful synchronization, several key parameters must be precisely controlled:

- *Frequency Alignment*: The frequency of the DG system should closely match the grid frequency. A minor deviation is permissible, typically around 0.33 Hz [13] higher than the grid frequency. This slight excess ensures that the DG unit is slightly ahead of the grid, facilitating a smoother connection and reducing the likelihood of oscillations.
- *Voltage Magnitude Consistency*: The voltage of the DG system must align with the grid voltage. Ideally, the terminal voltage should be a few percentage points higher than the grid voltage. This higher voltage facilitates the initial power flow from the DG unit to the grid, helping to stabilize the connection.
- *Phase Sequence Uniformity*: The phase sequence of the DG system must be identical to that of the grid. Any discrepancy in phase sequence can lead to severe power quality issues, including unbalanced currents and potential damage to the equipment.
- *Phase Sequence Uniformity*: The phase sequence of the DG system must be identical to that of the grid. Any discrepancy in phase sequence can lead to severe power quality issues, including unbalanced currents and potential damage to the equipment.
- *Phase Angle Coordination*: The phase angle difference between the DG unit and the grid must be tightly controlled, typically within a 5° range [14]. Proper phase angle coordination ensures that the power flows in the correct direction and minimizes the risk of circulating currents that can cause losses and heating in the system.

Proper synchronization of DG systems with the grid is essential for maintaining power quality, stability, and reliability. By ensuring frequency alignment, voltage consistency, phase sequence uniformity, and precise phase angle coordination, DG systems can seamlessly integrate with the grid. Advanced inverters and synchronization circuit breakers are critical components that facilitate this process, providing the necessary control and protection mechanisms to achieve successful synchronization.

2.9.2 Shallow and Deep Connections

The DG integration is often characterized by the concepts of deep and shallow connections, which primarily focus on the distribution of costs associated with grid connection and infrastructure enhancements.

Shallow connection refers to a scenario where the DG owner is responsible only for the direct costs associated with connecting their generation facility to the grid. This typically includes the cost of connection equipment and minor grid upgrades necessary to accommodate the new source of power. The main advantage of shallow connections is that they lower the barrier for smaller or individual renewable energy projects by minimizing upfront expenses, thereby encouraging the proliferation of distributed renewable energy resources. Costs beyond the immediate connection are usually absorbed by the utility or distributed among all grid users. This method can promote a quicker expansion of DG installations as it minimizes the financial burden on individual generators. In contrast, a deep connection approach places a greater financial burden on the DG owner. Here, the owner must cover not only the direct costs of connection but also more substantial grid upgrades and enhancements that may be required far from the actual point of connection. This can include upgrading transmission lines or enhancing grid stability measures to handle the new input of power, which can be particularly costly if the DG is located far from existing grid infrastructure or in areas with older infrastructure. The choice between deep and shallow connections can significantly influence the economic feasibility of renewable energy projects. Shallow connections lower the cost of entry, potentially increasing the number of projects developed. Deep connections, while more costly, can incentivize strategic placement of new generation facilities and ensure that the necessary infrastructure upgrades are made to support the long-term integration of renewable energy.

The regulatory framework surrounding connection charges can also vary by region, influencing the decision-making process for potential developers. For example, some regions may offer incentives or subsidies for certain types of connections or specific technologies, altering the financial dynamics between shallow and deep connection strategies.

2.9.3 Connection Charge and Voltage Considerations

Connection charges are employed to recover the investment costs for grid integration. These fees are structured to encourage or discourage certain behaviors:

- *Low Voltage Connections*: Generally preferred by distributed generation developers, especially for installations up to 500 kW, due to lower costs and less complex integration processes.
- *High Voltage Connections*: Distribution network operators favor these for larger installations (above 500 kW) to ensure reliability and manage the network efficiently.

2.9.4 Network Impacts and Voltage Regulation

Integrating DGs impacts the electrical grid in several ways:

- *Voltage Variations*: DG can cause voltage fluctuations depending on their output and the existing load on the grid. During low demand periods, these variations can exceed safe operational limits. To manage voltage levels and maintain grid stability, several strategies are employed: *(a) On Load Tap Changing (OLTC) Transformers*: These adjust their voltage output in response to changing load conditions without

interrupting the supply. *(b) Line Drop Compensation (LDC)*: Adjusts the voltage based on current flow changes and line impedance, stabilizing the voltage across the network. *(c) Induction and Synchronous Generators*: These can either supply or absorb reactive power, thus influencing the voltage levels along the distribution network.
- *Power Quality and Protection*: DG can introduce harmonics and flicker due to intermittent production patterns. Proper protection schemes are crucial to safeguard both the generation equipment and the network. Additionally, maintaining power quality is essential to prevent damage to sensitive equipment and ensure compliance with regulatory standards.
- *Stability and Fault Levels*: DG contributions to fault currents can alter the network's fault levels, potentially requiring upgrades in protection equipment like switchgears. Strategies such as employing fault current limiters are necessary to manage these increases in fault levels without costly equipment upgrades.

The integration of DG into the electrical grid presents a complex array of technical, financial, and regulatory challenges. Shallow and deep connection policies, along with strategic voltage regulation and protection mechanisms, are essential for ensuring that DG contributes positively to grid stability and reliability. As distributed energy resources continue to proliferate, adapting grid management practices and regulatory frameworks will be critical to harnessing their full potential while maintaining the security and efficiency of the power system.

3 Energy Storage

Energy storage systems (ESS) play a pivotal role in modern energy infrastructure, offering a multitude of benefits that enhance the efficiency, reliability, and flexibility of power grids and local energy systems. These systems store energy in various forms–electrical, chemical, mechanical, or thermal–and release it when needed, thereby balancing supply and demand, stabilizing the grid, and increasing the integration of renewable energy sources. In the first section, we will talk about energy storage, from key functions, characters, to every single storage technology. We will also cover on how to compare and select these technologies in the real world applications.

3.1 Key Functions of Energy Storage

This part will introduce key functions of energy storage systems from the national wide power grids, local energy systems, all the way through to individual energy consumers. Then, we will go through some key characterization of energy storage technologies. These characters can help us compare and select storage technologies for certain applications.

3.1.1 Load Levelling

Load leveling, also known as energy arbitrage, is a fundamental function of energy storage systems designed to manage the supply and demand of energy over time efficiently. This process involves storing electrical energy during periods of low demand (when energy is cheaper) and releasing it during periods of high demand (when energy is more expensive). The primary goal of load leveling is to enhance grid stability, reduce energy costs, and increase the economic efficiency of power systems. The concept of energy arbitrage, a key driver for load leveling, involves taking advantage of the price variations in energy markets. As indicated in Fig. 20, during off-peak hours, energy prices typically decrease due to lower consumption rates. Energy storage systems capitalize on these lower prices by storing excess energy generated. Conversely, during peak demand times, when energy prices are high due to increased consumption, stored energy is released. This not only provides a financial benefit by capitalizing on the price differential but also aids in balancing the grid by supplementing the higher demand without the need for additional generation capacity.

Load leveling through energy storage represents a critical solution to the challenges of modern energy systems, offering both economic and operational benefits. By smoothing out the discrepancies between supply and demand, energy storage not only facilitates more stable grid operations but also plays a pivotal role in making energy systems more sustainable and cost-effective.

3.1.2 Renewable Energy Integration

Energy storage systems play a pivotal role in addressing two of the main challenges associated with renewable energy sources such as wind and solar: their intermittency and the misalignment of their output with actual energy demand. These challenges stem primarily from the nature of renewable resources, which are inherently dependent on environmental conditions–solar intensity and wind availability–which do not always coincide with peak energy demand periods.

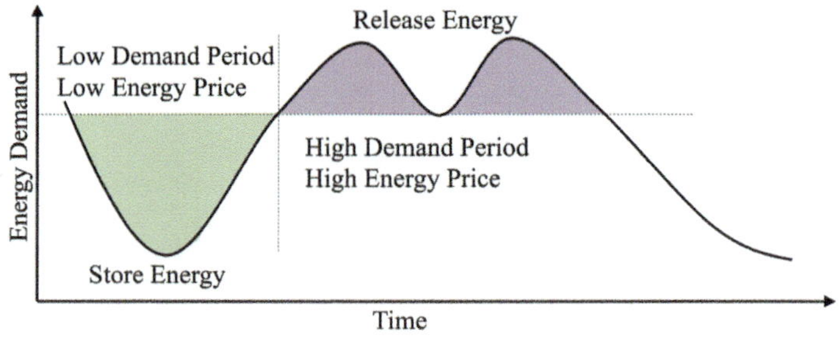

Fig. 20 Schematic illustration of energy arbitrage through using peak and off-peak pricing signals

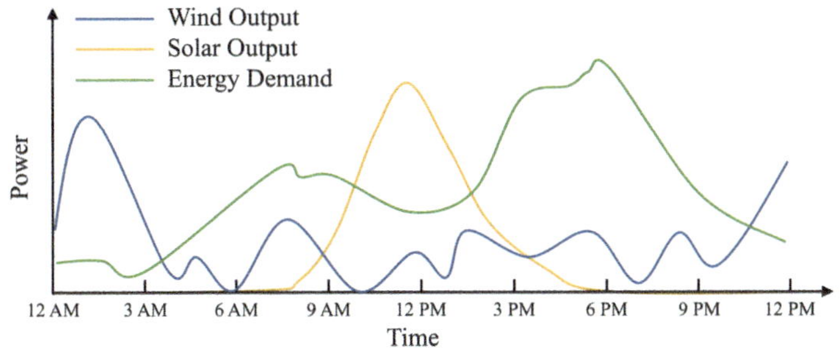

Fig. 21 Intermittency of wind and solar outputs and the misalignment with actual energy demand

Renewable energy sources are not constant; their production fluctuates based on day-to-day and seasonal changes in weather conditions. This variability presents significant challenges for energy grid managers who must ensure a constant balance between supply and demand to maintain grid stability. Energy storage technologies can mitigate these fluctuations by storing excess energy produced during peak production times and then releasing this energy during periods of low production or high demand.

Figure 21 illustrates a typical scenario where solar energy peaks at midday, which may not align with peak energy consumption, typically in the evening. Similarly, wind energy might peak at night when demand is low. By using energy storage, energy managers can 'shift' this energy to times of higher demand. This capability not only ensures a more reliable supply but also maximizes the economic value of the generated renewable energy by selling it at peak price times.

The technical implementation of energy storage for renewable integration can be summarised as follows:

- *Smoothing*: Energy storage acts as a buffer to smooth out the output of renewable energy sources. By continuously charging and discharging, storage systems can dampen the abrupt changes in energy generation caused by varying wind speeds or solar insolation, leading to a more stable energy output.
- *Dispatching*: With advanced predictive controls, energy storage systems can release stored power based on anticipated demand and generation conditions. This 'dispatchable' characteristic of energy storage is crucial for integrating renewables, as it provides flexibility in when and how energy is released back into the grid.
- *Increasing Grid Resilience*: By providing a reliable alternative source of energy during periods when renewable generation is low, energy storage systems help maintain grid resilience and prevent outages, which are more likely when relying on variable energy sources alone.

Energy storage promotes a higher penetration of renewable energy into the grid by making renewable sources more reliable and economically viable. This transition supports the broader goals of reducing reliance on fossil fuels, decreasing greenhouse gas emissions, and promoting sustainable energy landscapes.

3.1.3 Peak Shaving

Peak shaving is a crucial strategy used in energy management to reduce the demand on the electrical grid during peak usage times. This practice not only helps in stabilizing the grid but also reduces dependence on less efficient and more expensive peak power plants. Traditionally, peak demands have been managed by utilizing fast-ramping power plants that operate on fossil fuels or natural gas. However, these plants, while effective in meeting immediate energy needs, are often costly and contribute significantly to environmental pollution.

Large capacity energy storage devices offer a more sustainable and efficient solution for managing peak demand. By storing energy during periods of low demand and releasing it during peak periods, energy storage systems help to flatten out the demand curve, thus reducing the need for peak power plants. This capability not only ensures a more efficient use of energy but also contributes to significant cost savings and environmental benefits. Energy storage technologies such as batteries, pumped hydro storage, and compressed air energy storage are employed to absorb excess energy generated during off-peak hours. This stored energy is then discharged back into the grid when the demand rises, thereby "shaving" the peaks of energy usage. This process is crucial for maintaining grid stability and ensuring continuous power supply without the need to resort to quick-start power generation options.

The economic and environmental advantages of peak shaving can be summarised as follows:

- *Cost-Effectiveness*: Operating peak power plants is inherently more expensive due to higher fuel costs and the inefficiencies associated with rapid cycling. Energy storage systems, once installed, have lower operational costs and can offer a quicker return on investment through energy arbitrage.
- *Efficiency Enhancements*: Peak power plants typically operate below their maximum efficiency due to their sporadic and short-term usage. Energy storage systems can release energy consistently at optimal efficiency, thereby enhancing the overall efficiency of the energy system.
- *Reduction in Emissions*: By reducing the reliance on fossil-fuel-based peak power plants, energy storage systems contribute significantly to the reduction of greenhouse gas emissions and other pollutants. This transition is vital for meeting global climate targets and reducing environmental impact.
- *Enhanced Grid Reliability*: Energy storage provides a more reliable energy supply during peak times without the risk of outages or the need for frequent maintenance, which is common with traditional peaker plants.

While the benefits of peak shaving through energy storage are clear, several challenges need to be addressed for its broader adoption:

- *High Initial Investment*: The upfront cost of large-scale energy storage systems can be prohibitive, although this is gradually changing with technological advances and economies of scale.
- *Regulatory and Market Barriers*: In many regions, the energy markets and regulatory frameworks are not fully adapted to integrate energy storage solutions effectively. Policies that recognize and financially incentivize the benefits of energy storage are needed to support its adoption.
- *Technology and Capacity Limitations*: The capacity and efficiency of energy storage technologies are continually improving but still need to match the requirements for handling peak demands in larger grids.

Peak shaving through energy storage is a transformative approach that aligns with the global shift towards more sustainable and efficient energy systems. By integrating advanced storage solutions, energy systems can become less reliant on inefficient fossil-fuel plants, leading to a cleaner, more economical, and resilient energy landscape. As technology progresses and costs decline, the adoption of energy storage for peak shaving is expected to accelerate, marking a significant step forward in modern energy management practices.

3.1.4 Grid Decentralisation

Grid decentralization represents a transformative shift from traditional centralized power systems to a more distributed energy network. In conventional power systems, electricity is generated in large-scale, centralized facilities such as coal, nuclear, or hydroelectric power plants. This electricity is then transmitted over long distances and distributed to end-users across extensive networks. While effective, this system often leads to inefficiencies including significant energy losses during transmission and distribution, and heightened vulnerability to system failures or disruptions.

The modern approach to energy management increasingly favors decentralized systems, characterized by the widespread deployment of distributed energy resources such as solar panels, wind turbines, and energy storage systems. These resources are typically installed at or near the point of consumption–be it in residential, commercial, or industrial settings.

Energy storage is pivotal in the decentralization of the grid for several reasons:

- *Enabling Dispatchability of Localised Renewable Resources*: As renewables such as solar and wind are intermittent by nature, energy storage solutions provide a way to store surplus energy generated during peak production times. This stored energy can later be dispatched when production drops or demand increases, ensuring a consistent and reliable power supply.
- *Facilitation of Microgrids and Virtual Power Plants*: Microgrids are localized grid systems that can operate independently or in conjunction with the main power grid. Energy storage systems within microgrids help balance load and generation, enhancing resilience and allowing for more flexible energy management. Virtual

Power Plants (VPPs) integrate several types of energy producers and storage systems to operate as a single power-generating unit. Energy storage is crucial in both scenarios to provide real-time energy balancing and grid stabilization.
- *Reduction of Centralized Grid Dependency*: By allowing local generation and storage of energy, decentralized systems reduce dependency on large power plants and the central grid. This shift not only enhances energy security for end-users but also reduces transmission losses and lowers the environmental impact associated with large-scale energy production.

The typical scale for storage in decentralized systems ranges from small batteries in residential applications to larger systems used in industrial contexts, generally measured in kilowatt-hours (kWh). These systems must be capable of quick energy release to effectively respond to rapid fluctuations in demand and generation, a capability that is crucial for maintaining grid stability and ensuring efficient operation.

While the potential for energy storage in decentralized grids is vast, several challenges remain:

- *Economic Barriers*: High initial costs for storage technologies can be a barrier for widespread adoption.
- *Regulatory Issues*: The integration of DERs and storage solutions often faces regulatory hurdles that can impede their effective deployment.
- *Technical Limitations*: The technological development of energy storage must continue to advance to meet increasing demands for capacity, efficiency, and reliability.

As the energy landscape evolves towards more sustainable and efficient systems, the role of energy storage in enabling grid decentralization becomes increasingly critical. By facilitating the integration of renewable energy sources, enhancing grid resilience, and reducing dependency on centralized power systems, energy storage stands at the forefront of the transition to a decentralized energy future.

3.1.5 Power Quality

In modern electrical grids, maintaining high power quality is essential for the stability and efficiency of energy systems. Responsive energy storage systems play a critical role in achieving this by managing rapid fluctuations in power demand and generation. These systems are uniquely capable of immediate adjustments in power output, crucial during events such as switching operations, short outages, or the startup phases of power plants.

Responsive energy storage serves multiple functions within the power grid, primarily focusing on frequency regulation and voltage support–two critical aspects of power quality.

Fig. 22 Relationship between frequency and supply-demand balance. If the generation exceeds demand, the frequency rises 50 Hz, and vice versa

For frequency regulation:

- *Operational Frequency Stability*: The standard frequency for grid operation in Great Britain 50 Hz. Maintaining this frequency ensures the stability and efficiency of the grid.
- *Supply-Demand Balance*: Energy storage can quickly balance generation and consumption discrepancies. As shown in Fig. 22, if the generation exceeds demand, the frequency rises 50 Hz, and vice versa, storage systems can either charge (absorb power) or discharge (release power) to correct these imbalances.
- *Rapid Response*: The capability to instantaneously adjust power output allows these systems to swiftly address frequency deviations, thereby stabilizing the grid within seconds.

For voltage support:

- *Voltage Level Management*: Energy storage systems also help in maintaining voltage within safe and efficient operational limits across the grid.
- *Reactive Power Control*: Through the use of inverters, which convert DC power from storage to AC power for the grid, energy storage systems can manage the phase angle between current and voltage. This capability allows them to either absorb or inject reactive power, depending on the grid's needs, which helps in voltage stabilization.

The integration of advanced inverter technology is crucial for the effective operation of energy storage systems in providing reactive power support. Inverters adjust the phase angle between current and voltage to effectively manage power flows, which can be expressed in the following equations:

$$P = VI\cos(\phi), \tag{6}$$

$$Q = VI\sin(\phi). \tag{7}$$

These equations represent the fundamental relationship between voltage (V), current (I), and the phase angle (ϕ), determining how energy storage systems can influence both active and reactive power flows to support voltage levels.

Responsive energy storage systems are pivotal in modern energy infrastructure, addressing key challenges in power quality management through frequency regulation and voltage support. Their ability to rapidly respond to grid dynamics makes them invaluable for ensuring grid stability, integrating renewable energy, and supporting decentralized energy systems. The continued advancement in storage technology and inverter functionality will further enhance their role in meeting future energy demands and sustainability goals.

3.1.6 Automotive Applications

Energy storage systems, particularly in the form of lithium-ion batteries, are revolutionizing the automotive industry by powering electric vehicles (EVs) and enhancing hybrid electric vehicles (HEVs). This shift towards electric mobility is a crucial part of the broader transition from traditional combustion engines to more sustainable, efficient, and environmentally friendly transportation options.

Energy storage plays multiple roles in the automotive sector, each contributing significantly to advancements in vehicle technology and grid interaction. In EVs, lithium-ion batteries provide the primary source of power for electric motors, replacing the internal combustion engines of traditional vehicles. These vehicles offer a quieter, smoother ride and, most importantly, produce zero direct emissions, contributing to a significant reduction in urban air pollution. HEVs combine energy storage systems with conventional engines. The battery in an HEV is typically used for low-speed driving conditions, while the combustion engine is used for higher speeds and longer distances. By utilizing energy storage, HEVs can significantly reduce fuel consumption and emissions compared to their non-hybrid counterparts. Energy storage systems enable the engine to operate at optimal efficiency, reducing wasteful energy expenditure. Both EVs and HEVs feature regenerative braking systems that capture kinetic energy during braking and convert it into electrical energy, which is then stored in the vehicle's battery. This recovered energy is later used to power the vehicle, enhancing overall efficiency and extending the driving range without additional energy consumption.

Vehicle-to-Grid (V2G) technology allows EVs and HEVs to connect to the power grid to feed stored electrical energy back into the grid during peak demand times. Vehicles act as mobile energy reservoirs that can help stabilize the grid by providing ancillary services such as frequency regulation and voltage support. Participating in V2G programs can offer financial incentives to vehicle owners. By selling back energy to the grid, they can offset the cost of charging the vehicle, potentially turning an under-utilized asset into a revenue-generating one.

The integration of energy storage into the automotive sector is poised to transform transportation infrastructure and energy systems. As battery technology advances, the cost of energy storage is expected to continue to decline, making electric and hybrid vehicles increasingly accessible. Moreover, as more renewable energy sources are integrated into the grid, the role of EVs and HEVs in energy management and grid stability will become more critical.

3.2 Characterization of Energy Storage

This section delves into the core characteristics of energy storage systems, highlighting their critical role in the transition towards sustainable energy systems. Energy storage systems vary widely in type, capacity, and performance, each characterized by several key parameters that define their suitability for specific applications.

3.2.1 State of Charge (SoC) and Depth of Discharge (DoD)

The State of Charge is an indicator of the remaining energy in a storage device, represented as a percentage of its total capacity. As indicated in Fig. 23, a fully charged battery, with a SoC of 100%, is primed for maximum output, whereas a fully depleted battery shows a SoC of 0%. Complementing SoC, the Depth of Discharge describes the fraction of energy that has been extracted from the battery, again in percentage terms. Understanding these metrics helps in optimizing the operational strategy for energy storage, particularly in prolonging battery life and maximizing efficiency.

3.2.2 Self-discharge Time

Even when disconnected from a load or external circuit, batteries naturally lose some of their stored charge over time, a phenomenon known as self-discharge. The rate of self-discharge is crucial for applications where the battery may go long periods without use. For instance, as shown in Fig. 24, nickel-based batteries might lose 10–15% of their capacity in the first 24 h and more over an extended period, influencing the storage system's design and maintenance schedules.

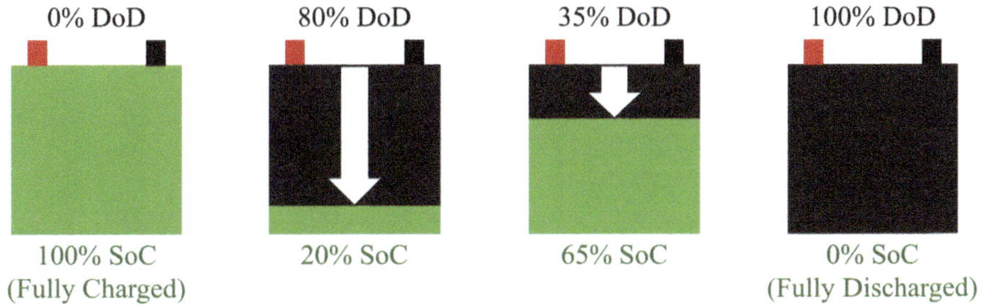

Fig. 23 Schematic illustration of state of charge (SoC) and depth of discharge for a battery at different states

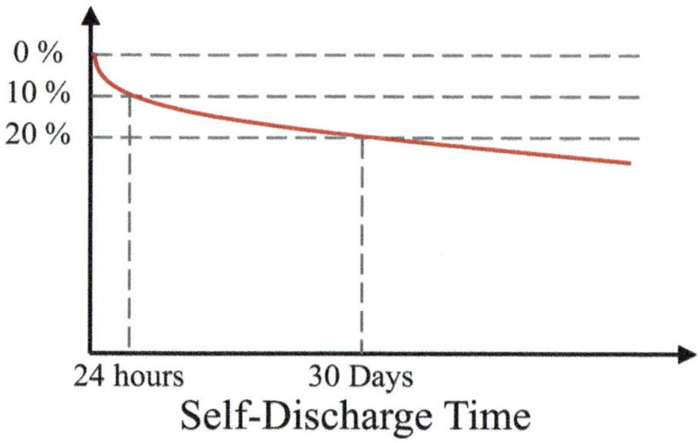

Fig. 24 Example of self-discharge time for nickel-based batteries

3.2.3 Efficiency

Efficiency is pivotal in determining the economic viability of energy storage. It is defined by the ratio of the energy delivered by the battery to the energy used to charge it. As an example in Fig. 25, an efficiency of 90% means that 10% of the energy is lost during the charge and discharge processes, typically due to heat generation, resistance, and other inefficiencies. High-efficiency levels are crucial for applications where energy costs and conservation are paramount.

3.2.4 Cycle Life

The cycle life of a battery indicates its durability and how many complete charge and discharge cycles it can undergo before its capacity degrades to a specific threshold, typically around 80% of its original capacity. The cycle life is affected by the depth of discharge; deeper discharges usually result in shorter battery lifespans. Manufacturers

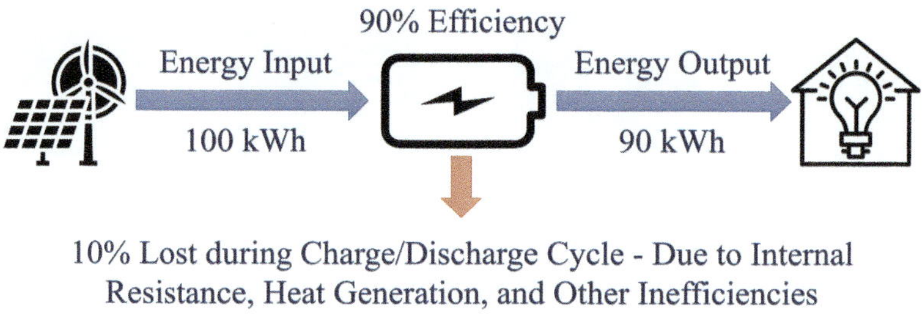

Fig. 25 Energy flow through a battery from input to output, including energy losses during charging and discharging cycles

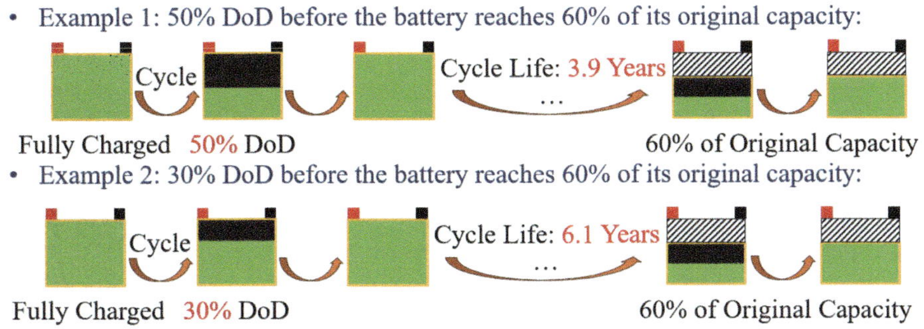

Fig. 26 Comparison of the cycle life of a battery under two different depth of discharge

often specify the expected cycle life at various depths of discharge to guide users in managing batteries for extended service life.

Figure 26 shows a comparison of two examples. If in every cycle, the battery is discharged to 50% of DoD, the cycle life will be 3.9 years until the battery gets to 60% of original capacity. By contrast, if in every cycle, the battery is discharged to 30% of DoD, the cycle life will be longer as 6.1 years until the battery gets to the same 60% of original capacity.

3.2.5 Specific Energy and Energy Density

These characteristics determine the usability of energy storage in portable and stationary applications. Specific energy, expressed in watt-hours per kilogram, measures the amount of energy stored per unit mass. Higher specific energy is advantageous for mobile applications such as electric vehicles and portable electronics where weight is a constraint. Energy density, measured in watt-hours per liter, indicates how much energy a storage system stores per unit volume, influencing how compact the storage solution can be. High energy density is critical for minimizing the space occupied by the storage system in densely packed environments.

3.3 Energy Storage Technologies

In this section, we will explore a diverse array of energy storage technologies, each distinguished by the form of energy it retains. For instance, batteries convert and store energy chemically, while capacitors directly store electrical energy. Pumped hydro systems capitalize on gravitational potential energy, flywheels maintain energy in the form of kinetic motion, and superconducting magnetic energy storage systems harness electromagnetic energy. Each technology offers unique advantages and challenges, reflecting its mechanism of energy conservation and release.

3.3.1 Battery

The first storage technology we will discuss in this chapter is batteries, fundamental devices that store and release electrical energy by capitalizing on the electrochemical potential of metals. This potential is essentially a propensity of metals to lose electrons and participate in electrochemical reactions. Understanding the operation of batteries requires delving into their fundamental principles and distinguishing between the types and efficiencies of different battery systems.

Batteries operate on the principle of electrochemical potential, a measure of the tendency of a metal within the battery to lose electrons. When two metals with different electrochemical potentials are used within a battery, it leads to an electrochemical reaction that creates an external flow of electrons, commonly known as electricity. This section explores the electrochemical potentials of various metals, noting that metals like lithium (Li) have a higher potential, making them more efficient at electron donation and thus excellent candidates for battery materials.

Batteries are broadly categorized into two types based on the reversibility of their electrochemical reactions:

- *Primary Batteries (Single-Use)*: These batteries operate on a non-reversible electrochemical reaction. Once the chemical substances within the battery are exhausted, the battery ceases to produce electricity and cannot be recharged. These are typically used in applications requiring high energy density and reliability over short durations, such as in remote controls, smoke detectors, and emergency devices.
- *Secondary Batteries (Rechargeable)*: Unlike primary batteries, secondary batteries are designed with reversible electrochemical reactions. This allows them to be recharged and reused multiple times. Recharging is achieved by applying a direct current that reverses the electrochemical reaction that occurs during discharge. Common examples include lithium-ion batteries used in smartphones, laptops, and electric vehicles.

The performance of batteries can also be measured in terms of efficiency, primarily categorized into one-way and round-trip efficiencies:

- *One-Way Efficiency*: This metric refers to the efficiency of a single operational phase–either charging or discharging. For instance, during the charging phase, it measures how much of the electrical energy used to charge the battery is actually stored as chemical energy. Typically, the one-way efficiency of batteries is between 85 and 90% [15].
- *Round-Trip Efficiency*: This is a critical measure for rechargeable batteries, representing the efficiency over a complete discharge and recharge cycle. It assesses the percentage of energy put into the battery during charging that can be retrieved during discharging. The round-trip efficiency for most batteries ranges from 70 to 80% [16].

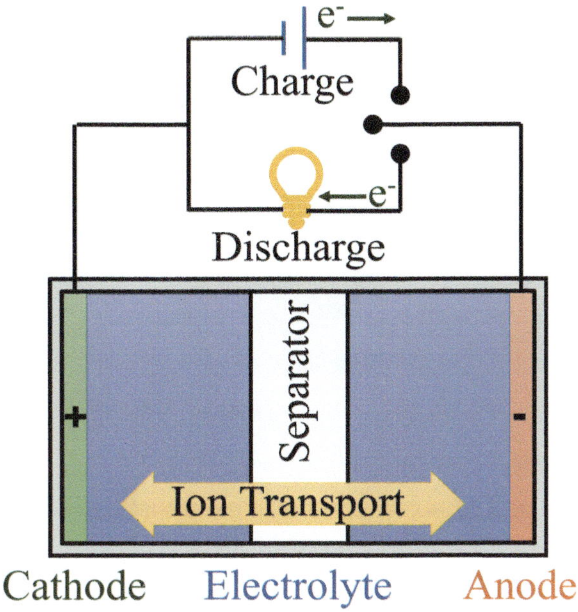

Fig. 27 Typical structure of a batter cell

As shown in Fig. 27, a typical battery consists of several electrochemical cells, each featuring an anode, cathode, and an electrolyte that facilitates ion transport. During discharge, the anode undergoes oxidation, losing electrons, while the cathode undergoes reduction, gaining electrons. This flow of electrons generates electrical energy used to power devices.

Batteries are characterized by various ratings that define their capacity and operational efficiency:

- *Capacity Rating (Ampere-Hours, Ah)*: Indicates the total charge a battery can store.
- *Voltage Rating*: Represents the electrical potential difference the battery can provide. For instance, lithium-ion batteries typically offer 3.7 V, whereas car batteries provide 12 V.
- *C-Rate (Ampere)*: Describes the rate at which a battery can be charged or discharged relative to its maximum capacity. For example, a discharge rate of C/2 for a 100 Ah battery means it would be fully depleted in 2 h if discharged at 50 amps.
- *Cycle Life and Energy Density*: These parameters indicate the longevity and energy storage efficiency of the battery.
- *Temperature Range*: Specifies the operational climate range of the battery, highlighting its robustness in different environmental conditions.

First, lead Acid (Pb-Acid) batteries, one of the most ubiquitous types of batteries, are highlighted for their distinct chemical properties and reactions which are fundamental in understanding basic battery technology. As illustrated in Fig. 28, these batteries utilize an electrolyte composed of one-third sulfuric acid and two-thirds

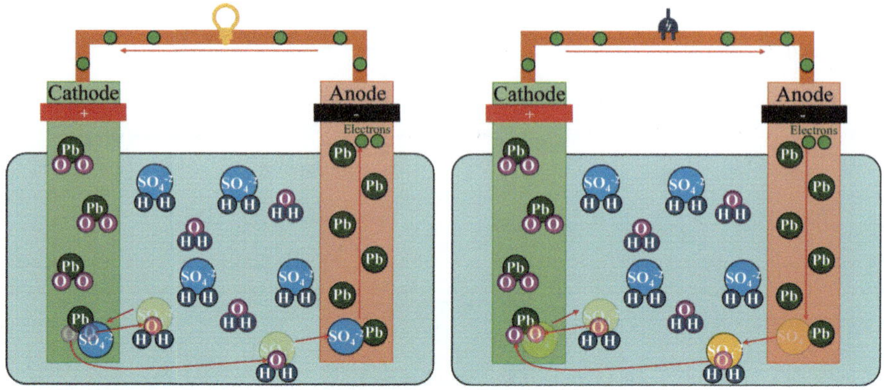

Fig. 28 Comparison of discharge and recharge of a Lead Acid (Pb-Acid) battery

water, with a cathode of lead oxide and an anode of pure lead. At the cathode, lead oxide reacts with sulfate ions from the electrolyte, forming lead sulfate. This reaction also liberates oxygen ions which then combine with hydrogen ions within the electrolyte to produce water. Concurrently, the anode's lead reacts with sulfate ions to form more lead sulfate while releasing electrons that move towards the positive terminal, completing the circuit externally. As the battery discharges, the sulfuric acid in the electrolyte becomes diluted and the accumulation of lead sulfate on both electrodes makes them chemically similar, thereby impeding further reactions. Recharging the battery involves reversing the discharge reactions. Providing an external power source allows electrons to re-enter the anode, breaking down the lead sulfate back into lead and sulfate ions. These sulfate ions return to the electrolyte and react to regenerate the electrolyte's acidic strength and recreate lead oxide at the cathode. Understanding these reactions is critical for effectively using and maintaining lead acid batteries. The reversible nature of these chemical processes allows for multiple cycles of discharge and recharge, making lead acid batteries suitable for a variety of applications, from automotive to energy storage systems. However, the efficiency and life span of these batteries can be significantly affected by the depth of discharge and the maintenance of proper charging cycles.

Second, lithium-ion (Li-ion) batteries are pivotal in modern technology, powering devices from smartphones to electric vehicles and laptops. The cathode of the Li-ion is composed of lithium metal oxide, which provides stability due to its robust chemical structure. The anode is made from graphite, characterized by a layered structure that efficiently stores lithium ions due to its loosely bonded layers. The interface between the lithium metal oxide cathode and graphite anode involves a liquid electrolyte that facilitates ion transfer while creating a passivating film. This film is crucial as it allows lithium ions to pass through while blocking electrons, ensuring ions move through the electrolyte and electrons travel through the external circuit only. As shown in Fig. 29, upon connecting to a power source, electrons are drawn away from the lithium metal oxide, moving through the external circuit towards the graphite anode. Concurrently,

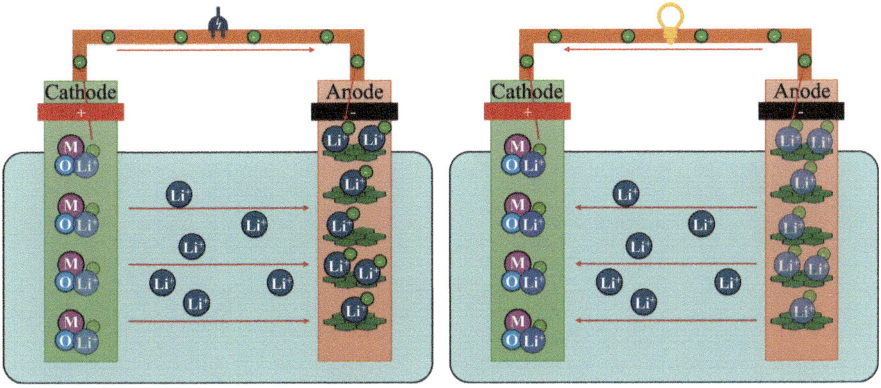

Fig. 29 Comparison of discharge and recharge of a lithium-ion (Li-ion) battery

lithium ions, now positively charged, migrate through the electrolyte towards the anode. During the discharging process, the stored energy is released as lithium ions move back towards the cathode, driving the electrons through the external circuit to power the connected device. Li-ion batteries boast high energy density, which allows for a smaller battery size while providing the same energy as larger batteries made from other materials. They maintain a relatively constant voltage throughout the discharge cycle, enhancing device performance. However, these batteries require sophisticated control mechanisms to prevent damage from overcharging, which can lead to overheating and potential hazards. The battery's chemistry and structure necessitate integrated protective measures to ensure safety and longevity.

Furthermore, lithium polymer (Li-Poly) batteries represent an innovative advancement in battery technology, particularly in the realm of portable electronics. This section will delve into the structure, functionality, and advantages of Li-Poly batteries to provide a detailed understanding appropriate for academic and technological contexts. Li-Poly batteries utilize a solid polymer electrolyte instead of the traditional liquid electrolyte found in Lithium-Ion batteries. This solid state not only simplifies the battery's structure by eliminating the need for a separate separator but also enhances the battery's overall energy capacity and stability. The solid polymer acts both as an electrolyte and a separator, which significantly contributes to the battery's higher specific energy. This means that Li-Poly batteries can store more energy for their size compared to many other battery types. One of the standout features of Li-Poly batteries is their safety profile. The solid polymer electrolyte minimizes the risk of leakage and is more resistant to puncture and other physical damages, which reduces the risk of fire and chemical spills. Li-Poly batteries are known for their low self-discharge rate, which makes them particularly suitable for applications where long-term reliability and minimal maintenance are crucial, such as in medical devices and safety equipment.

Next, Nickel Cadmium (Ni–Cd) batteries incorporate a design where the cathode is made of nickel hydroxide and the anode from cadmium, separated by a nylon

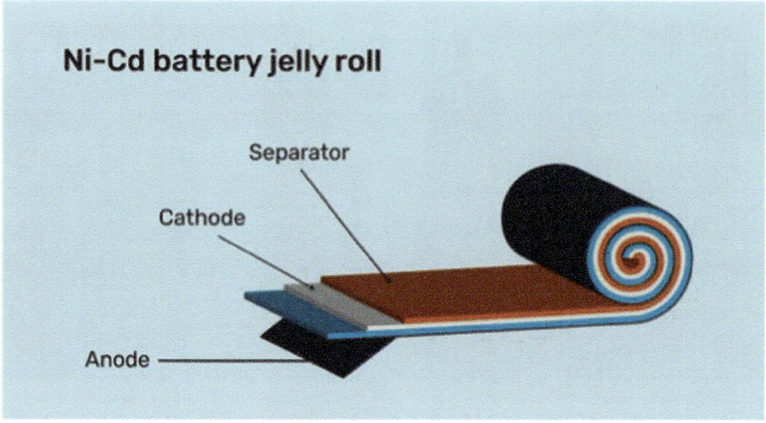

Fig. 30 roll structure of electrodes and separator of a Nickel Cadmium (Ni–Cd) battery [17]

material that prevents direct contact while allowing ionic flow. The electrodes and separator are tightly rolled into a "jelly roll" structure as shown in Fig. 30, which conserves space and increases efficiency. These batteries are prized for their ability to be recharged multiple times, maintaining a robust cycle life, and their durability across a wide range of temperatures. However, Ni–Cd batteries suffer from the memory effect, where they may only recharge to the frequent discharge point, potentially losing maximum capacity. Additionally, the use of cadmium, a toxic heavy metal, raises environmental and health concerns, requiring careful disposal and recycling protocols to mitigate negative impacts. This balance of high functionality with significant environmental considerations presents both opportunities and challenges for the use of Ni–Cd batteries in various applications.

The memory effect in batteries, particularly notable in Nickel–Cadmium (Ni–Cd) types, occurs when a battery "remembers" the discharge level at which it is frequently recharged, leading to decreased capacity. Initially, a new battery holds a full charge and functions optimally. However, if frequently recharged after using only a partial capacity–say, recharging after only 80% as illustrated in Fig. 31, discharge–crystalline formations may develop within the battery's structure. These crystals occupy space within the battery, effectively reducing the volume available for storing charge and thus diminishing the battery's total capacity. To mitigate the memory effect, it is advisable to deplete the battery to a low charge level before recharging. Moreover, removing the battery from the charger once it is fully charged helps prevent this condition by ensuring that the battery cycles through its full capacity, thereby avoiding the formation of memory-impairing crystals. This approach not only prolongs the battery's effective life but also maintains its performance closer to its original state.

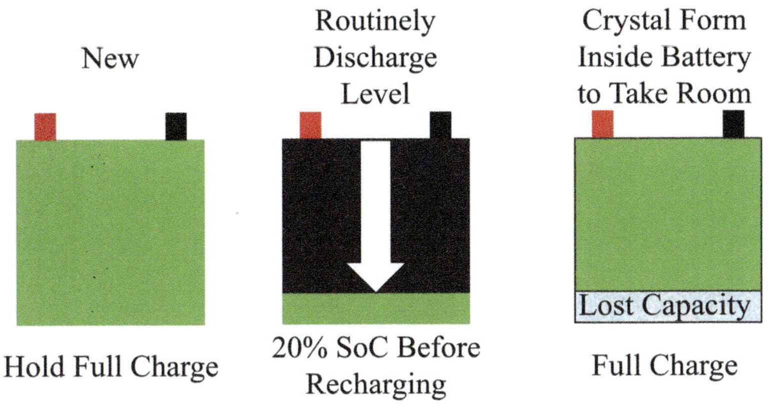

Fig. 31 Memory effect of a battery when frequently recharged after using 80% depth of discharge

Nickel Metal Hydride (Ni–MH) batteries represent a significant advancement over Nickel Cadmium (Ni–Cd) batteries, primarily addressing environmental and performance concerns. The substitution of cadmium with a metal hydride alleviates environmental risks, while also diminishing the memory effect and enhancing the energy density. Despite these improvements, Ni–MH batteries are more costly, exhibit higher self-discharge rates, and are more vulnerable to damage from overcharging compared to their Ni–Cd counterparts. Predominantly used in electric vehicles, their potential for cost-effectiveness increases with anticipated advancements in large-scale production, which could make them a more viable option in future energy storage solutions.

The last type of battery we will introduce is the Zinc-air batteries. Zinc-air batteries distinguish themselves with a unique design and chemistry. The cathode, constructed from a porous carbon material coated with a catalyst, facilitates air passage, essential for its operation. The anode consists of zinc, and potassium hydroxide serves as the electrolyte. During discharge, zinc's oxidation at the anode releases electrons that travel through an external circuit to the cathode where they reduce oxygen from the air. The key advantages of zinc-air batteries include exceptionally high energy density and cost-effectiveness, attributes derived from the abundant and environmentally benign nature of zinc, coupled with the atmospheric availability of oxygen. However, these batteries typically exhibit limited rechargeability and depend heavily on air flow, which complicates their design and can limit their application scope. Such characteristics position zinc-air batteries uniquely in the battery technology landscape, especially for applications where high energy density and environmental sustainability are prioritized.

As we explore the realm of battery technologies and their applications, an understanding of how to model these systems is crucial for any electrical engineer. The

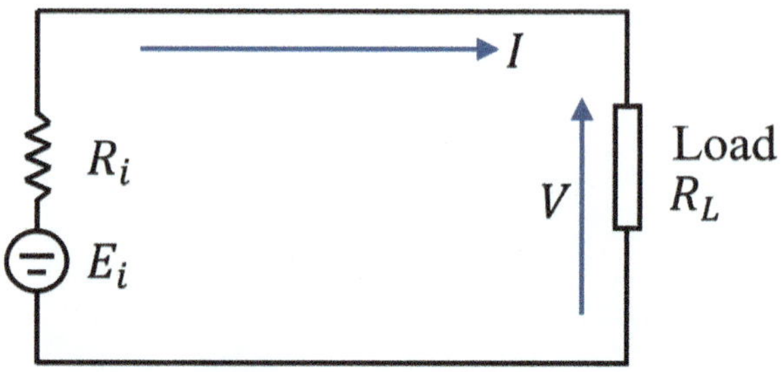

Fig. 32 Steady state equivalent circuit used to represent a battery's behavior under load

diagram provided in Fig. 32 illustrates a steady state equivalent circuit used to represent a battery's behavior under load as

$$E_i = E_0 - K_1 Q_d, \tag{8}$$

$$R_i = R_0 + K_2 Q_d, \tag{9}$$

where E_i is the open circuit voltage, R_i is the internal resistance of a battery, E_0 and R_0 are open circuit voltage and internal resistance of a battery at the fully charged state, Q_d is the ampere-hour discharged energy, and K_1 and K_2 are constants derived by curve fitting data. This model is crucial for predicting battery behavior under various operating conditions and is instrumental in designing more efficient battery management systems. It allows engineers to simulate and analyze the performance, efficiency, and lifespan of batteries in a controlled and predictable manner.

To understand the efficiency and thermal behavior of batteries, it is essential to consider their operational characteristics under different SoC. Efficiency in this context is defined as the ratio of power output (the power delivered to the external load, i.e., $I^2 R_L$) to power input (the total power supplied including losses due to internal resistance, i.e., $I^2(R_L + R_i)$ as

$$\begin{aligned} \eta &= \frac{I^2 R_L}{I^2(R_L + R_i)} \\ &= \frac{R_L}{R_L + R_i}. \end{aligned} \tag{10}$$

From Eq. (9), the internal resistance R_i of a battery increases when it is partially discharged (lower SoC). This increase in internal resistance directly impacts efficiency; higher R_i means a greater proportion of the input energy is lost as heat rather than being usefully transferred to the load. Consequently, a battery at a lower SoC not only becomes less efficient but also generates more heat during operation. This inverse relationship between SoC and internal resistance is crucial for designing battery management systems to optimize performance and longevity.

In exploring maximum power delivery from a battery, the Maximum Power Transfer Theorem is instrumental. This theorem posits that maximum power is transferred when the source's internal impedance matches the load's impedance. For direct current (DC) systems where impedance is purely resistive, this condition is met when the load resistance (R_L) equals the internal resistance of the battery (R_i). Therefore, to calculate maximum power delivery to a DC load, we have

$$
\begin{aligned}
P^{\max} &= I^2 R_L \\
&= \left(\frac{E_i}{R_i + R_L}\right)^2 R_L \\
&= \left(\frac{E_i}{2R_i}\right)^2 R_i \\
&= \frac{E_i^2}{4R_i}.
\end{aligned} \tag{11}
$$

At maximum power, the resistance of the load matches the internal resistance, optimizing the power output by reducing energy lost to internal resistance.

3.3.2 Supercapacitor

Supercapacitors, also known as ultracapacitors, offer a distinct advantage over traditional batteries due to their ability to charge and discharge energy very rapidly. This attribute stems from their design, where energy is stored electrostatically by polarizing an electrolytic solution, rather than through a chemical reaction as in batteries. While supercapacitors excel in delivering quick bursts of energy and endure millions of cycles without degrading significantly, they do not match batteries in terms of energy density. Batteries, with their chemical-based storage mechanism, can hold much more energy per unit volume, making them more suitable for applications requiring prolonged energy supply, such as mobile phones and electric vehicles. However, the rapid charging capability of supercapacitors makes them ideal for functions that require immediate power delivery, such as in regenerative braking systems in vehicles or for power stabilization in renewable energy systems, where they can quickly absorb and then release energy, smoothing out fluctuations in power generation.

Let us revisit how a capacitor stores energy. As shown in Fig. 33, a capacitor stores energy electrostatically by creating a voltage differential across its electrodes. These electrodes are separated by a material known as a dielectric, which prevents them from coming into contact. When a voltage source is connected to these electrodes, electrons accumulate on one plate, creating a potential difference and thus storing energy. When the voltage source is replaced by a load, the stored energy is released, powering the load through discharge.

Fig. 33 Charging and discharging processes of a capacitor

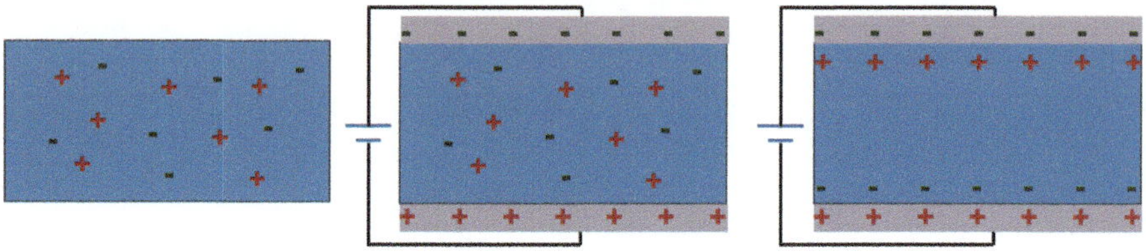

Fig. 34 Mechanism of Electrostatic Double Layer Capacitor (EDLC)

Advancements in capacitor technology have led to the development of supercapacitors, which offer significantly increased energy density. Supercapacitors utilize an electrolyte and a thin insulator instead of the traditional dielectric, which contributes to their enhanced performance. There are three main types of supercapacitors:

- *Electrostatic Double Layer Capacitor (EDLC)*: As shown in Fig. 34, the device uses an electrolyte solution containing both positively and negatively charged ions. When voltage is applied across the capacitor's electrodes, one electrode becomes positively charged and the other negatively charged. This results in electrostatic forces driving the attraction of opposite charges: negative ions move towards the positive electrode and positive ions towards the negative electrode. This migration forms a double layer of charges at each electrode interface, effectively storing electrical energy in an electrostatic field. This mechanism allows EDLCs to achieve rapid charging and discharging cycles compared to chemical energy storage options like batteries. In EDLCs, the electrodes are enveloped by a tightly bound layer of ions, which significantly reduces the distance between charged species across the electrode-electrolyte interface. This configuration leverages the physical proximity of opposite charges, enhancing the capacitor's ability to store electrical energy by minimizing the effective thickness of the dielectric barrier. Although this arrangement allows EDLCs to achieve a higher energy density than traditional capacitors by utilizing thinner layers, their energy density still remains substantially lower than that of batteries. This characteristic shapes the application scope of EDLCs, particularly in systems where rapid charge and discharge cycles are more critical than high energy storage capacity.
- *Pseudocapacitors*: Pseudocapacitors differ from EDLCs in their energy storage mechanism, employing porous materials that facilitate swift redox reactions between the electrode and the electrolyte. This capability allows them to store

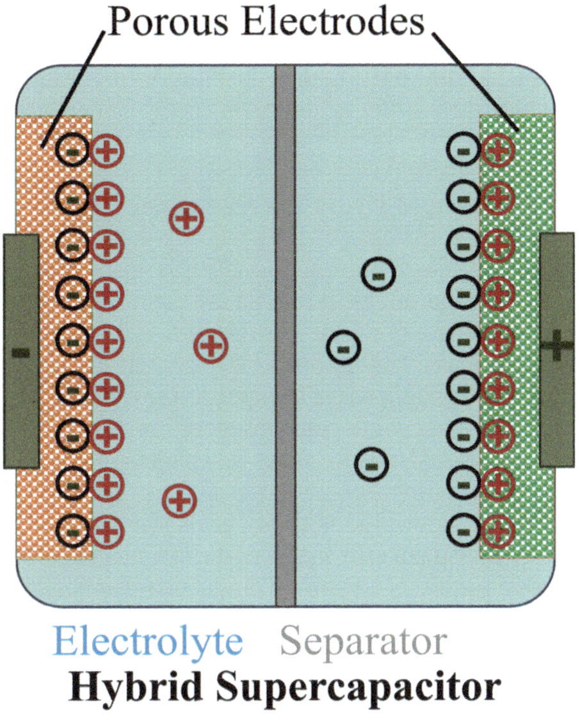

Fig. 35 Structure of a hybrid supercapacitors storing energy both electrostatically and electrochemically

energy electrochemically, similar to batteries, enabling higher energy storage capacities and energy density.

- *Hybrid Supercapacitors*: As indicated in Fig. 35, hybrid supercapacitors merge the qualities of both EDLCs and pseudocapacitors. They store energy both electrostatically and electrochemically, offering the rapid charging and discharging capabilities of EDLCs alongside the higher energy density typical of pseudocapacitors. This combination makes them highly versatile for applications requiring both high power and energy density.

These supercapacitors are distinguished by their ability to charge and discharge rapidly, making them particularly useful in applications requiring quick energy exchanges but they still lag behind batteries in terms of overall energy storage capacity.

The energy storage capability of a supercapacitor is quantified by

$$E = \frac{1}{2} C V^2, \tag{12}$$

where C represents the capacitance, and V is the voltage across it. Supercapacitors are preferred over batteries in applications requiring rapid energy exchanges because they can be charged and discharged at high speeds and endure over 100,000 cycles–significantly more than the typical 100–10,000 cycles a battery can handle. However, a notable drawback is their inability to maintain charge over prolonged periods, limiting their utility in long-term energy storage applications.

3.3.3 Pumped Hydro Energy Storage

Pumped hydro energy storage (PHES) operates using two water reservoirs situated at distinct elevations. It plays a critical role in balancing electrical grids by storing excess electricity during periods of low demand and generating electricity during peak demand. During off-peak hours, electricity, which is often cheaper, powers motors that pump water from the lower reservoir to the upper reservoir, converting electrical energy into potential energy as shown in Fig. 36. Conversely, during high-demand periods, water flows downhill through turbines, transforming this stored potential energy back into electrical energy, ready for use on the grid as shown in Fig. 37. This method not only helps stabilize grid operations by matching supply with demand but also enhances the efficiency of other renewable energy sources by storing surplus energy when production exceeds consumption.

The stored energy in a pumped hydro system is calculated by

$$E_s = V \times d \times g \times h, \tag{13}$$

where V is the the volume of water lifted, d is the density of water, g is the acceleration due to gravity, and h is the elevation difference between the reservoirs. This energy is measured in joules. However, because not all stored energy is converted back into

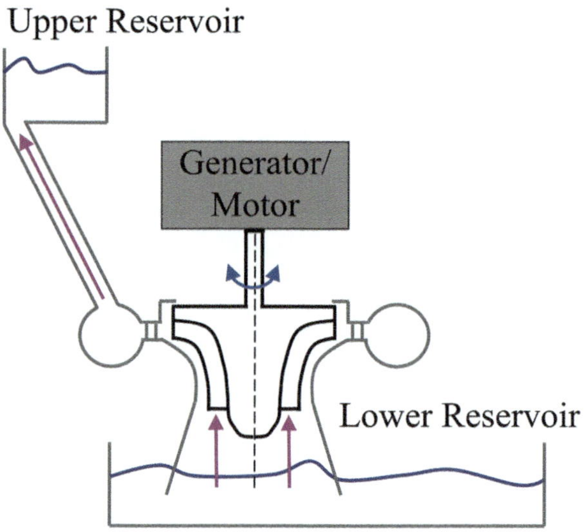

Fig. 36 Schematic illustration of energy storing phase of a pumped hydro energy storage

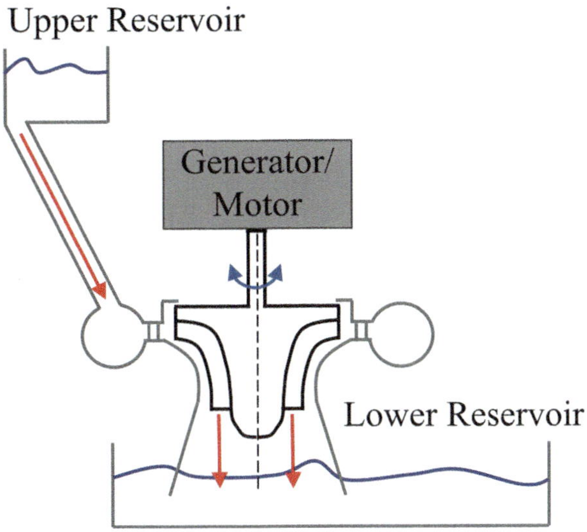

Fig. 37 Schematic illustration of energy realising phase of a pumped hydro energy storage

electrical energy due to losses in pumping, turbine operation, and evaporation, the actual usable energy is

$$E_{\text{usable}} = E_s \times \eta, \tag{14}$$

where η is the round-trip efficiency of the system.

PHES plays a crucial role in energy management on a large scale. It is particularly effective for stabilizing the power grid, managing the variability of demand during peak and off-peak hours, and integrating intermittent renewable energy sources like solar and wind. This method currently contributes to about 3% of the global electricity generation capacity, representing over 160 GW [18]. Despite its benefits in grid management and renewable integration, PHES systems are marked by high initial costs and extended construction periods. However, once constructed, they are highly efficient, with new facilities achieving about 80% efficiency in energy conversion. These characteristics underscore the importance and potential for further adoption of PHES systems as the demand for sustainable and reliable energy solutions grows.

3.3.4 Superconducting Magnetic Energy Storage

Superconducting magnetic energy storage (SMES) systems utilize the principles of superconductivity to store and release electrical energy with high efficiency and minimal energy loss. Superconductivity is a quantum mechanical phenomenon characterized by the complete disappearance of electrical resistance in certain materials when they are cooled below a specific temperature known as the critical temperature.

This critical temperature varies significantly among materials and typically lies at very low values, necessitating sophisticated cooling techniques using liquid helium or liquid nitrogen.

In the energy storage phase, direct current (DC) is supplied to a coil made of superconducting material. Since the coil exhibits zero resistance, it can maintain the current flow indefinitely without any energy loss. This persistent current generates a stable magnetic field capable of storing substantial amounts of energy. The energy is stored in the magnetic field generated by the current flowing through the coil. When the stored energy is needed, the flow of current is interrupted using switches designed to operate at low temperatures and high magnetic fields. This interruption causes the magnetic field to begin collapsing, converting the stored magnetic energy back into electrical energy. The system includes converters and inverters that transform this energy into alternating current (AC), which is compatible with the power grid. The ability to rapidly switch from storage to release mode makes SMES systems particularly valuable for applications requiring high power for short durations, providing stability to power grids during peak load demands.

The operation of a SMES system involves a series of procedures designed to efficiently store and release electrical energy using the properties of superconductors. The system's working process can be divided into three primary stages: initial power reception, conversion to DC, and energy storage and release.

- *Initial Power Reception*: The first step in the operation of a SMES system is receiving electrical power from the grid, which is typically in AC form. The initial power reception involves sophisticated equipment designed to handle high voltage AC power efficiently and safely. This equipment is also responsible for providing the first level of regulation and filtering to ensure that the power is suitable for further processing within the SMES system.
- *Conversion to DC*: Once the AC power is received, it must be converted into DC. This conversion is crucial because superconducting coils, which are the heart of the SMES system, require DC to maintain a stable and constant magnetic field. DC is preferred over AC in this application because an AC-induced magnetic field would continually change direction, leading to instability and potential inefficiencies. The conversion is achieved using a power conversion system that typically consists of rectifiers and associated control electronics. These components are specifically designed to handle high power levels with minimal losses, ensuring that the efficiency of the SMES system is maximized. The quality of the DC power is critical, as any fluctuations or impurities could affect the performance of the superconducting coil.
- *Storage and Release of Energy*: The core of the SMES system is the superconducting coil where the energy is stored. Once the DC is supplied to the coil, it generates a magnetic field proportional to the amount of current flowing through it. This magnetic field represents the stored energy. Due to the superconducting nature of the coil (zero resistance), the current will continue to flow indefinitely without power input, thereby storing the energy with negligible losses. When the stored energy is needed, the current flow is interrupted, which begins the process of

collapsing the magnetic field. This change in the magnetic field induces a current in the opposite direction, which can then be converted back to AC and fed into the grid. The ability to rapidly release the energy makes SMES systems particularly useful for applications requiring quick responses, such as in stabilizing the power grid during transient faults or fluctuations.

In practice, the design and operation of SMES systems involve addressing several challenges, such as ensuring the reliability of the power conversion electronics and managing the thermal loads associated with the cooling of superconducting coils. The economic and practical viability of SMES systems also depends on advances in superconductor technology and the availability of cost-effective and efficient cooling solutions. A SMES system is composed of several critical components that enable its efficient operation. Each component plays a unique role in managing the energy flow, maintaining operational conditions, and ensuring system stability and safety. As shown in Fig. 38, this section elaborates on four main components: the superconducting coil, cryogenic system, power conditioning system, and the control and protection system.

- *Superconducting Coil*: The superconducting coil is the heart of the SMES system where electrical energy is converted into magnetic energy and stored. Made from materials that exhibit superconductivity at low temperatures, these coils can carry a DC indefinitely without energy loss due to their zero electrical resistance. The magnetic field created by the DC flowing through the coil is where the energy is

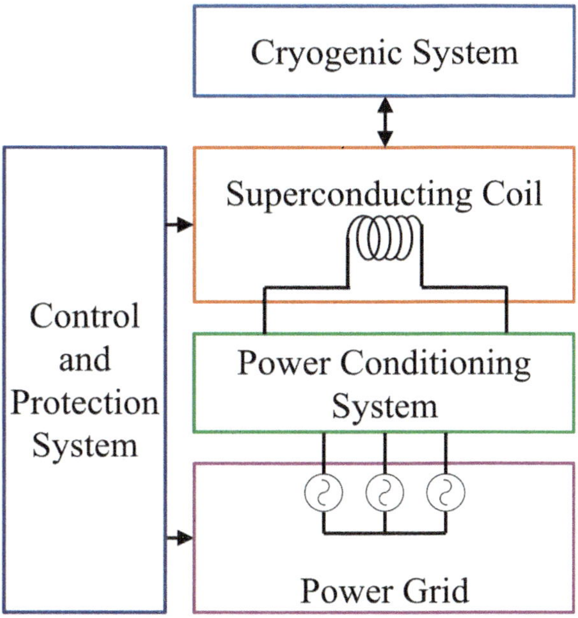

Fig. 38 Components of a superconducting magnetic energy storage

actually stored. The design of the coil, including its size and the material used, directly influences the amount and duration of energy that can be stored.

- *Cryogenic System*: The cryogenic system is crucial for maintaining the superconducting coil at its critical temperature, below which the material exhibits superconductivity. Most SMES systems utilize liquid helium to cool the coils, as helium remains in liquid form at extremely low temperatures conducive to achieving and maintaining superconductivity. The efficiency of the cryogenic system impacts the overall energy efficiency of the SMES system, as maintaining low temperatures requires significant energy. Advances in cryogenics and the development of high-temperature superconductors are key to reducing these cooling demands.
- *Power Conditioning System*: This system is responsible for the conversion of AC from the power grid into DC to be stored in the superconducting coil and then converting it back to AC when the energy is discharged. It includes rectifiers, inverters, and other conditioning equipment to ensure the quality and stability of the energy conversion process. The power conditioning system must handle high currents and voltages efficiently, with minimal energy losses during both conversion processes. This system also plays a vital role in synchronizing the discharged energy with the power grid's frequency and phase.
- *Control and Protection System*: The control and protection system provides comprehensive management of the SMES system. It regulates voltage, manages power quality, and protects against overloads and faults. This system includes sensors and switches to monitor and control the operational parameters of the SMES, ensuring that it functions within safe and optimal conditions. It is also instrumental in integrating the SMES system with the broader power grid, ensuring that it responds correctly to grid demands and disturbances.

The ability to calculate the energy stored in a SMES system is crucial for understanding its efficiency and designing systems that can meet specific energy requirements. Stored energy in a SMES system can be calculated using two different approaches: through the energy density related to the magnetic field surrounding the superconducting coil, and by considering the properties of the superconducting coil itself.

- *Energy Density Related to the Magnetic Field*: The first method involves calculating the energy density in the space around the superconducting coil. The energy density E_s in joules per cubic meter (J/m^3) is given by the equation:

$$E_s = \frac{1}{2} B^2 \mu, \tag{15}$$

where B is the magnetic flux density in Teslas (T), and μ is the magnetic permeability of the medium in Henry per meter (H/m). Magnetic flux density B is a measure of the strength of the magnetic field produced by the coil and is influenced by the electrical current flowing through the coil (I) and the number of turns (N) in the coil. It is crucial for determining how much energy the magnetic field can store. Magnetic permeability μ is a property of the medium surrounding the coil that indicates how much the medium can be magnetized. It is a constant for a given material and plays a significant role in how the magnetic field affects the surrounding space.

- *Properties of the Superconducting Coil*: The second method focuses on the superconducting coil's properties, specifically its inductance L, and the current I flowing through it. The total energy E_s stored in the coil can be calculated by:

$$E_s = \frac{1}{2}LI^2, \tag{16}$$

where L is the inductance of the coil in Henrys (H), and I is the current through the coil in amperes (A). Inductance L measures the coil's ability to store energy in its magnetic field, which is significantly influenced by the coil's geometry, the number of turns of the wire, and the core material. The design of the coil, therefore, directly affects its energy storage capacity. The current I flowing through the coil not only influences the magnetic flux density but also directly impacts the total energy stored. Higher currents can store more energy but require the coil to be capable of handling such intensities without losing its superconducting properties.

SMES systems offer several significant advantages that make them highly effective for specific applications in power systems:

- *High Round-Trip Efficiency*: SMES systems exhibit exceptionally high round-trip efficiencies, approximately 95%, due to the superconducting state of the coil which operates without electrical resistance. This efficiency is critical for reducing energy losses during storage and retrieval, making SMES economically and energetically favorable compared to other storage technologies.
- *Longer Lifespan*: The superconducting coil in a SMES system typically has fewer moving parts, resulting in minimal mechanical wear and tear. This characteristic extends the operational lifespan of SMES systems to as much as 30 years, significantly longer than many chemical battery technologies.
- *Fast Response Time*: One of the standout features of SMES systems is their ability to discharge and charge energy within milliseconds. This capability allows them to deliver large amounts of power almost instantaneously, which is particularly valuable in stabilizing power grids and managing peak load demands.

The operational characteristics of SMES systems make them suitable for a variety of applications in modern power grids. SMES systems can quickly release energy to stabilize the grid during short-term power fluctuations and outages. This rapid response capability is vital for maintaining uninterrupted power supply and preventing the cascading failures that can occur in interconnected power systems. By compensating for dips and surges in power supply, SMES systems help maintain consistent power quality. This is crucial in industrial settings where sensitive equipment requires stable power for optimal operation. SMES systems can store excess energy generated from renewable sources such as wind and solar during periods of low demand. This stored energy can later be released to the grid during peak demand periods, thereby enhancing the utilization of renewable energy resources and reducing reliance on fossil-fuel based power generation.

3.3.5 Compressed Air Energy Storage

Compressed Air Energy Storage (CAES) is a technology that stores energy in the form of compressed air in underground caverns or above-ground tanks, providing a viable solution for balancing grid demand and supply. This technology leverages off-peak, low-cost electricity to compress and store air, releasing it during peak demand to generate electricity. As illustrated in Fig. 39, the storage phase of CAES involves several critical components and processes:

- *Motor and Compressor*: During off-peak hours when electricity prices are lower, a motor powered by electricity drives a compressor. The compressor intakes ambient air and compresses it, significantly increasing the air's pressure and temperature. The process of air compression is inherently energy-intensive but crucial for the storage of energy in a compact form.
- *Heat Management*: As air is compressed, its temperature increases due to the compression process's thermodynamic properties. To enhance efficiency and ensure safety, the heat generated during compression is often removed through cooling systems. This heat, if not managed, could reduce the system's overall efficiency and pose safety risks due to the increased pressure and temperature.
- *Storage in Caverns or Tanks*: The compressed air is then stored under high pressure in specially engineered underground caverns or sealed above-ground tanks. These storage facilities are designed to hold large volumes of high-pressure air without leakage, maintaining the energy potential for extended periods.

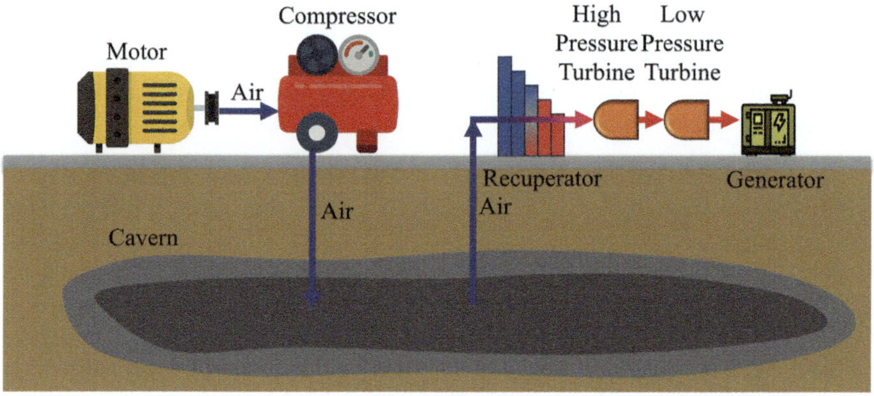

Fig. 39 Mechanism of energy storage and release of a compressed air energy storage (CASE) system

When the demand for electricity rises, the stored compressed air is released and utilized to generate power:

- *Heating and Expansion*: Initially, the high-pressure air is heated to increase its volume and potential energy. This heating phase can be achieved using external energy sources such as natural gas, although some modern CAES systems use adiabatic processes where the heat generated during compression is stored and reused, thus enhancing the system's efficiency.
- *Turbine Operation*: The heated, high-pressure air expands through a series of turbines. Different turbines are designed to operate optimally at various stages of expansion.The high-pressure turbines handle air at higher pressures and lower temperatures. These turbines are robust, designed to withstand the significant mechanical stresses of compressed air at high pressures. Low-pressure turbines are optimized for lower pressures and higher temperatures. The design considerations include tolerance to thermal expansion and lower mechanical stresses compared to high-pressure turbines.

As the air expands through the turbines, it spins the turbine blades, converting the mechanical energy into rotational energy. This rotational energy drives generators, converting it into electrical energy, which is then supplied back to the grid.

Understanding the thermodynamic principles behind CAES is essential for optimizing system design and efficiency. The fundamental variables in these calculations include pressure (P), volume (V), and temperature (T), which relate to each other through the gas laws. In the energy storage phase of CAES, air is compressed from an initial state characterized by pressure P_1 and volume V_1 to a final state with pressure P_2 and volume V_2. The work done on the air, which equals the energy stored (assuming no losses), is described by the formula:

$$E_s = \frac{n(P_2 V_2 - P_1 V_1)}{n - 1}. \tag{17}$$

Here, n represents the polytropic index, which varies depending on the thermodynamic process assumed for the compression. For the isentropic process, an ideal process with no heat transfer, n equals 1.4. For the polytropic process, A more realistic process accounting for heat transfer, typically n is around 1.3.

The relationship between pressure, volume, and temperature during compression follows the polytropic process, where:

$$PV^n = \text{Constant}. \tag{18}$$

This equation indicates that the product of pressure and volume raised to the power of n remains constant if the temperature is managed appropriately. If P_1, V_1, P_2, and n are known, V_2 can be calculated using this law.

Compressing air increases its temperature. The temperature change from T_1 to T_2 can be estimated using the following relationship, derived from the polytropic process:

$$\frac{T_2}{T_1} = \left(\frac{P_2}{P_1}\right)^{\frac{n-1}{n}}. \tag{19}$$

This expression helps predict the final temperature T_2 after compression, essential for designing safety and efficiency measures in CAES systems.

During compression, managing the increased heat is crucial for maintaining efficiency and safety. Advanced CAES systems use methods like intercooling between compression stages to reduce temperature, thus enhancing the overall energy efficiency of the storage process. In addition to storing energy, the heat generated during compression can be captured and reused during air expansion to improve system efficiency. This approach is central to adiabatic CAES, where nearly all the compressive heat is stored and reused, significantly increasing the system's efficiency.

A crucial aspect of CAES systems is how they manage the variables of pressure and volume during the storage process, leading to two distinct types: constant volume CAES and constant pressure CAES. In constant volume CAES systems, the volume of the air storage reservoir is held fixed. Typical storage facilities include underground caverns, depleted oil or gas fields, and abandoned mines. These are environments where the structural integrity allows for a constant volume but varying pressure levels. When air is compressed and stored, the pressure increases. As energy is extracted and the volume of air decreases, the pressure within the cavern also drops. This change in pressure affects the amount of energy that can be extracted. For instance, 1 million cubic feet of air stored at 600 PSI can provide up to 0.25 million kWh of energy. The direct relationship between pressure and the amount of stored energy is governed by the gas law in Eq. (18). A significant drawback of the constant volume system is that the energy output decreases as the stored air is depleted. This is because the pressure of the air decreases with volume, reducing the system's ability to do work. Constant pressure CAES systems, in contrast, maintain the pressure of the air in the storage reservoir at a constant level. This is typically achieved by adjusting the volume of the storage space, which may involve the use of overground variable volume tanks or flexible underground aquifers. As air is released or stored, the physical size of the reservoir changes to compensate for the volume changes, thereby keeping the pressure steady. Despite maintaining constant pressure, the energy density in constant pressure systems is often lower than in constant volume systems. For example, 1 million cubic feet of air at 600 PSI in a constant pressure system provides about 0.1 million kWh, which is less than half of what is offered by a constant volume system under similar conditions. The main advantage of maintaining constant pressure is the consistent power output as air is drawn from the system. However, the requirement for mechanically adjustable storage volumes can introduce complexity and potential reliability issues.

The integration of CAES systems with wind turbines represents a significant advancement in renewable energy technology. This combination harnesses the variable and intermittent nature of wind energy, enhancing grid stability and increasing the efficiency of power generation systems. In regions with significant wind resources, wind turbines convert kinetic energy from wind into electrical energy. Traditionally, the intermittency of wind can pose challenges for grid management and consistent power supply. By integrating CAES systems, the electricity generated during periods of high wind speeds is utilized to compress air into storage reservoirs, such as underground caverns or above-ground tanks. This stored energy is then available for release during periods of low wind activity or peak demand, thereby smoothing out the supply and ensuring a more stable energy output. CAES systems can rapidly respond to grid demands, making them particularly valuable in maintaining grid stability. The ability of CAES to quickly release stored energy means that it can act as a buffer against sudden drops in wind energy production, thus maintaining a steady flow of electricity to the grid. This rapid dispatchability is crucial for balancing supply with fluctuating demand, enhancing the reliability of the grid, and reducing the need for backup generation from fossil fuels. Integrating CAES with wind turbines can also provide economic advantages. By maximizing the utilization of wind turbines and reducing energy wastage, operators can achieve better returns on investment. Additionally, the ability to store and release energy on demand can help energy providers better manage energy prices, taking advantage of lower prices during off-peak periods for storage and selling energy during peak demand at higher prices.

While the integration of CAES with wind energy offers numerous benefits, there are challenges that need addressing. The initial cost for setting up CAES systems, particularly the infrastructure for air storage, can be substantial. Furthermore, the efficiency of the CAES process itself is dependent on the thermodynamic properties of the air compression and expansion cycles, which currently entail some energy loss. Ongoing research and technological advancements aim to improve the efficiency and cost-effectiveness of CAES systems. Innovations in materials science and engineering are expected to enhance the performance of compressors and expanders, reduce energy losses, and lower the environmental impact of these systems.

3.3.6 Flywheel Energy Storage

Flywheel energy storage systems (FESS) represent a sophisticated technology designed for high-performance energy storage. Utilizing the kinetic energy of a rotating flywheel, FESS systems are particularly valued for their high power capacity, rapid response times, and longevity.

As shown in Fig. 40, key components of a FESS are detailed as follows:

- *Flywheel*: The flywheel, typically a solid cylinder made from advanced composite materials such as carbon fiber, is the core component of the FESS. It has a large mass and is capable of rotating at very high speeds–often tens of thousands of

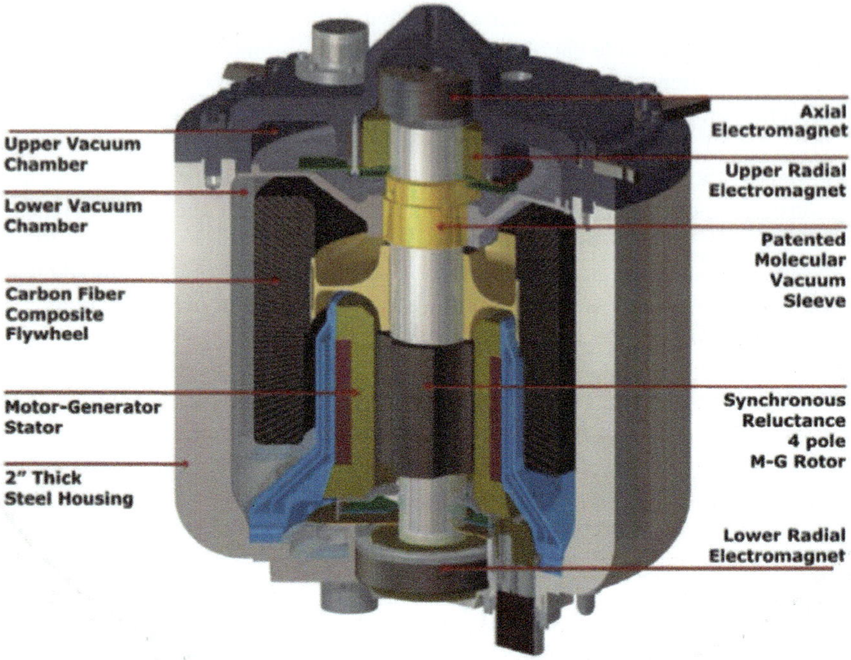

Fig. 40 Structure and components of a flywheel energy storage system [19]

revolutions per minute (RPM). The ability of the flywheel to store kinetic energy is proportional to the square of its angular velocity, making high-speed operation crucial.

- *Rotor*: The rotor encompasses the flywheel and is the primary rotating part of the system. It is directly connected to a motor-generator unit, allowing it to function both as a recipient of energy when acting as a motor and as a source of energy when acting as a generator.
- *Motor/Generator*: This dual-function component is pivotal in both the storage and release phases of operation. During the storage phase, the motor uses electrical energy from the grid to accelerate the flywheel to its operating speed. Conversely, during the energy release phase, the kinetic energy of the flywheel is converted back into electrical energy by the generator.
- *Bearings*: To minimize frictional losses and wear, flywheels are often supported by magnetically levitated bearings. These bearings facilitate nearly frictionless rotation, significantly enhancing the system's efficiency and reducing maintenance requirements.
- *Containment and Vacuum Chamber*: The flywheel and its associated components are housed within a robust containment unit to ensure safety and operational integrity. The containment is designed to withstand extreme forces should the

flywheel disintegrate. Additionally, the system often operates within a vacuum chamber to minimize air resistance and further reduce energy losses.

During discharge, the stored kinetic energy is converted back into electrical energy by reversing the process: the flywheel slows down, and the kinetic energy reduction is converted by the generator into electricity.

Next, we will explore the fundamental principles behind the storage of energy in flywheels, the crucial role of the flywheel's moment of inertia, and how design considerations affect the system's energy capacity and efficiency. The kinetic energy stored in a flywheel is mathematically expressed as:

$$E_s = \frac{1}{2} I \omega^2, \tag{20}$$

where I is the moment of inertia of the flywheel in kilogram square meters ($\mathrm{kg} \cdot \mathrm{m}^2$), and ω is the angular velocity in radians per second (rad/s).

This formula highlights that the energy stored in the flywheel is proportional to the square of its angular velocity and directly proportional to its moment of inertia. The moment of inertia of a flywheel is a measure of how difficult it is to change its rotational speed. It is calculated as:

$$I = k \times m \times r^2, \tag{21}$$

where k is the inertial constant, which depends on the shape and mass distribution of the flywheel, m is the mass of the flywheel in kilograms (kg), and r is the radius of the flywheel in meters (m).

From the above equations, a heavier flywheel typically has a higher moment of inertia, allowing it to store more energy at a given angular velocity. Increasing the mass, while keeping the radius constant, increases the energy storage capability. Similarly, increasing the radius of the flywheel disproportionately increases its moment of inertia, as the radius is squared in the moment of inertia formula. Larger radii are thus highly effective in enhancing energy storage capacity. The energy stored is also proportional to the square of the angular velocity. Higher rotational speeds exponentially increase the energy stored, highlighting the importance of material strength and engineering to manage centrifugal forces at high speeds. The design must also consider safety due to high rotational speeds and the potential for catastrophic failure if the flywheel disintegrates. Efficient energy transfer mechanisms and robust containment structures are essential.

FESS offers a range of benefits that make them a compelling choice for energy storage applications, especially where high power and rapid discharge are required. Firstly, flywheels in modern FESS are often designed with advanced magnetic levitation technology that significantly reduces mechanical contact and thus, wear and tear. The absence of physical bearings eliminates a major source of friction, leading to lower heat generation and reduced deterioration over time. Housing the flywheel in a vacuum chamber minimizes air resistance, which can otherwise lead to energy losses. This design choice further reduces the mechanical stress on the system, prolonging its operational life. These technological advancements contribute to a low maintenance requirement for FESS. With fewer mechanical breakdowns and less need for regular servicing, the typical lifespan of a flywheel can extend to about 20 years or more, making it an economically viable option over the long term.

Secondly, flywheels are constructed from environmentally friendly materials such as carbon fiber composites, which offer high strength-to-weight ratios and are recyclable. This choice of materials reflects a commitment to sustainability, minimizing the environmental footprint of the storage system. Unlike chemical batteries, flywheels do not involve hazardous materials or produce toxic wastes, which can complicate disposal and recycling processes. Their clean operation makes them an attractive choice for projects aiming to reduce environmental impact.

Thirdly, FESS is particularly well-suited for applications requiring quick bursts of energy. They can accelerate from a standstill to their maximum speed in a matter of seconds, delivering substantial power instantaneously. This capability makes flywheels ideal for critical power quality applications, such as grid frequency regulation and voltage support. By quickly absorbing or injecting energy, flywheels help stabilize the grid during fluctuations in demand or supply.

Fourth, the efficiency of energy conversion in flywheel systems is typically in the upper 90% range, contingent on the reduction of frictional losses. This high efficiency is due to the effective design of the motor-generator units and the minimized energy dissipation in the levitation and vacuum systems. Although the initial setup costs can be significant, the high efficiency and low operational costs of flywheels render them cost-effective in scenarios requiring frequent charge and discharge cycles over their operational life.

Acknowledgements This work was supported by the EPSRC Innovation Launchpad Network + Researchers in Residence Scheme through the project of "Digitally-Enabled Flexibility Assessment of Multi-Energy Systems Toward Net-Zero Transition" (RIR35231118-1), and The British Academy through the project of "Assembling Digital Solutions for Inclusive and Net-Zero Transition of Community Energy Systems (ADvaNCE)" (PPHE2400207).

References

1. Méndez Quezada VH, Rivier Abbad J, Gomez San Roman T (2006) Assessment of energy distribution losses for increasing penetration of distributed generation. In: IEEE transactions on power systems, vol 21, no 2, pp 533–540
2. Adefarati T, Bansal RC (2017) Reliability assessment of distribution system with the integration of renewable distributed generation. Appl Energy 185:158–171
3. Hua W, Chen Y, Qadrdan M, Jiang J, Sun H, Wu J (2022) Applications of blockchain and artificial intelligence technologies for enabling prosumers in smart grids: a review. Renew Sustain Energy Rev 161:112308
4. Sadeeq M, Zeebaree S (2021) Energy management for Internet of Things via distributed systems. J Appl Sci Technol Trends 2(02):80–92

5. Herbert G, Iniyan S, Sreevalsan E, Rajapandian S (2007) A review of wind energy technologies. Renew Sustain Energy Rev 11(6):1117–1145
6. Introduction to gears and gearboxes in wind turbines. Windpower Engineering & Development. https://www.windpowerengineering.com/gears-gearboxes-101/. Accessed April 2024
7. How many volts does a single solar cell produce? Quora. https://www.quora.com/How-many-volts-does-a-single-solar-cell-produce. Accessed May 2024
8. FAQs: Your guide to fuel cells. Intelligent Energy. https://www.intelligent-energy.com/product-support/faqs-your-guide-to-fuel-cells/. Accessed June 2024
9. Fuel cell types and working. Electrical A2Z. https://electricala2z.com/renewable-energy/fuel-cell-types-working/. Accessed June 2024
10. Encyclopædia Britannica. Encyclopædia Britannica, Inc. https://www.britannica.com/. Accessed June 2024
11. Microturbine technology matures. Power Magazine. https://www.powermag.com/microturbine-technology-matures/. Accessed June 2024
12. Stirling engine. San Diego State University. http://solar.sdsu.edu/. Accessed June 2024
13. Smith J, Brown A (2020) Grid synchronization techniques in distributed generation systems. IEEE Trans Power Syst 35(4):1234–1245. https://doi.org/10.1109/TPWRS.2020.3042345
14. ASCO Power Technologies (2020) Basic power source synchronization and paralleling. https://www.ascopower.com/us/en/resources/technical-briefs/basic-power-source-synchronization.jsp
15. Marsh J (2023) Lithium-ion vs. Lead acid batteries. EnergySage [Online]. https://www.energysage.com. Accessed 27 July 2024
16. MIT OpenCourseWare (2020) Lecture #11: Batteries & energy storage [Online]. https://ocw.mit.edu/courses/. Accessed 27 July 2024
17. Best uses for Nickel-Cadmium batteries. SolarReviews (2023). https://www.solarreviews.com/blog/best-uses-for-nickel-cadmium-batteries. Accessed 11 Aug 2023
18. International Renewable Energy Agency (2024) Global pumped hydroelectric storage [Online]. https://irena.org. Accessed on 2024-08-12. Data indicating pumped hydro storage contributes about 3% to global electricity generation capacity, representing over 160 GW
19. Power-through. http://power-thru.com/

Key Technologies in the Digitalization of Local Energy Systems

Minzhen Li, Elnaz Azizi, Xiao-Ping Zhang, and Weiqi Hua

Abstract The advancement of digital technologies has become a pivotal factor in the transformation of local energy systems. As we strive towards creating sustainable, reliable, and efficient energy networks, the adoption of cutting-edge technologies like smart grids, blockchain, Internet of Things (IoT), digital twins, cybersecurity measures, and artificial intelligence (AI) with machine learning (ML) capabilities, is reshaping how energy is produced, distributed, and consumed. This chapter briefly introduces each of these technologies, exploring their roles and implications within the context of local energy systems. The details of each technologies and their applications will be discussed in the following chapters of this book.

1 Internet of Things (IoT): Connecting the Energy Ecosystem

The integration of Internet of Things (IoT) technologies within energy systems interconnects various elements, ranging from generation sources to consumption devices. This interconnectedness allows IoT-enabled devices to provide critical data, facilitating significant improvements in several key areas:

- Improved Asset Management and Operational Efficiency: Real-time data collection helps in optimizing the performance and management of energy assets.

M. Li · X.-P. Zhang · W. Hua (✉)
School of Engineering, University of Birmingham, Birmingham, UK
e-mail: w.hua@bham.ac.uk

M. Li
e-mail: mxl1125@student.bham.ac.uk

X.-P. Zhang
e-mail: x.p.zhang@bham.ac.uk

E. Azizi
Department of Engineering Science, University of Oxford, Oxford, UK
e-mail: elnaz.azizi@eng.ox.ac.uk; david.wallom@oerc.ox.ac.uk

W. Hua et al. (eds.), *Digitalisation of Local Energy Systems*, Green Energy and Technology, https://doi.org/10.1007/978-3-031-77833-9_3

- Enhanced Demand-Side Management: Through real-time data on energy consumption patterns, energy providers can better manage and predict demand, leading to more efficient energy use.
- Predictive Maintenance: By addressing potential equipment failures preemptively, IoT systems reduce downtime and repair costs.

IoT has revolutionized energy production and consumption by making renewable energy sources more efficient, adaptable, and attainable. Furthermore, it enhances consumer trust in energy practices, promoting a more sustainable and efficient energy system. One of the key benefits of IoT technology in energy systems is its ability to automate processes to minimize energy wastage. IoT enables energy providers to monitor consumer energy usage and make informed strategic decisions. This capability allows for precise control of voltage, load switching, and adjustments in network configurations. For instance, in power grids, smart switches can detect and isolate areas experiencing issues, thereby enhancing grid reliability and efficiency.

The concept of the Internet of Energy (IoE) represents a distributed energy system that links various elements to optimize energy production and distribution. Key components of the IoE include energy producers, distribution utilities, and grid operators, all working together to improve the efficiency of energy infrastructure and minimize waste. The IoE serves as a prime platform for innovations in energy management and distribution.

As we delve deeper into the IoT's role in energy systems, it is essential to understand the fundamental aspects of IoT functionality and how it transforms industries by enabling interconnected devices to communicate and collaborate effectively.

1.1 The Basics of IoT Functionality

The Internet of Things (IoT) is transforming industries by enabling interconnected devices to communicate and collaborate. This section outlines the essential aspects of IoT functionality, focusing on energy management and smart industries as shown in Fig. 1.

IoT systems comprise key components that gather, analyze, and respond to data. Sensors in data capturing modules collect real-time information on energy consumption, temperature, and humidity [1, 2]. This information is contextualized through a semantic framework, facilitating meaningful analysis [2]. An action engine then uses predictive models and rules to suggest ways to enhance energy efficiency [2]. Interconnectivity and interoperability are critical elements of IoT systems. Standard information models can ensure smooth data exchange across various domains, enabling effective communication. IoT communication frameworks, utilizing protocols such as IPv6 (Internet Protocol version 6), CoAP (Constrained Application Protocol), and 6LoWPAN (IPv6 over Low-Power Wireless Personal Area Networks), improve compatibility and data transmission [3]. Effective communication is essential. The 6LoWPAN protocol adapts IPv6 for low-power networks, facilitating

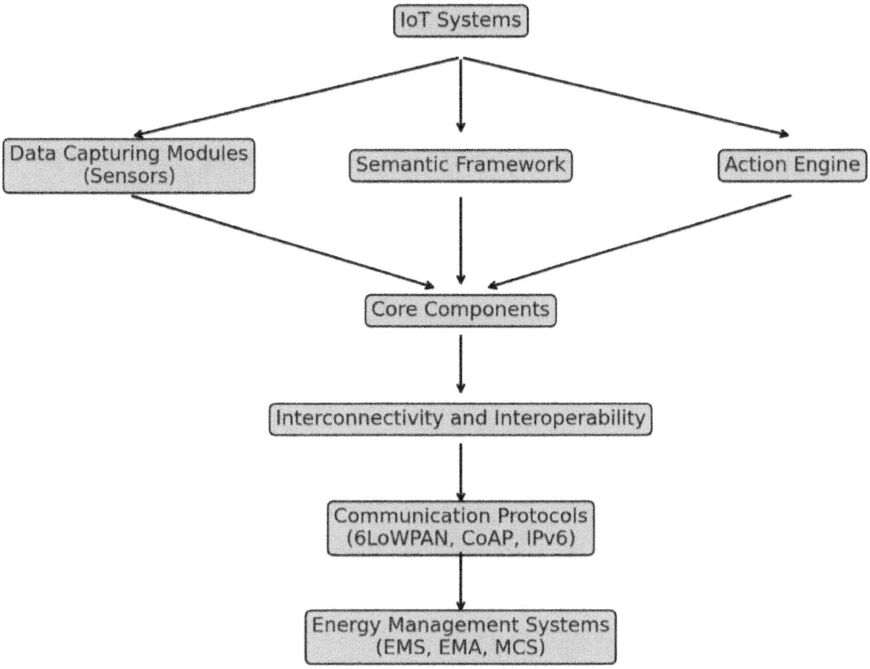

Fig. 1 IoT functionality

efficient communication for small devices. CoAP enables resource-constrained devices to communicate by simplifying HTTP integration with web technologies [3]. IPv6 provides benefits such as a larger addressing space and enhanced security.

IoT systems manage energy consumption through demand response (DR) programs, which adjust energy usage based on electricity prices or incentives [3]. By scheduling tasks during off-peak times, IoT systems can lower costs and reduce grid strain. Integrating distributed energy resources (DERs) like solar panels and storage systems offers greater flexibility and sustainability in energy management [3]. Comprehensive energy management platforms include the Energy Management System (EMS), Energy Manager Agent (EMA), and Monitoring and Control System (MCS). The EMS supervises energy management, running algorithms to optimize usage. The EMA focuses on monitoring and controlling electricity consumption for specific tasks, while the MCS automates device monitoring and control [4]. The rise of IoT-based energy management systems is driven by factors such as modern manufacturing needs, increasing energy costs, and environmental concerns. These systems play a vital role in achieving green energy goals and managing energy supply and demand effectively. Practical applications in smart industrial planning and wireless power transfer illustrate IoT's potential to optimize energy usage and achieve substantial savings [2, 4].

1.2 Streamlining Asset Management and Operations with IoT

The integration of the Internet of Things (IoT) into asset management and operations significantly enhances efficiency, accuracy, and decision-making capabilities. Through real-time monitoring, data collection, and analytics, IoT systems optimize performance and reduce costs [3, 5].

IoT systems use various sensors and devices to capture real-time data from diverse sources, including energy consumption, environmental conditions, and equipment status. These sensors provide continuous monitoring, allowing for quick detection of anomalies and operational inefficiencies. For example, in energy management, sensors track parameters such as electricity usage, indoor temperature, and humidity to optimize energy consumption in buildings and industrial facilities [2, 3]. One notable advantage of IoT in asset management is predictive maintenance. By analyzing data from sensors, IoT systems can anticipate potential failures and maintenance needs before they occur. This proactive approach minimizes downtime and extends the lifespan of equipment. In industrial settings, IoT systems can detect performance irregularities and schedule maintenance to prevent costly repairs or replacements [2–4].

IoT-enabled energy management systems (EMS) offer advanced tools for monitoring and controlling energy usage. These systems integrate data from various sources, such as smart meters and renewable energy modules, to provide a comprehensive view of energy consumption. Utilizing IoT, facilities can implement demand response strategies, optimize energy use during off-peak hours, and integrate distributed energy resources like solar panels and energy storage systems [3, 4]. The success of IoT in streamlining operations relies heavily on the interconnectivity and interoperability of devices and systems. IoT platforms often use standardized communication protocols and common information models to ensure seamless data exchange between different systems. This interconnected framework allows for better coordination and more efficient management of resources across various domains, such as energy, transportation, and industrial operations [3].

IoT systems not only collect data but also analyze it to provide actionable insights. Advanced analytics and machine learning algorithms process the vast amounts of data generated by IoT devices to identify patterns and trends. These insights support decision-making processes, helping organizations optimize operations, reduce costs, and improve overall performance. For instance, in smart energy management, IoT analytics can predict peak load times and suggest measures to balance demand and supply, enhancing grid stability and efficiency.

Through these technologies and methods, IoT significantly enhances various aspects of asset management and operations, driving more efficient and reliable industrial practices [4].

1.3 Leveraging IoT for Real-Time Energy Demand Insights

The integration of the Internet of Things (IoT) in energy management offers significant benefits, particularly in generating real-time insights into energy demand. By using advanced sensors, networks, and data analytics, IoT systems can greatly improve the understanding and management of energy consumption patterns [2, 3].

IoT systems use various sensors to monitor real-time data on energy consumption, environmental conditions, and equipment status. These sensors continuously collect data on electricity usage, indoor temperature, and humidity, which is essential for optimizing energy use in buildings and industrial facilities. The constant stream of data allows for the immediate detection of anomalies and inefficiencies, enabling timely interventions to prevent energy wastage. The semantic framework within IoT systems integrates and contextualizes data from multiple sources. Using Semantic Web technologies, these frameworks provide comprehensive and meaningful analysis of energy data [2]. This integration offers a holistic understanding of energy demand by considering various factors that influence consumption patterns. Advanced analytics and machine learning algorithms process the vast amounts of data collected by IoT devices. These technologies analyze data to identify trends, patterns, and anomalies in energy consumption. Machine learning models can predict future energy demands based on historical data and current usage patterns, enabling proactive energy management and accurate demand forecasting.

One key application of IoT in energy management is demand response (DR) programs. IoT-enabled DR programs adjust energy consumption in real-time based on electricity prices or grid demands. By shifting energy usage to off-peak periods and integrating distributed energy resources (DERs) like solar panels and energy storage systems, these programs help balance the load, reduce energy costs, and enhance grid stability. Moreover, IoT systems enhance energy management through their interoperability and connectivity. Using standardized communication protocols and common information models, IoT platforms ensure seamless data exchange and integration across various devices and systems [3]. This connectivity enables comprehensive monitoring and control of energy consumption, leading to more efficient and sustainable energy practices.

1.4 Enhancing Equipment Longevity Through Predictive Maintenance

Predictive maintenance is a method that uses data analysis and detection of performance anomalies to predict equipment failures. Internet of Things (IoT) devices offer valuable insights into equipment status through real-time monitoring and data analysis, facilitating predictive maintenance.

Predictive maintenance employs equipment like sensors and data analytics to monitor the operational status and performance metrics of equipment. These sensors gather data on various parameters such as vibration, temperature, pressure, and current. By analyzing trends and anomalies in this data, IoT systems can predict when equipment is likely to fail, allowing for maintenance to be conducted before a failure occurs. Compared to traditional preventive and reactive maintenance, predictive maintenance offers significant advantages. Firstly, it extends the lifespan of equipment. By detecting and addressing potential issues early, equipment can be maintained before minor problems escalate into major failures, preventing extensive damage. Secondly, predictive maintenance reduces downtime. Since maintenance is scheduled based on predictions and planned ahead, it can be carried out during non-peak hours, minimizing disruption to production. Lastly, predictive maintenance lowers maintenance costs. Timely maintenance prevents major failures, reducing expensive repair costs and the need for replacement parts.

In the energy sector, IoT devices have successfully implemented predictive maintenance with impressive results. For example, Amazon published a AWS IoT system to help power companies use IoT sensors to monitor and predict the renewable energy [6]. By analyzing vibration and temperature data, these companies can identify abnormal operating conditions and perform maintenance before failures occur. This approach has not only extended the lifespan of equipment but also significantly reduced downtime and maintenance costs. Specifically, some companies have reported a 30% decrease in equipment failure rates, a 25% reduction in maintenance costs, and a 20% increase in average equipment uptime. These case studies demonstrate that applying IoT technology in predictive maintenance improves equipment reliability and lifespan while significantly lowering maintenance costs and downtime, providing substantial economic benefits and operational advantages. By continuing to advance IoT technology in predictive maintenance, various industries can further enhance their equipment management, achieving more efficient and sustainable operational models.

2 Artificial Intelligence and Machine Learning: The Intelligent Core

AI and ML are transformative forces within local energy systems, capable of:

- Analyzing vast datasets to extract meaningful patterns and insights [7].
- Forecasting energy demand and generation, particularly from intermittent renewable sources, to optimize grid balance [8].
- Automating the operation of grid components, improving efficiency, and reducing human error [9].

AI algorithms can predict equipment failures, while ML models can accurately forecast short-term and long-term energy loads, facilitating more reliable and efficient energy distribution [10]. The integration of smart grids, IoT, blockchain, digital twins, cybersecurity, and AI with ML heralds a new era for local energy systems. These technologies collectively enhance the intelligence, agility, and resilience of the energy infrastructure, aligning it with sustainability goals and adaptability to changing energy landscapes. The future of energy is not just about the energy itself but also about the innovative digital frameworks that will govern its flow and utilization [11].

2.1 Transformative AI Techniques for Real-Time Energy Management

The integration of Artificial Intelligence (AI) and Machine Learning (ML) into IoT-based energy management systems is revolutionizing data analysis and utilization, leading to significant improvements in energy efficiency and reliability [12]. These advanced algorithms can process vast datasets to uncover meaningful patterns and insights, which are crucial for optimizing energy consumption and forecasting future energy requirements like (1). IoT systems deploy a wide array of sensors to capture real-time data from diverse sources, including energy consumption metrics, environmental conditions, and equipment status [13]. This continuous monitoring enables swift detection of anomalies and inefficiencies. For instance, sensors track parameters such as electricity usage, indoor temperature, and humidity to optimize energy consumption in buildings and industrial facilities.

AI and ML algorithms, particularly Artificial Neural Networks (ANNs) and Generative Adversarial Networks (GANs), are employed to analyze the collected data and identify patterns [14]. These algorithms can predict energy demand and recognize trends that aid in maintaining grid balance. Predictive maintenance is another critical application of AI in energy management. By analyzing data from IoT sensors, AI systems can forecast potential equipment failures before they occur, thereby minimizing downtime and extending equipment lifespan. Advanced modeling techniques like GANs are utilized to simulate power grid systems and generate synthetic feeder models. This approach enhances our understanding and prediction of potential failure points and maintenance needs under various conditions. For example, the FeederGAN framework employs GANs to generate synthetic distribution feeders, improving the ability to test and optimize grid operations without disrupting the actual system [9].

IoT systems not only collect data but also analyze it to provide actionable insights. Advanced analytics and machine learning algorithms process vast amounts of information to identify patterns and trends. These insights support decision-making processes, helping organizations optimize operations, reduce costs, and improve overall performance [7]. AI-based techniques have been successfully applied in various energy management case studies. One example involves using a probabilistic approach with ANNs to predict daily power consumption for an industrial region, improving prediction accuracy and optimizing maintenance scheduling based on forecasted energy consumption [13]. Another application is the use of AI to integrate distributed energy resources such as solar panels and energy storage systems, enabling demand response strategies and enhancing grid stability [11].

In conclusion, the synergy between AI, ML, and IoT in energy management systems is driving significant advancements in energy efficiency, predictive maintenance, and grid optimization. As these technologies continue to evolve, they promise to play an increasingly crucial role in shaping the future of sustainable and intelligent energy systems.

$$L_{\text{forecast}}(t) = f(H, C) \tag{1}$$

where $L_{\text{forecast}}(t)$ represents the load forecast at time t, H represents historical data, and C represents current conditions.

$$E(t) = \sum_{i=1}^{n} D_i(t) \times P_i(t) \tag{2}$$

where $E(t)$ represents the energy consumption at time t, $D_i(t)$ is the activity of device i at time t, and $P_i(t)$ is the power consumption of device i at time t.

2.2 AI-Driven Data Analysis and Pattern Recognition in Energy Management

AI and ML algorithms play a critical role in forecasting energy demand and generation, particularly from intermittent renewable sources such as wind and solar power. These forecasts are essential for optimizing resource allocation and ensuring grid stability [10]. AI algorithms use historical data and real-time inputs to predict future energy demand. For instance, ML models like Artificial Neural Networks (ANNs) and Support Vector Machines (SVMs) analyze patterns in energy consumption data to generate accurate forecasts. These predictions help to anticipate peak demand periods and adjust energy production accordingly [14].

One key application of AI in this domain is predicting the output from renewable energy sources. Given the variability of solar and wind power, accurate forecasting is vital. Advanced AI techniques, such as Recurrent Neural Networks (RNNs) and Long Short-Term Memory (LSTM) networks, are particularly effective in handling time-series data, providing precise short-term and long-term forecasts of energy generation (3). This capability allows grid operators to better integrate renewable energy into the grid, balancing supply with demand more effectively [8].

$$G_{\text{forecast}}(t) = g(W, H_g) \tag{3}$$

where $G_{\text{forecast}}(t)$ represents the energy generation forecast at time t, W represents weather data, and H_g represents historical generation data.

Optimizing resource scheduling based on these forecasts is another crucial benefit. AI-driven predictions enable grid operators to allocate resources more efficiently, minimizing waste and ensuring a reliable supply of energy. For example, during periods of high renewable energy generation, excess power can be stored in batteries or other storage systems for later use. Conversely, during low generation periods, the grid can rely on stored energy or alternative power sources to meet demand [12].

The practical application of these predictive models extends to improving the accuracy and reliability of forecasts which is shown as (4). By continuously refining their algorithms with new data, AI systems can reduce prediction errors and enhance the robustness of their models. Techniques such as ensemble learning, which combines multiple models to improve prediction accuracy, are commonly used. Furthermore, continuous monitoring and updating of these models ensure they adapt to changing conditions and maintain their efficacy over time [11].

$$E_{\text{pred}} = \frac{1}{n}\sum_{i=1}^{n}(A_i - F_i)^2 \tag{4}$$

where E_{pred} represents the prediction error, A_i is the actual value for instance i, and F_i is the forecasted value for instance i.

2.3 Advanced Automation in Grid Operations Through AI and ML

The integration of AI and ML in energy systems is revolutionizing grid operations through advanced automation and real-time insights [7]. These technologies enable the analysis of vast datasets to extract meaningful patterns and provide accurate forecasts for energy demand and generation. Specifically, they are crucial in optimizing the balance of the grid by predicting the intermittent outputs from renewable energy sources like solar and wind power.

AI algorithms excel in predictive maintenance by analyzing operational data to foresee equipment failures, thereby reducing downtime and extending the lifespan of grid components [14]. ML models, meanwhile, are proficient in forecasting both short-term and long-term energy loads, facilitating more reliable and efficient energy distribution [13]. Additionally, AI and ML can enhance the intelligence of grid operations by continuously learning from new data and improving their predictive accuracy. These models can identify inefficiencies and suggest optimizations in real-time, ensuring a more resilient and adaptable energy grid. For example, advanced AI techniques can adjust power flows to prevent overloads and manage the integration of renewable energy sources seamlessly. Cybersecurity measures, bolstered by AI, are integral to protecting the integrity of these interconnected systems. AI-driven security protocols can detect and respond to threats faster than traditional methods, ensuring robust defense against cyber attacks [12].

The synergy of AI and ML in energy systems is driving significant advancements in smart grid, enhancing the reliability and sustainability of energy infrastructure.

3 Digital Twin: The Virtual Reflection for Real-Time Insight

A "digital twin" is a virtual representation that serves as the real-time digital counterpart of a physical object or process. This concept extends beyond traditional simulation by bridging the physical and digital worlds. By creating a digital mirror of the physical object, complete with its dynamics and nuances, digital twins enable simulations that can reflect real-world conditions in real time.

3.1 Concept of Digital Twins

To illustrate the differences between a digital twin and conventional simulation, let's consider the example of punting as shown in Fig. 2. Imagine a punting scenario where the boat, the punter, and the environment are recreated in a digital environment. Sensors placed on an actual punt gather data such as the vessel's position, the force exerted by the punter, water currents, and prevailing weather conditions. This live data feed allows the digital twin to simulate various scenarios in real time, such as making turns, accelerating, or decelerating.

The digital twin allows the punter, in this case, to explore a multitude of actions before taking them, assessing the outcomes of each scenario. This immediate insight

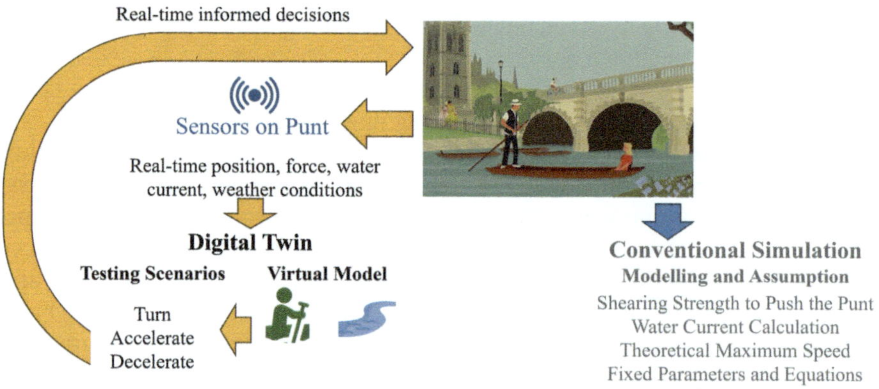

Fig. 2 Comparison between digital twins and conventional simulation model in an example of punting

provides a significant advantage, as it enables informed decision-making that takes into account the current state of the environment and the punt.

Key components of a Digital Twin include:

- *Physical Entity*: The actual system or process being modeled, such as a local energy network.
- *Digital Model*: The virtual representation of the physical entity.
- *Data*: Real-time and historical data collected from sensors, meters, and other monitoring devices.
- Connectivity: The communication infrastructure that enables real-time data flow between the physical and digital entities.
- Analytics and Visualization: Tools and techniques that analyze the data and provide actionable insights.

In contrast, a conventional simulation does not have a real-time data connection. It relies on static or semi-dynamic models built on general knowledge, with assumptions and fixed parameters. For the punting scenario, a conventional simulation might use predefined equations to calculate the shear strength needed to push the punt or to estimate water currents. These calculations would be based on theoretical maximum speeds and fixed physical parameters that may not account for the real-time variability of the actual conditions.

The following subsections will detail how digital twin can be implemented to assist requirements of local energy systems.

3.2 Digital Twins for Local Energy System Planning

Local energy systems require meticulous planning to ensure they are capable of meeting community energy demands while effectively integrating renewable energy sources, managing demand fluctuations, and ensuring overall reliability. Digital Twins are instrumental in performing scenario analysis, which is crucial for the effective planning of local energy systems. By simulating various scenarios, such as fluctuations in energy demand, integration of new renewable energy sources, infrastructure planning, or the impact of extreme weather events, planners can evaluate the potential outcomes of different strategies and optimize system design and operation accordingly. For instance, digital twins can simulate the effects of adding a new solar power installation to a local grid and analyze how this would interact with existing infrastructure and energy demand patterns.

To implement a digital twin for local energy system planning, the following steps need to be considered.

- *First*, objectives of the digital twin needs to be clearly defined, such as improving energy efficiency, integrating renewables, or enhancing resilience. Also the scope of the digital twin needs to be identified.

- *Second*, data needs to be gathered from various sources, including sensors, smart meters, weather forecasts, and historical records. The data needs to be integrated into a cohesive framework that can be used by the digital twins.
- *Third*, based on collected data and defined scopes, we can develop the digital model of the local energy system and ensure it accurately represents the physical system. The model should be able to simulate various scenarios and incorporate real-time data.
- *Fourth*, simulations will be run using the digital twins to validate the model against real-world conditions, and examine if the digital twin model can accurately predict system behavior under different scenarios.
- *Fifth*, the digital twin will be deployed in the planning process and used to inform decision-making. Continuously updating and refining the digital twins as new data becomes available and as the energy system evolves will also be crucial.

While Digital Twins offer significant potential for improving local energy system planning, there are several challenges that need to be addressed. First, the effectiveness of a digital twin depends on the availability and quality of data. In many cases, data may be incomplete, inconsistent, or unavailable, making it difficult to develop an accurate model. Second, local energy systems are often composed of diverse and interconnected subsystems, each with its own data sources, technologies, and stakeholders. Integrating these into a single digital twin can be complex and resource-intensive. Third, as digital twins rely on real-time data and connectivity, they are vulnerable to cyber-attacks. Therefore, ensuring the security and integrity of the digital twin is critical. Fourth, developing a digital twin that can scale with the growing complexity of local energy systems is a significant challenge. The digital twins must be adaptable to accommodate new technologies, data sources, and regulatory requirements. Last, the development and maintenance of a digital twin can be costly, particularly for small communities with limited resources. The cost-benefit analysis of implementing a digital twin needs to be carefully considered.

3.3 Digital Twins for Operational Control

One of the most powerful features of digital twins is their ability to provide real-time data on the operation of local energy systems. This real-time monitoring capability allows power system operators to continuously optimize the performance of local energy systems, adjusting strategies dynamically in response to changing conditions. For example, during periods of high demand, a digital twin can identify potential bottlenecks in the system and suggest operational adjustments, such as redistributing load or activating energy storage systems, to maintain stability and efficiency.

Operational control in local energy systems involves managing energy generation, distribution, and consumption within a localized network, often integrating various renewable energy sources, storage solutions, and demand-response strategies. Traditional control methods may struggle to accommodate the dynamic nature of local energy systems, where energy flows can be highly variable due to factors like weather conditions and fluctuating demand. Effective operational control must ensure reliability, efficiency, and sustainability in energy supply while balancing the needs of multiple stakeholders.

Digital Twins can revolutionize operational control in local energy systems by providing a virtual representation of the physical energy system that mirrors its real-time operations. The key roles digital twins play include:

- *Real-time Monitoring*: Digital twins continuously monitor the status of energy assets, such as solar panels, batteries, and grid components. This real-time visibility allows for quick identification and resolution of issues, ensuring that the system operates within optimal parameters.
- *Predictive Maintenance and Control*: Digital twins analyze historical and real-time data to predict potential failures or inefficiencies, enabling proactive maintenance and adjustments to control strategies. For example, a digital twin could forecast the degradation of battery storage systems and optimize their charging cycles to extend lifespan.
- *Optimization of Distributed Energy Resources*: Digital twins help in optimizing the integration and operation of distributed energy resources within the local energy system. By simulating different scenarios, such as varying weather conditions or load demands, digital twins assist in planning and operating the system more effectively, maximizing the use of renewable resources while maintaining grid stability.
- *Enhanced Decision Support*: Digital twins provide decision-makers with detailed simulations of various operational strategies, helping to evaluate the impact of different choices on system performance. This capability is particularly valuable in complex local energy system where multiple energy sources and loads must be balanced.

Implementing a Digital Twin for local energy system operational control involves several key steps:

- *First*, develop a comprehensive digital model that represents all physical components of the local energy systems, including generation units, storage systems, distribution networks, and loads. This model should be capable of reflecting real-time operations and adapting to changes in the system.
- *Second*, collect and integrate data from various sources, including sensors, smart meters, weather forecasts, and historical performance data. Ensuring seamless data flow between the physical system and its digital counterpart is critical for maintaining the accuracy of the digital twins.
- *Third*, use the digital twins to simulate different operational scenarios, such as peak load conditions, equipment failures, or the addition of new energy assets. Scenario analysis helps identify the most effective operational strategies under varying conditions.

- *Fourth*, develop and integrate advanced control algorithms that leverage the digital twin's insights to optimize system performance. These algorithms should be capable of real-time adjustments based on the digital twin's predictions and simulations.
- *Fifth*, regularly update and refine the digital twin based on new data and evolving system dynamics. Continuous learning and adaptation are essential for maintaining the effectiveness of the digital twin in controlling local energy systems.

3.4 Digital Twin for Asset Management

Asset management is a critical component of maintaining the efficiency, reliability, and sustainability of local energy systems. With the increasing complexity of local energy systems, which integrate various distributed energy resources, energy storage systems, and digital infrastructure, traditional asset management approaches are often inadequate. Digital twin technology offers a transformative solution by providing a virtual representation of both physical and non-physical assets, enabling real-time monitoring, predictive maintenance, and optimized asset utilization. This section explores the implementation of digital twins for asset management in local energy systems, focusing on methodologies, benefits, and challenges.

Asset management in local energy systems involves the systematic process of operating, maintaining, and upgrading physical assets such as generation units, storage systems, and distribution networks, as well as non-physical assets like data, software systems, and contractual agreements. The goal is to ensure that all assets perform optimally throughout their lifecycle while minimizing costs and risks. Effective asset management is essential for sustaining the long-term viability and resilience of local energy systems.

Digital Twins play a crucial role in enhancing asset management by creating a dynamic, real-time model of both physical and non-physical assets. Key roles include:

- *Real-time Monitoring and Diagnostics*: Digital twins continuously monitor the condition of physical assets, such as transformers, solar panels, and batteries, providing real-time data on their performance. This enables early detection of issues, reducing downtime and maintenance costs.
- *Predictive Maintenance*: By analyzing historical and real-time data, digital twins can predict when assets are likely to fail or degrade. This allows for proactive maintenance, which can extend the lifespan of assets and prevent costly failures.
- *Lifecycle Management*: Digital twins assist in managing the entire lifecycle of assets, from installation to decommissioning. They provide insights into how assets age and perform under different conditions, helping operators plan for upgrades or replacements.
- *Non-physical Asset Management*: Digital twins can also model non-physical assets such as software systems, data flows, and contractual obligations. This includes

monitoring the performance of control algorithms, managing data integrity, and ensuring compliance with regulatory requirements.
- *Optimization of Asset Utilization*: Digital twins enable operators to optimize the use of assets by simulating different operational scenarios. For example, they can help determine the most efficient way to use battery storage or optimize the output of renewable energy sources in response to changing demand.

Implementing digital twins in asset management comes with several challenges:

- *Data Quality and Integration*: The success of a digital twin depends on the quality and completeness of the data it uses. Integrating data from diverse sources and ensuring its accuracy can be challenging, particularly in complex local energy system with multiple stakeholders.
- *Modeling Complexity*: Developing accurate and comprehensive models of both physical and non-physical assets requires significant expertise. The models must be detailed enough to capture the nuances of asset behavior while remaining flexible to accommodate changes over time.
- *Cybersecurity and Data Privacy*: As digital twins rely on continuous data exchange, they are vulnerable to cybersecurity threats. Protecting the integrity of the data and ensuring compliance with data privacy regulations are critical considerations.
- *Cost and Resource Requirements* Developing and maintaining a digital twin for asset management can be resource-intensive, requiring investments in technology, expertise, and ongoing support. Smaller local energy system operators may find it challenging to justify these costs.

3.5 Digital Twins for Stakeholder Engagement

The visual and interactive nature of digital twins makes them an effective tool for engaging with stakeholders involved in the planning and operation of local energy systems. By providing a clear and accessible visualization of complex energy systems, digital twins facilitate better communication with stakeholders, allowing them to understand the impact of different planning decisions and participate more actively in the decision-making process.

Local energy systems involve multiple stakeholders, each playing a distinct role in the planning, development, operation, and management of energy resources. The key stakeholders include:

- *Utility Companies*: Responsible for the generation, distribution, and billing of energy within the local energy systems. They are key players in ensuring the reliability and efficiency of the energy supply.
- *Local Governments and Regulators*: These entities oversee the implementation of energy policies, regulations, and standards. They are also involved in permitting and monitoring compliance with environmental and safety regulations.

- *Community Members and Consumers*: The residents and businesses that rely on the local energy systems for their energy needs. They are directly impacted by the cost, reliability, and sustainability of the energy supply.
- *Renewable Energy Developers*: Companies or organizations that develop and operate renewable energy projects within the local energy systems, such as solar farms, wind turbines, and battery storage systems.
- *Investors and Financial Institutions*: Entities that provide the capital necessary to develop and maintain energy infrastructure. Their primary concern is the financial return on investment.
- *Technology Providers*: Companies that supply the hardware and software necessary for the operation of the local energy systems, including sensors, control systems, and digital twin platforms.
- *Environmental Groups and NGOs*: Organizations focused on promoting sustainable energy practices and ensuring that local energy system development aligns with environmental goals.
- *Academic and Research Institutions*: These stakeholders are involved in the development of new technologies, strategies, and policies related to local energy systems.

Digital twins can play a transformative role in engaging these stakeholders by offering a platform for real-time interaction, data sharing, and scenario analysis. First, digital twins provide a transparent view of the local energy systems, allowing stakeholders to see how the system operates in real-time. This transparency fosters trust and facilitates better communication between stakeholders, particularly in situations where decisions must be made collectively. Second, digital twins enable stakeholders to simulate various scenarios, such as changes in energy demand, integration of new renewable resources, or impacts of regulatory changes. This capability helps stakeholders to understand the potential outcomes of different decisions and to collaborate on finding the best solutions. Third, for community members and consumers, digital twins offer an accessible way to understand how their energy system works and to participate in decision-making processes. By visualizing energy flows, costs, and environmental impacts, digital twins can make complex energy systems more understandable to non-experts. Fourth, digital twins allow for real-time monitoring of energy assets and system performance, providing stakeholders with up-to-date information on the status of the local energy systems. This capability is particularly valuable for utility companies, regulators, and renewable energy developers who need to respond quickly to changes in the system. Fifth, for local governments, utility companies, and renewable energy developers, digital twins offer a platform for collaborative planning and development. By sharing data and insights, these stakeholders can work together to optimize the design and operation of the local energy systems, ensuring that it meets the needs of the community while aligning with regulatory requirements. Sixth, investors and financial institutions can use digital twins to assess the risks associated with local energy system projects. By simulating different financial and operational scenarios, digital twins help investors to understand potential risks and returns, making them more confident in their investment decisions. Last, environmental groups and regulators can use digital twins to assess the environmental impact of local energy system projects. By simulating different scenarios, digital twins can help to identify potential environmental risks and to develop strategies for mitigating those risks.

4 Blockchain: Securing Transactions and Enabling Decentralization

Blockchain technology offers a secure and transparent platform for energy transactions within the local energy market. This section will introduce the main functions of blockchain and smart contract and how they can filfull the requirements of local energy systems.

4.1 Blockchain and Smart Contract

Blockchain is a decentralized, distributed ledger technology that records transactions across a network of computers in a way that ensures security, transparency, and immutability. Each block in the blockchain contains a list of transactions, and once a block is added to the chain, it is nearly impossible to alter the information without altering all subsequent blocks, which requires consensus from the network participants. This decentralized nature removes the need for a central authority or intermediary, making blockchain particularly useful in scenarios where trust and transparency are critical.

As shown in Fig. 3, a blockchain is composed of a series of blocks, each of which contains a list of transactions. Each block has three key components:

- *Data*: This includes the actual transaction information, such as the sender, receiver, and amount in the case of financial transactions, or any other data relevant to the blockchain's purpose.
- *Hash*: A unique identifier for the block, generated by processing the block's data through a cryptographic algorithm. The hash ensures the integrity of the block; any change in the block's data will produce a different hash.
- *Previous Block Hash*: This links the current block to the previous one, forming a chain. It ensures that blocks are chronologically ordered and that the chain cannot be altered retroactively without altering all subsequent blocks.

The blocks are linked together in a linear, chronological order, forming what is known as the blockchain. This chain of blocks is stored across a distributed network of nodes, with each node maintaining a copy of the entire blockchain.

Unlike traditional databases that are controlled by a single entity (like a bank or a government), blockchain is decentralized. This means that no single party has control

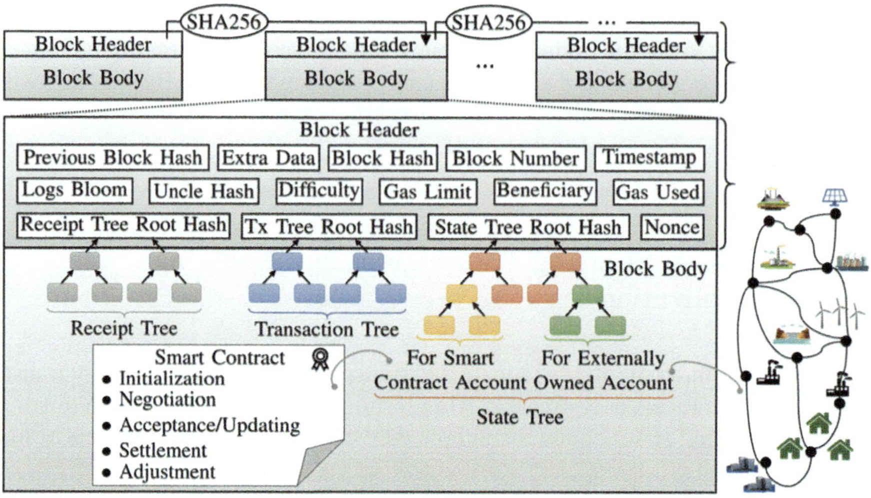

Fig. 3 Structure of a block and its integration into a blockchain network

over the entire blockchain. Instead, control is distributed across all the nodes in the network. Each node has equal authority and can validate and record transactions on the blockchain. Decentralization enhances security because there is no single point of failure. If one node is compromised, the rest of the network remains secure, making it extremely difficult for malicious actors to alter the blockchain's records.

One of the most significant features of blockchain is immutability. Once a block is added to the blockchain, it is nearly impossible to alter or delete the data it contains. This is because changing the data would require changing the hash of that block, which would then require changing the hash of every subsequent block. This process would require controlling over 50% of the network's nodes (in what is known as a 51% attack [15]), which is highly unlikely in a well-established and widely distributed network. Immutability ensures that all records on the blockchain are permanent and tamper-proof, making it ideal for applications where data integrity is crucial, such as financial records, supply chain tracking, and legal contracts.

Blockchain's transparency comes from its public nature. In public blockchains, anyone can view the transactions that have been recorded, although the actual identities behind those transactions are typically pseudonymous. This transparency is crucial for maintaining trust among participants in the network.

To maintain the integrity of the blockchain, all nodes in the network must agree on the validity of transactions before they are added to the blockchain. This agreement process is achieved through consensus mechanisms. The most common consensus mechanisms are:

- *Proof of Work (PoW)* [16]: Used by Bitcoin and several other blockchains, PoW requires nodes (often called miners) to solve complex mathematical puzzles to validate transactions and create new blocks. The first node to solve the puzzle gets to add the block to the blockchain and is rewarded with cryptocurrency.

- *Proof of Stake (PoS)* [17]: In PoS, nodes (validators) are chosen to validate transactions and create new blocks based on the number of tokens they hold and are willing to "stake" as collateral. This method is less energy-intensive than PoW and is used by blockchains like Ethereum 2.0 [18].
- *Other Consensus Mechanisms*: These include Proof of Authority (PoA) [19], Delegated Proof of Stake (DPoS) [19], and Byzantine Fault Tolerance (BFT) [20], each with its own approach to ensuring consensus while balancing security, scalability, and energy efficiency.

Blockchain is inherently secure due to its use of cryptographic techniques and decentralized consensus mechanisms. Each transaction is verified and recorded by multiple nodes, making it nearly impossible for a single entity to alter the record without detection.

Furthermore, smart contracts are self-executing agreements with the terms of the contract directly written into code. These contracts automatically execute and enforce the terms of an agreement when predefined conditions are met, without the need for intermediaries such as lawyers or notaries. The concept of smart contracts was first proposed by Nick Szabo in the 1990s [21], but it gained widespread attention and implementation with the advent of blockchain technology, particularly with the Ethereum platform.

A smart contract is essentially a program that runs on a blockchain. It contains rules and logic written in code, defining how the contract will behave. For instance, a simple smart contract for a payment might stipulate that "if Party A sends 10 ETH to the contract, then Party B will receive 10 ETH." Smart contracts operate on a decentralized blockchain network, meaning that once they are deployed, they are distributed across all nodes in the network. This decentralization ensures that no single party can alter or interfere with the contract once it is in operation, making the system trustless–participants do not need to trust each other or a central authority to execute the contract correctly. When the conditions coded into a smart contract are met, the contract executes automatically. This means that the execution is deterministic, with the outcome predetermined by the contract's logic, leaving no room for ambiguity or dispute over the terms. Once deployed, smart contracts are immutable, meaning they cannot be changed or tampered with. This immutability ensures the integrity of the contract and protects it from fraud or unauthorized modifications. However, this also means that any errors in the contract's code cannot be easily corrected without deploying a new contract.

The key features and advantages of smart contracts are summarised as follows:

- *Automation*: Smart contracts automate processes that would traditionally require manual intervention, reducing the need for intermediaries. This automation can lead to significant cost and time savings.
- *Transparency*: Since smart contracts operate on public blockchains, their code and transactions are transparent and can be audited by anyone. This transparency fosters trust among participants and reduces the risk of fraud.

- *Security*: Smart contracts benefit from the security features of blockchain technology, including cryptography and decentralized consensus mechanisms. This makes them highly resistant to tampering, fraud, and hacking.
- *Efficiency*: By removing intermediaries and automating processes, smart contracts can greatly increase the efficiency of various operations.
- *Enforcement*: Unlike traditional contracts, where enforcement relies on legal systems, smart contracts enforce themselves through code. This reduces the risk of non-compliance and speeds up the execution of contract terms.

4.2 Application of Blockchain Technologies in Peer-to-Peer (P2P) Energy Trading

P2P energy trading represents a novel approach that allows consumers to trade energy directly with each other, bypassing traditional energy utilities. Blockchain technology, with its decentralized, secure, and transparent nature, coupled with smart contracts, is well-suited to facilitate P2P energy trading. This section explores the implementation of blockchain and smart contracts in P2P energy trading systems.

P2P energy trading enables prosumers (entities that both produce and consume energy) to trade excess energy with other users within a localized network. This concept promotes the use of renewable energy sources and enhances grid resilience by reducing reliance on centralized power systems. The P2P model allows for more flexible and dynamic energy markets, where energy prices can be determined in real-time based on supply and demand. The decentralization nature of blockchain Removes the need for a central authority or intermediary. The transparency nature ensures that all transactions are visible and verifiable by all participants. The security nature Protects data and transactions from tampering and unauthorized access. The autonomy nature Allows participants to control their energy production, consumption, and trading activities.

Several blockchain platforms are suitable for P2P energy trading, including Ethereum [22], Hyperledger [23], and Binance Smart Chain [24]. The choice of platform depends on factors such as scalability, transaction speed, and the ability to support smart contracts.

- *Ethereum*: Known for its robust smart contract capabilities, widely used for decentralized applications (dApps).
- *Hyperledger*: A permissioned blockchain ideal for enterprise-level applications requiring higher security and privacy.
- *Binance Smart Chain*: Offers faster transaction times and lower costs, suitable for scalable energy trading systems.

In addition, in the context of P2P energy trading, smart contracts can automate various processes. For the energy trading, smart contracts can automatically match buyers and sellers based on predefined criteria, such as energy price and energy type.

For the payment, smart contract can facilitate instant payments once energy is delivered. For the compliance, smart contracts can ensuring that all transactions comply with regulatory requirements. For the dispute resolution, smart contracts can provide mechanisms to resolve disputes without external intervention. Smart contracts enhance the efficiency and reliability of P2P energy trading by reducing transaction costs and time, minimizing human error, and ensuring trust among participants.

When designing smart contracts for P2P energy trading, several considerations must be taken into account:

- *Energy Measurement*: Integrating with smart meters to accurately measure energy production and consumption.
- *Dynamic Pricing*: Implementing algorithms to calculate real-time energy prices based on supply and demand.
- *Security*: Ensuring the smart contract code is secure and free from vulnerabilities.
- *Interoperability*: Ensuring compatibility with different blockchain platforms and energy systems.

4.3 Application of Blockchain Technologies in Energy Tokenization, Certificates and Carbon Credits

Blockchain's decentralized, immutable, and transparent nature provides novel opportunities for addressing key challenges in energy tokenization, certificates, and carbon credits. In this section, we explore how blockchain and smart contracts can revolutionize these areas, offering new mechanisms for tracking energy production, ensuring transparency in renewable energy certifications, and facilitating the trade of carbon credits.

4.3.1 Energy Tokenization

Energy tokenization refers to the process of converting units of energy, typically from renewable sources, into digital assets, or tokens, that can be traded on a blockchain. These tokens represent ownership rights to specific amounts of energy produced by a particular source (e.g., solar, wind) and can be exchanged in real-time between producers, consumers, and traders. Tokenization can also encompass the creation of tradable financial products linked to energy production and consumption.

Blockchain's decentralized ledger enables the recording of energy production, distribution, and consumption in a manner that is secure, transparent, and immutable. By using smart contracts, energy token transactions can be automated, removing the need for intermediaries and reducing transaction costs. Smart contracts ensure that when energy is generated by a producer, the corresponding energy tokens are automatically created and delivered to the respective stakeholders based on predefined rules.

Energy tokenization introduces several advantages:

- *Liquidity*: Energy tokenization transforms energy into a liquid asset that can be easily traded, bought, or sold on blockchain-based platforms.
- *Transparency*: The blockchain ledger provides a transparent record of all energy transactions, making it easy to trace the origin and ownership of energy tokens.
- *Decentralization*: Tokenization enables energy markets to operate in a decentralized manner, reducing dependence on central authorities and utilities.
- *Empowerment of Consumers*: Consumers are empowered to buy renewable energy directly from producers, potentially reducing costs and fostering green energy adoption.

While energy tokenization offers great promise, several challenges remain. Regulatory uncertainty, the scalability of blockchain platforms, and the volatility of energy prices can hinder widespread adoption. Additionally, technical and cybersecurity risks associated with smart contracts must be addressed. Nevertheless, with ongoing advancements in blockchain technology and clearer regulatory frameworks, energy tokenization could become a mainstream solution in the future energy market.

4.3.2 Energy Certificates

Energy certificates, such as Renewable Energy Certificates (RECs) and Guarantees of Origin (GOs), are tradable instruments that certify that a specific amount of energy was generated from renewable sources. These certificates play a crucial role in renewable energy markets, allowing consumers to support clean energy generation by purchasing certified green energy, even if they do not consume the energy directly.

Blockchain technology provides an efficient way to track the issuance, trading, and redemption of energy certificates. Currently, the verification and trading of certificates involve multiple intermediaries, which can lead to inefficiencies, fraud, and double-counting of certificates. Blockchain addresses these issues by providing a transparent and immutable record of certificate ownership and transactions. Smart contracts can automate the issuance and transfer of certificates. For example, when renewable energy is generated and verified by a trusted entity, smart contracts can automatically issue energy certificates to the producer's blockchain wallet. These certificates can then be traded between participants without the need for intermediaries, reducing transaction costs and increasing market efficiency.

The advantages of the blockchain based energy certificates are summarized as follows:

- *Transparency*: Blockchain provides a public ledger that allows participants to verify the authenticity and ownership of certificates.
- *Efficiency*: The automation of certificate issuance and trading via smart contracts reduces administrative overhead and transaction costs.
- *Fraud Prevention*: Blockchain's immutable nature prevents the double issuance or falsification of certificates, enhancing trust in the system.

Adopting blockchain-based energy certificates faces several challenges, including interoperability with existing certificate systems, regulatory approval, and the complexity of integrating blockchain with current energy infrastructure. However, as blockchain technology matures and gains broader acceptance, it is likely to play an increasingly central role in energy certification markets.

As an instance, the work in [25] introduces an innovative trading platform for RECs using a Directed Acyclic Graph (DAG) blockchain system. The proposed platform addresses critical challenges in the existing REC markets, including issues related to scalability, security, and privacy. By incorporating a reputation-based consensus mechanism, the system not only reduces the computational load and energy consumption typically associated with traditional blockchain technologies but also enhances the security and privacy of REC transactions. The DAG structure enables more efficient transaction processing, while the use of smart contracts automates the trading process, reducing the need for intermediaries and minimizing transaction costs. Additionally, the privacy-preserving features ensure that sensitive user information remains secure during the trading process, promoting broader participation in the renewable energy market. The study's findings suggest that this blockchain-based approach can significantly improve the efficiency, transparency, and scalability of REC markets, making it a viable solution for global adoption and contributing to the broader goals of renewable energy adoption and climate change mitigation.

4.3.3 Carbon Credits

Carbon credits are tradable permits that allow an entity to emit a certain amount of carbon dioxide or other greenhouse gases. They are a key component of cap-and-trade systems, which aim to reduce global greenhouse gas emissions. Carbon credits can be bought, sold, or traded, and they play an important role in mitigating climate change.

Blockchain can transform carbon credit markets by providing greater transparency and traceability. Today's carbon credit markets suffer from several issues, including fraud, lack of transparency, and inefficiencies in credit verification. Blockchain offers a way to verify the authenticity of carbon credits, track their ownership, and ensure that credits are retired once they have been used. Smart contracts can automate the issuance and trading of carbon credits. When a project achieves a verified reduction in carbon emissions, a smart contract can automatically issue carbon credits to the project owner. These credits can then be traded on a blockchain-based platform, where all transactions are recorded on an immutable ledger. Smart contracts can also enforce rules that prevent the double-counting or re-use of credits, further enhancing the integrity of carbon credit markets.

The benefits of implementing blockchain in carbon credit markets are summarized as follows:

- *Transparency and Traceability*: Blockchain provides an immutable record of carbon credit transactions, ensuring that credits are authentic and have not been double-counted or reused.
- *Efficiency*: By automating the issuance and trading of credits via smart contracts, blockchain reduces the need for intermediaries and increases market efficiency.
- *Trust*: Blockchain's transparent nature enhances trust among market participants, including regulators, businesses, and consumers.

Analogously, the adoption of blockchain in carbon credit markets faces several hurdles, including regulatory challenges, interoperability with existing markets, and the need for widespread industry acceptance. However, as carbon markets grow and climate change mitigation becomes an increasingly urgent priority, blockchain has the potential to play a significant role in ensuring the integrity and efficiency of these markets.

As an instance, the work in [26] proposes a decentralized trading system that combines energy trading with carbon allowance exchanges using blockchain technology. This framework enables prosumers, who both produce and consume energy, to trade energy and carbon allowances directly with each other through a peer-to-peer network. The system utilizes smart contracts to automate transactions, ensuring transparency, security, and efficiency in trading. The proposed framework aims to incentivize prosumers to modify their energy consumption and generation behaviors, thereby promoting regional energy balance and reducing carbon emissions. Case studies demonstrate that this blockchain-enabled system outperforms traditional centralized and aggregator-based trading models by achieving better energy balance and greater carbon savings, illustrating its potential as an effective tool for energy and carbon markets.

5 Cybersecurity: Protecting the Energy Infrastructure

As energy systems become increasingly digitized and interconnected, they also face greater vulnerabilities in terms of cybersecurity. Energy infrastructure, being a critical component of modern society, requires robust protection to ensure continuous operation, safeguard sensitive data, and maintain the integrity of energy markets. Any disruption in these systems can have significant implications for national security, economic stability, and public trust. Therefore, implementing comprehensive cybersecurity measures is crucial to protect energy consumers and other stakeholders. This section will examine strategies to enhance the security of energy infrastructure across three key areas: protecting the privacy of energy consumers, securing the infrastructure of local energy systems, and ensuring trusted energy trading practices.

5.1 *Cybersecurity for Protecting Privacy of Energy Consumers*

As energy systems become increasingly digitized, the importance of safeguarding consumer privacy has grown significantly. The deployment of smart meters and advanced metering infrastructure (AMI) has transformed the way energy consumption data is collected and managed. However, these technological advancements also present substantial privacy risks. For example, smart meters gather detailed energy usage data, which can potentially reveal sensitive information about consumers' daily routines, habits, and even the presence or absence of individuals in their homes [27].

To address these risks, it is crucial to implement robust cybersecurity measures that protect consumer privacy. One key approach is the encryption of data collected by smart meters. By encrypting this data at the source, it becomes much more difficult for unauthorized parties to access and misuse it. In addition, access controls should be put in place to ensure that only authorized personnel can view or manage consumer data [28]. Another important strategy involves the use of privacy-enhancing technologies, such as data anonymization. By anonymizing energy consumption data before it is stored or shared, consumer identities can be protected while still allowing for broader analysis of usage patterns. This approach is particularly effective when combined with data minimization principles, which advocate for the collection of only the data necessary for specific purposes, thereby reducing the risk of privacy breaches.

Furthermore, integrating blockchain technology into energy systems offers a promising solution for enhancing consumer privacy. Blockchain's decentralized and immutable nature ensures that energy transactions are secure, transparent, and less vulnerable to cyberattacks. By utilizing cryptographic algorithms and consensus mechanisms, blockchain can provide secure identity management and data protection, giving consumers greater control over their personal information [29].

As energy systems continue to evolve, protecting consumer privacy must remain a top priority. Implementing encryption, access control, data anonymization, and exploring innovative technologies like blockchain are all essential steps towards safeguarding consumer data in the increasingly interconnected energy landscape.

5.2 *Cybersecurity for Infrastructure of Local Energy Systems*

The infrastructure of energy systems has expanded significantly as more local energy systems adopt digital technology. While this advancement brings numerous benefits, it also introduces potential vulnerabilities. Traditional grids were not designed with cybersecurity in mind, and with the rise of the Internet of Things (IoT), local energy systems are now exposed to increased cyber threats.

The first challenge facing local energy systems is availability. Ensuring the availability of these systems is paramount, as any cybersecurity issue that compromises

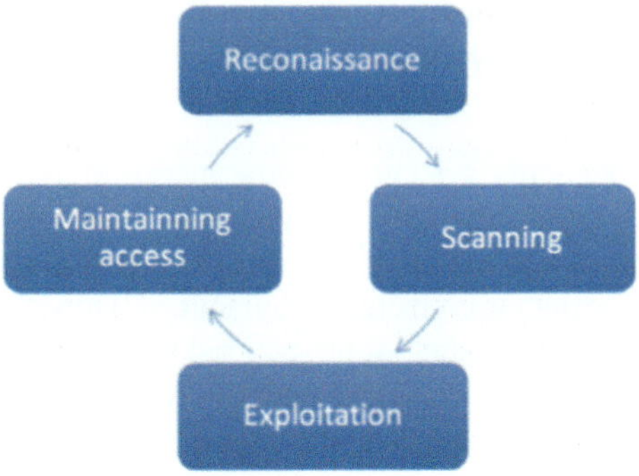

Fig. 4 Attacking cycle followed by hackers to get control over a system

stability could have severe societal and economic consequences. The second challenge is decentralization. An increasing number of individuals and corporations are participating in energy generation, particularly through renewable sources like solar power. However, the cybersecurity defenses of these new entrants are often inadequate, making the entire system more susceptible to attacks. Lastly, legacy systems pose a significant problem. Many existing grid components are outdated and were not designed to counter contemporary cyber threats. Upgrading these systems to meet modern security standards is challenging due to the high costs associated with replacing components and the long operating cycles required [30].

To further illustrate the vulnerability of local energy systems to cyber threats, Fig. 4 demonstrates the typical attack cycle that malicious actors use to compromise a system. This cycle includes four key steps: reconnaissance, scanning, exploitation, and maintaining access. Understanding this cycle is crucial in developing robust defenses against potential attacks. For instance, during the reconnaissance and scanning phases, attackers gather information and identify vulnerabilities within the system. Exploitation then follows, where attackers use identified weaknesses to gain control, and finally, they establish methods to maintain their unauthorized access over time [31].

To solve these challenges, local energy systems must implement comprehensive cybersecurity measures at each stage of this attack cycle. Preventative actions, such as regular security audits, intrusion detection systems (IDS), and secure communication protocols, can significantly reduce the risk of successful attacks. Additionally, updating legacy systems and incorporating advanced technologies like blockchain can enhance the overall resilience of local energy infrastructure [32].

5.3 Cybersecurity for Trusted Energy Trading

In distributed energy trading, cybersecurity is fundamental to ensuring trust in the marketplace and the security of transactions. With the digital transformation of energy markets and the proliferation of smart grids, more and more transactions are taking place through online platforms, exposing systems to increased security threats.Cybersecurity for Trusted Energy Trading is a key area of focus in protecting these markets, aiming to ensure the confidentiality, integrity, and availability of transactions. Through the use of advanced encryption technologies, authentication mechanisms, and smart contracts, cybersecurity measures prevent data breaches, malicious attacks and fraudulent behaviour, thus creating a secure and trustworthy trading environment.

Against this backdrop, distributed energy trading markets such as Local Energy Markets (LEMs) face additional security challenges while providing flexibility and decentralisation benefits. Maintaining the security of these markets is not only to prevent potential economic losses, but also to ensure the stable operation of the grid. Therefore, cybersecurity plays a crucial role in building a trustworthy energy trading system. In distributed energy trading markets like Local Energy Markets (LEMs), safeguarding the security and reliability of transactions is crucial. As these markets grow more decentralized and complex, they encounter several significant cybersecurity challenges that need to be effectively managed to maintain stability and bolster participant confidence. A key concern in distributed energy trading is the risk of malicious activities. Participants with ill intentions might engage in actions such as misreporting energy usage or production, failing to fulfill energy delivery or payment obligations, falsely claiming non-receipt of energy or payment, or submitting fraudulent bids or offers. Such behaviors can destabilize the market, cause imbalances in the grid, and potentially lead to widespread power outages. Therefore, it is essential to implement strong mechanisms to detect and prevent these activities, ensuring the market's integrity.

Despite the decentralized design of LEMs, intended to reduce the risk of Single Points of Failure (SPOF), the complexity of the system can still introduce such risks. These vulnerabilities may stem from critical software or hardware components, which, if compromised, can affect the entire system. Addressing these weak points is vital for enhancing the resilience and security of distributed energy markets [33]. Smart contracts, commonly used to facilitate Distributed Ledger Technology (DLT) transactions in these markets, also pose significant challenges. While their immutability prevents tampering, it also restricts the ability to address vulnerabilities after deployment. Malicious actors could exploit these weaknesses, potentially disrupting market operations, making the security of smart contracts a crucial focus.

Data privacy protection is another major concern in LEMs, where transactions often occur between prosumers–users who both produce and consume energy. While blockchain technology enhances transaction security and transparency, it can also introduce risks related to data leaks and privacy breaches. Balancing efficient energy trading with the protection of sensitive information is a challenge that must be carefully managed. To tackle these cybersecurity challenges, a trust evaluation model has been proposed to assess participant reliability in real time. By analyzing historical behavior and contextual data, such as weather conditions, the model predicts participant actions in future transactions. Trust levels are dynamically adjusted based on observed behavior, helping to identify and exclude potentially malicious actors from the market [34].

Acknowledgements This work was supported by the EPSRC Innovation Launchpad Network + Researchers in Residence Scheme through the project of "Digitally-Enabled Flexibility Assessment of Multi-Energy Systems Toward Net-Zero Transition" (RIR35231118-1), and The British Academy through the project of "Assembling Digital Solutions for Inclusive and Net-Zero Transition of Community Energy Systems (ADvaNCE)" (PPHE2400207).

References

1. WebbyLab (2024) How IoT can help with energy managemen. WebbyLab Blog [Online]. https://webbylab.com/blog/how-iot-can-help-with-energy-management/#::text=An%20IoT%2Dbased%20energy%20management,implement%20more%20resource%2Defficient%20strategies. Accessed 21 May 2024
2. Marinakis V, Doukas H (2018) An advanced IoT-based system for intelligent energy management in buildings. Sensors 18(2):610
3. Liu Y, Yang C, Jiang L, Xie S, Zhang Y (2019) Intelligent edge computing for IoT-based energy management in smart cities. IEEE Netw 33(2):112–117
4. Saleem MU, Usman MR, Shakir M (2021) Design, implementation, and deployment of an IoT based smart energy management system. IEEE Access 9:59649–59664
5. Li X, Luo F, Li C (2024) Multi-agent deep reinforcement learning-based autonomous decision-making framework for community virtual power plants. Appl Energy 360:122813
6. Amazon Web Services (2024) Protecting renewable energy systems using AWS IoT. AWS IoT Blog [Online]. https://aws.amazon.com/cn/blogs/iot/protecting-renewable-energy-systems-using-aws-iot/
7. Baba A (2021) Advanced AI-based techniques to predict daily energy consumption: a case study. Expert Syst Appl 184:115508
8. Wang Q, Zhao M, Liu Y (2024) Forecasting electricity demand using machine learning techniques. Appl Energy 321:119329
9. Kulkarni V, Sahoo SK, Thanikanti SB, Velpula S, Rathod DI (2021) Power systems automation, communication, and information technologies for smart grid: a technical aspects review. TELKOMNIKA Telecommun Comput Electron Control 19(3):1017–1029
10. Smith J, Jones P, Brown A (2023) Renewable energy integration in smart grids: challenges and solutions. Energy Rep 9:567–580
11. Johnson K, Lee H, Kim M (2024) Energy storage systems for electric vehicles: a technical review. J Power Sour 492:229654
12. Chen R, Zhu T, Sun L (2022) Carbon emission reduction in urban areas: a comprehensive review. Sustain Cities Soc 85:104150
13. Li H, Wei Z, Liu X (2016) Application of life cycle assessment in residential buildings: a case study in China. Renew Sustain Energy Rev 69:379–388
14. Zhang Y, Wang L, Li W (2023) Improving energy efficiency in industrial systems: a review. J Clean Prod 387:131968
15. Bitcoin Wiki. Majority attack. Bitcoin Wiki [Online]. https://en.bitcoin.it/. Accessed 17 Aug 2024
16. Nakamoto S (2008) Bitcoin: a peer-to-peer electronic cash system [Online]. https://bitcoin.org/bitcoin.pdf. Accessed 17 Aug 2024
17. Buterin V (2013) Ethereum whitepaper [Online]. https://ethereum.org/en/whitepaper/. Accessed 17 Aug 2024
18. Ethereum Foundation (2020) Ethereum 2.0: proof of stake. Ethereum Foundation Blog [Online]. https://blog.ethereum.org/. Accessed 17 Aug 2024
19. BlockRum (2021) Proof-of-authority vs. proof-of-stake explained [Online]. https://blockrum.com/proof-of-authority-vs-proof-of-stake. Accessed 17 Aug 2024

20. Ethereum Foundation (2017) Proof of stake: How I learned to love weak subjectivity. Ethereum Foundation Blog [Online]. https://blog.ethereum.org/. Accessed 17 Aug 2024
21. Szabo N (1997) The idea of smart contracts [Online]. https://www.fon.hum.uva.nl/rob/Courses/InformationInSpeech. Accessed 17 Aug 2024
22. Wood Gavin (2014) Ethereum: a secure decentralised generalised transaction ledger. Ethereum Project Yellow Paper 151:1–32
23. Androulaki E et al (2018) Hyperledger Fabric: a distributed operating system for permissioned blockchains. In: Proceedings of the thirteenth EuroSys conference
24. Binance Academy (2021) BNB smart chain vs. ethereum: What's the difference? Binance Academy. https://academy.binance.com/en/articles/bnb-smart-chain-vs-ethereum-whats-the-difference
25. Liu WJ, Chiu WY, Hua W (2024) Blockchain-enabled renewable energy certificate trading: a secure and privacy-preserving approach. Energy 290:130110
26. Hua W, Jiang J, Sun H, Wu J (2020) A blockchain-based peer-to-peer trading framework integrating energy and carbon markets. Appl Energy 279:115539
27. McDaniel P, McLaughlin S (2009) Security and privacy challenges in the smart grid. IEEE Secur Privacy 7(3):75–77
28. Wang W, Lu Z (2013) Cyber security in the smart grid: survey and challenges. Comput Netw 57(5):1344–1371
29. Meydani A, Shahinzadeh H, Ramezani A, Gharehpetian GB (2024) Analysis of the cyber-security and privacy protection of blockchain-empowered Internet of Energy (IoE). In: 8th international conference on smart cities, Internet of Things and applications (SCIoT), pp 213–221
30. Xu H, Zhang B, Yao L, Wang C (2021) Cybersecurity in power grids: challenges and opportunities. Sensors 21(22), article 6225
31. El Mrabet Z, Kaabouch N, El Ghazi H, El Ghazi H (2018) Cyber-security in smart grid: survey and challenges. Comput Electr Eng 67:469–482
32. Wang Y, Chen Z, Ai M (2020) Cybersecurity in distributed power systems: a review. IEEE Trans Indus Inform 16(3):1893–1904
33. Andrade R, Wannous S, Pinto T, Praça I (2021) Extending a trust model for energy trading with cyber-attack detection. Electronics 10(16), article 1975
34. Sousa-Dias D, Amyot D, Rahimi-Kian A, Mylopoulos J (2023) A review of cybersecurity concerns for transactive energy markets. Energies 16(16), article 4838

Data Analytics and Energy Management of Local Energy Systems

Tong Wu

Abstract In the landscape of local energy systems, the integration of big data and machine learning emerges as a transformative force, driving efficiency and ensuring energy remains sustainable and reliable. However, navigating this frontier raises two predominant challenges. First, it is important to handle and deploy expansive, synchronized power system data, given its large volume, high velocity, and high dimensionality. Secondly, there is an urgent requirement for real-time fault event detection, where even minimal delays can lead to escalated issues. This chapter delves into these challenges by introducing the application of machine learning techniques in power system data analytics, emphasizing fault event detection and voltage stability assessment. This advanced data-driven technology improves our ability to monitor the power grid, revealing detailed information about the grid's performance.

1 Introduction

This chapter focuses on the crucial role of online security and stability analysis, which leverages streaming data from widely deployed phasor measurement units (PMUs). The high precision of synchrophasor readings empowers real-time evaluation of dynamic incidents, such as fault and oscillation events [1, 2]. However, efficiently managing and making use of the large amount of PMU data is a challenge due to its large volume, speed, and complexity [3]. Local energy networks are always changing, with their operational states and the way they are set up constantly shifting. Such variations present significant hurdles for learning-based algorithms, mainly because these algorithms typically rely on finite historical data for training and validation. As operating conditions change, the need arises to retrain these models with fresh data. This retraining is not only slow, especially when gathering sufficient samples, but also time-consuming, particularly in expansive power systems [4, 5].

T. Wu (✉)
Department of Electrical and Computer Engineering, University of Central Florida, Orlando, FL, USA
e-mail: tong.wu@ucf.edu

W. Hua et al. (eds.), *Digitalisation of Local Energy Systems*, Green Energy and Technology, https://doi.org/10.1007/978-3-031-77833-9_4

In electric power systems, there are two key scenarios that require consideration. The first pertains to fault event detection, an essential tool for identifying extreme conditions and system anomalies. The second, monitoring voltage stability, acts as a continuous preventative measure for the system's overall operational state. Within this context, the term "events" is used to describe atypical operations often caused by bus faults, line tripping, and various other disturbances [4]. In contrast, "low-quality data measurements" denote the inaccuracies observed in sensor readings, typically resulting from complications like sensing noise or data losses [6, 7]. The critical difference between these two is that events signal an abnormal state of the system, whereas with low-quality measurements, the system might be functioning typically, but the readings do not accurately reflect the current operating conditions.

Highlighting the prevalence of these issues, a report from the California Independent System Operation in 2011 revealed that low-quality data comprised between 10 and 17% of all measurements [8]. Even though these inaccuracies might seem small, they can significantly impact the overall stability of power systems. Specifically, voltage stability is crucial for the secure and reliable operation of these systems, as past instances of voltage instability have been responsible for major global blackouts [5, 9, 10]. This highlights the importance of monitoring static voltage stability, enabling power system operators to swiftly identify and respond to conditions that could lead to voltage collapse [11, 12]. Recent efforts have focused on harnessing the wide-area measurement system (WAMS) and PMUs for voltage stability monitoring [9, 13]. Compared with asynchronous nature of Supervisory Control and Data Acquisition (SCADA), the synchronized data from PMUs offers a timely and comprehensive insight into voltage stability parameters. Considering the high-speed and large volume of PMU measurements [14], there is an urgent need for novel online data analytics platforms which are capable of fast data processing.

1.1 Related Works

1.1.1 Power System Fault Detection

The increasing emphasis on swift event detection through synchrophasor data has led to a shift toward machine learning-based methods as the primary approach. However, the concurrent existence of both event and low-quality measurement data introduces two significant challenges for such learning methodologies: (1) The creation of an event detector that remains trainable with PMU data, even when mixed with low-quality measurements during its training phase. (2) The capability to accurately distinguish actual events from these low-quality measurements during testing. Our analysis reveals that many mainstream event detection techniques, including those based on reducing data dimensions and machine learning, often struggle to overcome these challenges [4, 6, 15, 16]. The presence of low-quality measurements in the training set often results in compromised detection performance [4]. Specifically, dimension-reduction-based detection methodologies become particularly susceptible to enhanced sensing noise, resulting from eigenvalue perturbations within Principal Component Analysis (PCA) [17]. Furthermore, many existing methods struggle to distinguish between real events and these low-quality measurements. This can lead to false alarms, where the system mistakenly identifies bad measurements as actual events. These misclassifications happen because unusual power system behavior and low-quality sensor data can sometimes look very similar in the data collected by synchrophasors (PMUs), which can confuse machine learning algorithms.

1.1.2 Power System Voltage Monitoring

Machine learning, particularly in recent times, has been harnessed to exploit PMU data for voltage stability monitoring [9, 10, 18–21]. Deep neural networks (DNNs), especially, have become the most popular approach for static security evaluations [10, 19, 21, 22]. As an illustration, Zhou et al. [10] offer a neural network architecture adept at rapidly estimating the long-term voltage stability margins, applicable in both standard and contingency scenarios. Sunitha et al. [19] see the application of artificial neural networks for instantaneous static security assessments, while the effectiveness of DNNs in real-time computational tasks and superior generalization within power system security assessments is studied in [23–25]. However, approaches based on DNNs often struggle because they need a lot of labeled data to train the models effectively [26, 27]. With PMU data growing rapidly and reaching the size of big data, manually labeling such large datasets is both expensive and challenging for tasks that need to be done in real time. This highlights the need to create voltage stability predictors that require very few labeled inputs for training in real time.

Moreover, network conditions and typologies are always changing. Many modern methods use decision trees to improve adaptability in real time [9, 20, 28, 29]. For example, Su and Liu [9] introduce an improved random forest model that adapts to changes by updating its trees. Maaji et al. [29] introduce a method called adaptive boosting, which builds a stronger classifier by adding a series of basic learners that get better with new PMU data. However, a major problem with many of these techniques is that they assume there is a lot of labeled data from new conditions available for training [9, 28, 29]. The significant effort and time needed to label numerous PMU datasets make these methods unrealistic for real-time changes. This situation creates an opportunity for developing a static voltage stability monitor that can adapt to changing situations using only a small amount of new labeled data.

2 Local Energy System Fault Detection

In this section, we explore the use of adaptive machine learning methodologies for fault event detection, specifically leveraging the adaptive isolation forests (*i*Forest). This ensemble learning method, which uses isolation trees, can quickly and accurately identify power system faults, even when the measurements are noisy [22, 30].

2.1 *i*Tree and *i*Forest

In this section, we delve into the concepts of *i*Tree, *i*Forest, and path length. These elements are fundamental for understanding the online isolation forest (*i*Forest)-based detection algorithm that we will discuss in the following section.

Anomalies can be regarded as rare data points that significantly differ from the standard instances. Here, an instance is described as a high-dimensional PMU reading, selected for its unique subsampled features [1, 22, 30]. In [30], we term these anomalies as event instances. Their unique properties make anomalies especially amenable to a process termed "isolation." Within a given random binary tree structure, there is a continuous effort to partition instances until each is distinct and isolated. This random method often leads to anomalies having much shorter paths. As illustrated in Fig. 1a and b, when there are fewer unusual events, there is less need to divide the data, leading to shorter paths. Additionally, instances with unique attributes often find themselves isolated in the earlier stages of the partitioning process. Drawing from the idea of isolation, we introduce the *i*Tree, which employs a binary tree design to separate instances [22]. We will proceed to define *i*Tree in the following.

Definition 1 (*Isolation Tree*) The *i*Tree, illustrated in Fig. 2, is a binary tree with a unique random design. Each node can have up to two children, but while every internal node has a pair, the external ones are childless. During its formation, each internal node selects a feature q and a corresponding threshold p at random. It then branches into two based on whether $q < p$. The tree develops in this manner until a node has identical values throughout or consists of a single entry.

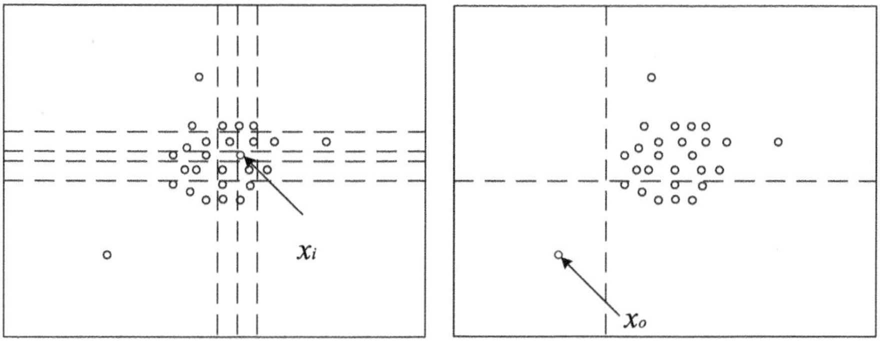

(a) Identifying anomalies. (b) Identifying normal observations.

Fig. 1 Isolation mechanism

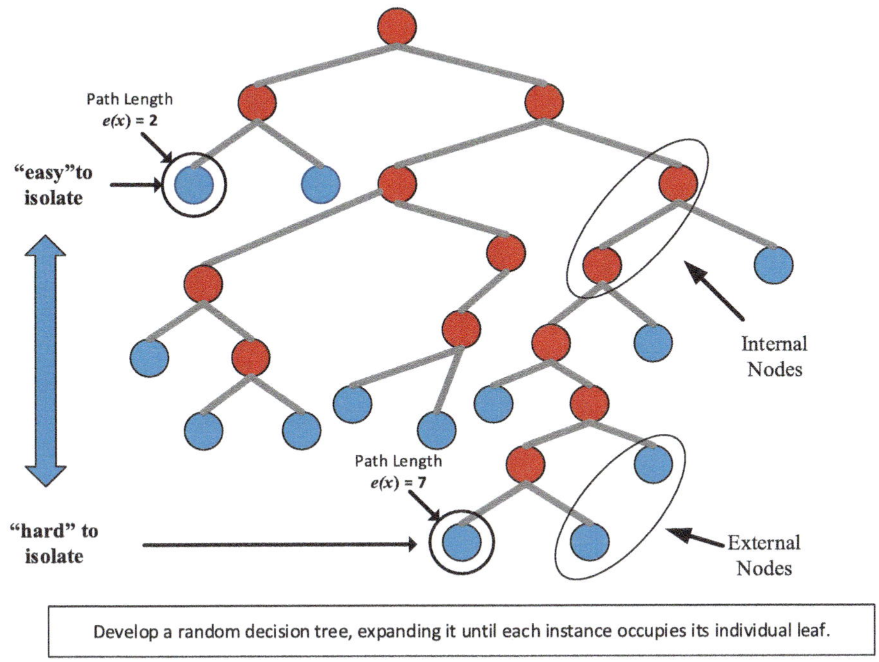

Fig. 2 Illustration of an *i*Tree

Consider a training dataset of N examples, symbolized as $\mathcal{X} = \{\boldsymbol{x}_1, \ldots, \boldsymbol{x}_N\}$. From this, we extract a subset $\hat{\mathcal{X}}$ comprising ψ examples, which forms the base for creating an *i*Tree. The creation of an *i*Tree involves breaking down $\hat{\mathcal{X}}$ into smaller segments. We emphasize that during *i*Tree creation, we always work with subsets of a fixed, smaller size, regardless of the overall dataset scale. This strategy guarantees the fast generation of each *i*Tree. Figure 2 illustrates that anomalies are typically closer to the root of the *i*Tree, leading to shorter paths. In contrast, normal data points are isolated into deeper nodes, resulting in longer paths. Hence, we can consider anomaly scores by considering these path lengths. We will explain exactly what we mean by "path length" in the next section.

Definition 2 (*Path Length*) In the *i*Tree, $e(\boldsymbol{x})$ denotes the count of edges that lie between the root node and the terminal node associated with the instance $\boldsymbol{x}$.

By using the dataset $\mathcal{X}$, we can create different *i*Trees. Each one varies because of the randomly selected features, division values, and a subset of the data used. Hence, anomaly scores for a particular instance might vary. To reduce the variability that comes from just one *i*Tree, we can put together an *i*Forest, which serves as a group of multiple *i*Trees. For local energy system fault detection, feature subspace selection is the first step. Begin by collecting a comprehensive training set, $\mathcal{X} = \{\boldsymbol{x}_1, \ldots, \boldsymbol{x}_N\}$, with features F. We term the fault event detector as *i*F.

2.2 Data Augmentation

Before diving into how the fault event detector works, we introduce data augmentation, a technique widely used in fields like speech recognition [31] and computer vision [32, 33], often involving the addition of synthetic data to training sets [33–35]. In this study, such a strategy bolsters the *i*Forest's capacity for generalization by promoting growth in its low-level leaves.

Real-world PMU data usually possess a limited number of event data, causing an imbalance between normal and anomalous data in the training set. If the recent training set lacks events, the *i*Trees may be dominated by deep-level leaves. Consequently, any subsequent event could be mistakenly assigned to a deep-level leaf, leading to detection errors. To address this, we propose a data augmentation approach that enriches the training set with more anomalous instances, ensuring *i*Trees exhibit a balanced blend of leaf levels. When an event arises, it's more aptly classified into a low-level leaf, ensuring accuracy. Here's how the augmentation works: For each feature in $\mathcal{X}$, we first determine its mean ($\boldsymbol{\mu}$) and standard deviation ($\boldsymbol{\sigma}$). To simulate real-world noise, we generate synthetic data that follows a normal distribution (Gaussian distribution) centered around the actual data's statistics. The amount of noise is controlled by a parameter called the augmentation coefficient λ. This coefficient scales the standard deviation ($\boldsymbol{\sigma}$) of the normal distribution, allowing us to adjust the level of noise added to the synthetic data (represented mathematically by $\boldsymbol{\lambda} \times \boldsymbol{\sigma}$). This synthetic data, denoted $\mathcal{X}_{fa}$, is then incorporated into the original training set.

To visualize the value of this synthetic data, consider Fig. 3 which plot two features: voltage magnitude (x-axis) and voltage angle (y-axis). In Fig. 3, we see training data clustered around normal PMU readings and readings during a bus fault event. This bias impacts anomaly score contours. For instance, a subsequent line tripping event could be misclassified, as shown in Fig. 3. Yet, by introducing synthetic data (Fig. 3), anomaly contours become more balanced (Fig. 3), leading to more accurate event detection.

2.3 Training Stage

The training process of the *i*Forest-based detection algorithm, described above, is summarized in Algorithm 2. Steps 7–15 correspond to the training of the *i*Forest. The training process of *i*Tree is shown in Steps 16–27.

2.4 Evaluation Stage

In the evaluation stage, we measure anomaly scores for each sample using the *i*F detectors. We use a normalization term $c(\psi)$ to compute the anomaly score. This

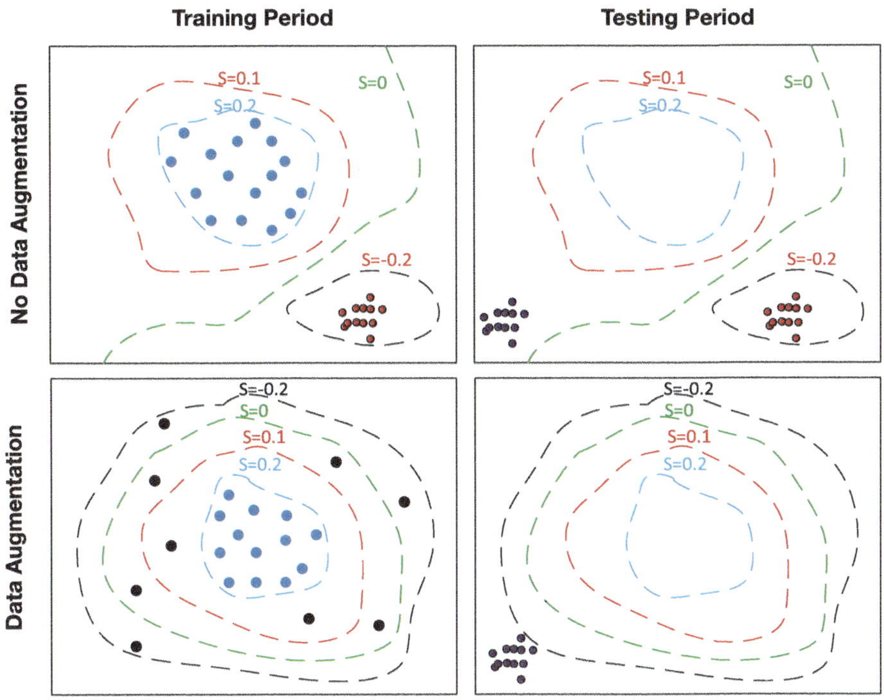

Fig. 3 The generalization performance of fake noisy data

term reflects the average path length for ψ samples from an iTree [22] and is given by:

$$c(\psi) = 2H(\psi - 1) - 2(\psi - 1)/\psi, \tag{1}$$

Here, ψ represents the size of the subsampled dataset. The harmonic number $H(\psi - 1)$ can be approximated as $\ln(\psi - 1) + 0.5772156649$, referencing Euler's constant. In this framework, $c(\psi)$ helps adjust the path length $e(\boldsymbol{x})$, forming the basis of the anomaly score s, which is

$$s(\boldsymbol{x}, \psi) = 0.5 - 2^{-\frac{E(e(\boldsymbol{x}))}{c(\psi)}}, \tag{2}$$

Here, $E(e(\boldsymbol{x}))$ denotes the average value of $e(\boldsymbol{x})$ over T iTrees for a given sample $\boldsymbol{x}$ and is found using:

$$E(e(\boldsymbol{x})) = \frac{1}{T}\sum_{t=1}^{T} e_t(\boldsymbol{x}). \tag{3}$$

Algorithm 1: Training Process of *i*Forest-based detection Algorithm

```
Input: X - data, ψ - subsampling size, λ - augmentation coefficient, T - number of trees
Output: iF
Select feature subspace to get X and F;
Calculate mean (μ) and standard deviation (σ) of X;
Generate noisy data: X_fa ~ N(μ, λ × σ);
Augment data: X ← X ∪ X_fa;
Construct forest: iF = iForest(X, T, ψ, F);
return iF;
Function iForest(X, T, ψ, F)
    Output: iForest
    Initialize iForest as empty;
    for t = 1 to T do
        Sample X̂_t from X with ψ instances;
        iT_t = iTree(X̂_t, F);
        Add iT_t to iForest;
    end
    return iForest;
end
Function iTree(X, F)
    Output: An iTree
    Initialize exnodes and innodes as empty;
    if X is indivisible then
        Set size of exnodes to |X|;
    else
        Randomly select q from F and split value p from range of q in X;
        Split X into left (X_l) and right (X_r) subtrees using q and p;
        Define nodes with left, right subtrees, and attributes q, p;
    end
    Combine exnodes and innodes to form iTree;
    return iTree;
end
```

Based on the score s, we infer the following: Positive scores indicate regular data, while negative scores suggest anomalies. Samples with $s_{L1} < 0$ are flagged as events.

2.5 Adaptive Learning Procedure

In the online detection phase, we define the training period as T_{trn}. The *i*Forest-based detection algorithm, with its adaptive learning process, is designed to adjust to the changing operational conditions of the power system. Here's a concise overview of the adaptive learning procedure:

1. In the absence of events, the *i*Forest is employed for swift event detection even with sub-optimal data readings. Specifically, we update the *i*Forest as $i\mathrm{F} \leftarrow i\mathrm{F} \cup i\mathrm{T}_{\mathrm{new}} - i\mathrm{T}_{\mathrm{old}}$. Here, each $i\mathrm{T}_{\mathrm{new}}$ is generated in real time.

2. When an event is identified and subsequently resolved by system operators, initial scores from the primary detector are negative. They revert to positive values once the issue is addressed, revealing that the system has transitioned to a new, stable operational state.
3. Upon entering this new operating condition, the entire *i*Forest is refreshed using real-time PMU samples.

In essence, our proposed detection approach is designed for real-time execution, leveraging live streaming data sourced from PMUs.

3 Local Energy System Voltage Stability Monitoring

While the fault event can be detected rapidly, a critical situation concerning voltage magnitudes is not necessarily a fault event. Instead, it closely approaches a voltage collapse. In this section, we will discuss how such situations can be monitored using a deep learning method.

3.1 *Voltage Stability Labeling*

First, we present the concept of voltage stability margin. Next, we frame the static voltage stability monitoring issue as a classification challenge.

One can evaluate voltage stability by examining the distance between the current operating point (OP) and the threshold of voltage collapse (VCP). Figure 4 showcases the P-V curve, illustrating voltage magnitude variations against different load

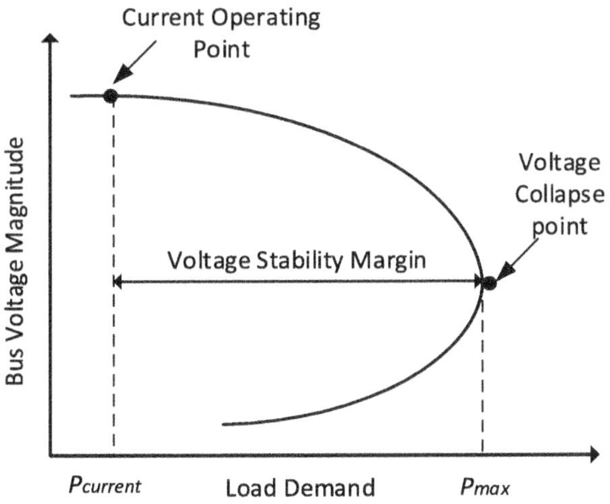

Fig. 4 VCP and VSM on load demand—voltage magnitude curve

demands. The P-V curve is commonly determined through the continuation power flow (CPF) method [36]. If the OP gets too close to the VCP, a voltage collapse is on the horizon. To gauge this, the voltage stability margin (VSM) is defined by:

$$VSM = \frac{P_{\max} - P_{op}}{P_{\max}}, \tag{4}$$

where $P_{\max}$ is the maximum achievable power and P_{op} is the base load demand at the active OP. Building on this, Wu et al. [37] segregate an OP into three potential states: stable, alert, or critical. This differentiation aids operators in taking proactive measures, notably if there's a drift toward a critical state, thus reducing voltage collapse chances. As described in [14], the method involves accumulating M OPs from the system's PMUs. The mean VSMs of these M OPs are given by $\mathbf{VSM}_m$, and the first quartile VSMs by $\mathbf{VSM}_q$. An OP exceeding $\mathbf{VSM}_m$ is labeled stable ($y = 1$). Those beneath $\mathbf{VSM}_m$ yet above $\mathbf{VSM}_q$ are considered alert ($y = 2$), while any falling under $\mathbf{VSM}_q$ are termed critical ($y = 3$).

In our research, every OP is characterized by synchronized measurements of voltage, current, and both active and reactive power injections, represented as magnitude and phase angles. A training sample is symbolized as $\boldsymbol{x} = \{\boldsymbol{V}, \boldsymbol{\theta}, \boldsymbol{P}, \boldsymbol{Q}\}$, where $\boldsymbol{V}$ covers voltages from V_1 to V_N, $\boldsymbol{\theta}$ encompasses phase angles from θ_1 to θ_N, and similarly for $\boldsymbol{P}$ and $\boldsymbol{Q}$. Here, N indicates the count of buses. For every training example $\boldsymbol{x}_l$, we employ the CPF approach to determine its corresponding VSM and deduce its label y_l. In this context, a labeled dataset is represented as $\mathcal{L} = \{(\boldsymbol{x}_l, y_l)\}_{l=1}^{l_s}$, comprising l_s labeled instances, and an unlabeled dataset is symbolized as $\mathcal{U} = \{(\boldsymbol{x}_u)\}_{u=1}^{u_t}$, with u_t unlabeled instances. A notable challenge is the extensive computational time required to determine the VSM for each OP using the CPF approach. This means generating labeled OPs becomes time-consuming, whereas collecting unlabeled OPs is relatively easy. Given these constraints, our goal is to introduce a learning technique that leverages a minimal subset of labeled data.

3.2 Semi-supervised Deep Learning for Voltage Stability Monitoring

In this section, we will explain the semi-supervised deep learning which is designed for monitoring voltage stability. Specifically, in this subsection, we present the tri-net DNN, a model that requires only a minimal amount of labeled data for training, significantly augmented by the utilization of unlabeled data. It is important to note that while unlabeled data are easy to obtain, collecting labeled data requires a lot of time.

3.2.1 Tri-Net DNN

The tri-net consists of a shared module, M_s, and three distinct modules: M_1, M_2, and M_t, as depicted in Fig. 5. The shared module, M_s, produces features common to the three separate networks. The roles of M_1 and M_2 are to provide complementary features and generate pseudo-labels to enhance the training set. On the other hand, M_t is primarily responsible for classifying features derived from M_s. During training, the discrepancies between the three modules, M_1, M_2, and M_t, are harnessed. Notably, when the complementary modules, M_1 and M_2, concur on the prediction for an unlabeled data point from the set $\mathcal{U}$, that data point is pseudo-labeled. This collection of pseudo-labeled samples is represented by $\mathcal{L}_l$. Subsequently, to enrich the training data, we combine the original labeled dataset, $\mathcal{L}$, with these pseudo-labeled samples, resulting in an augmented dataset, $\mathcal{PL} = \mathcal{L} \cup \mathcal{L}_l$.

3.2.2 Training and Implementation on Initial Systems

Loss function: For the modules M_1, M_2, and M_t, we define their respective loss functions as Loss(M_1), Loss(M_2), and Loss(M_t). The functions for the shared module, M_1-module, M_2-module, and the main module are represented by f_{M_s}, f_{M_1}, f_{M_2}, and f_{M_t}, in that order.

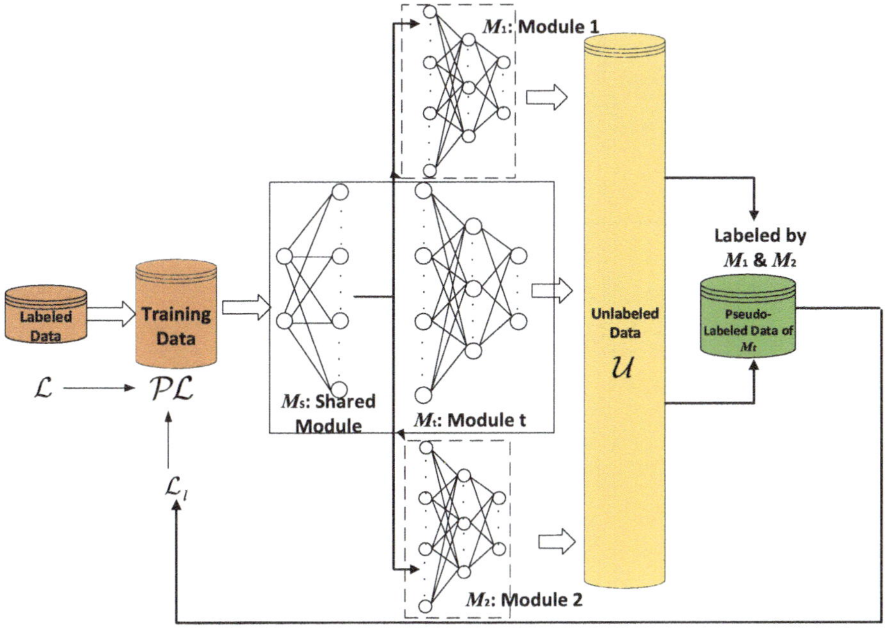

Fig. 5 The Tri-net DNN framework

Algorithm 2: Training Procedure

Input: PMU data: $\mathcal{L} = \{(\boldsymbol{x}_l, y_l)\}_{l=1}^{l_s}$, $\mathcal{U} = \{(\boldsymbol{x}_u)\}_{u=1}^{u_t}$, and initialized $\mathcal{L}_l = \emptyset$; U: Size of $\mathcal{U}$; σ: threshold for filtering pseudo-labels; N_{init}: initial count of unlabeled samples; **AssignLabels**: labeling function; *Niter*: iteration count.
Output: Trained DNN composed of M_S and M_t.
Initialize: Train M_S, M_1, M_2 using mini-batches from $\mathcal{L}$;
Set $N_t = N_{init}$;
$\mathcal{L}_l \leftarrow$ **AssignLabels**(M_S, M_1, M_2, $\mathcal{U}$, N_t, σ);
Construct augmented set: $\mathcal{PL} = \mathcal{L} \cup \mathcal{L}_l$;
for $t = 1$ *to* T **do**
  for $i = 1$ *to Niter* **do**
    Train M_S, M_1, M_2 using mini-batches from $\mathcal{PL}$;
    Train M_S, M_t using mini-batches from $\mathcal{PL}$;
  end
  Adjust N_t: $N_t \leftarrow \min(t/20 \times U, U)$;
  Clear augmented set: $\mathcal{PL} = \emptyset$;
  $\mathcal{L}_l \leftarrow$ **AssignLabels**(M_S, M_1, M_2, $\mathcal{U}$, N_t, σ);
  Update augmented set: $\mathcal{PL} = \mathcal{L} \cup \mathcal{L}_l$;
end
Remove modules M_1 and M_2;
return M_S and M_t;
Function ***AssignLabels***(M_S, M_1, M_2, $\mathcal{U}$, N_t, σ)
  Initialize labeled set: $\mathcal{L}_l = \emptyset$;
  for $i = 1$ *to* N_t **do**
    Determine predictions and confidences: $\bar{y}_1 \leftarrow \arg\max_y f_{M_1}(f_{M_s}(y|\boldsymbol{x}_i))$;
    $\bar{y}_2 \leftarrow \arg\max_y f_{M_2}(f_{M_s}(y|\boldsymbol{x}_i))$;
    $\sigma_1 \leftarrow f_{M_1}(f_{M_s}(\bar{y}_1|\boldsymbol{x}_i))$;
    $\sigma_2 \leftarrow f_{M_2}(f_{M_s}(\bar{y}_2|\boldsymbol{x}_i))$;
    if *confidence criteria met:* $\sigma_1 \geq \sigma$, $\sigma_2 \geq \sigma$, *and* $\bar{y}_1 = \bar{y}_2$ **then**
      Add to labeled set: $\mathcal{L}_l \leftarrow \mathcal{L}_l \cup \{(\boldsymbol{x}_i, y_i = \bar{y}_1)\}$;
    end
  end
  return $\mathcal{L}_l$;
end

$$\text{Loss}(M_1) = \frac{1}{n}\sum_{i=1}^{n}\{L_y(f_{M_1}(f_{M_s}(\boldsymbol{x}_i)), y_i)\} + \alpha_1(\|W_1\|_2^2 + \|W_S\|_2^2), \tag{5}$$

$$\text{Loss}(M_2) = \frac{1}{n}\sum_{i=1}^{n}\{L_y(f_{M_2}(f_{M_s}(\boldsymbol{x}_i)), y_i)\} + \alpha_2(\|W_2\|_2^2 + \|W_S\|_2^2), \tag{6}$$

$$\begin{aligned}\text{Loss}(M_t) = &\frac{1}{n}\sum_{i=1}^{n}\{L_y(f_{M_t}(f_{M_s}(\boldsymbol{x}_i)), y_i)\} + \lambda|W_1(1)^T W_2(1)| \\ &+ \alpha_3(\|W_t\|_2^2 + \|W_S\|_2^2),\end{aligned} \tag{7}$$

where let L_y represent the softmax cross-entropy loss function. To mitigate the risk of overfitting, we incorporate an l_2-norm regularization term, such as $\| W_1 \|_2^2$, $\| W_2 \|_2^2$, or $\| W_S \|_2^2$. Initially, we train the paired complementary modules M_1 and M_2, along with the shared module M_s, aiming to minimize the combined result of Eqs. (5) and (6). Subsequently, we focus on the primary modules M_t and M_s, minimizing according to Eq. (7). This iterative training methodology can be observed in Steps 7–8 of Algorithm 2. In Eq. (7), we encounter the term $|W_1(1)^T W_2(1)|$, wherein $W_1(1)$ and $W_2(1)$ represent the initial layer weights of M_1 and M_2, in that order. Given that both M_1 and M_2 are trained on identical datasets [38], there's an inherent propensity for them to exhibit likeness. To counteract this and promote diversity, the term $|W_1(1)^T W_2(1)|$ is integrated into our objective function, specified in Eq. (7). Consequently, M_1 and M_2 adapt to distinct features, empowering them to confidently label samples from $\mathcal{U}$.

Training Process: Algorithm 2 illustrates the training procedure. Initialization is depicted in Step 1, where the loss functions of M_S, M_1, and M_2 are refined as per Eqs. (5) and (6). The **Labeling** actions in Steps 3 and 12 can be described in this manner: Trained modules M_1 and M_2 predict samples from $\mathcal{U}$, producing pseudo-labels $\bar{y}_1$ and $\bar{y}_2$. If both $\bar{y}_1$ and $\bar{y}_2$ concur and their highest posterior probabilities surpass the threshold σ, it signifies classifier consensus. Accordingly, the entity $\{(\boldsymbol{x}_l, y_l = \bar{y}_1 = \bar{y}_2)\}$ gets incorporated into the augmented dataset $\mathcal{PL}$ as seen in Step 4. Steps 5 through 14 detail the training utilizing mini-batch instances. The alternating training of M_1, M_2, and M_t is portrayed in Steps 7 and 8. Step 10 methodically augments the pool of unlabeled instances prepared for labeling, captured as $N_t \longleftarrow \min(t/20 \times \mathrm{U}, \mathrm{U})$, with U representing the utmost count of unlabeled instances. To avert model over-reliance on pseudo-labels, $\mathcal{PL}$ is cleared in Step 11, followed by a new sampling and labeling from $\mathcal{U}$ in Step 12.

Online Implementation: Once the tri-net DNN completes its training, we retain only the modules M_s and M_t, omitting the complementary modules M_1 and M_2 as outlined in Steps 15–16. Subsequently, we assess voltage stability by introducing incoming OPs to the trained DNN. This process allows us to determine the associated operating conditions, which can be classified as stable, alert, or critical.

4 Case Studies

In this section, we present the application of the *i*Forest algorithms for local energy system fault detection and delve into the semi-supervised deep learning techniques for voltage stability monitoring. We validate the effectiveness of the proposed algorithm using PMU data generated by Siemens PSS/E [38], realistic data from ISO New England [39], and actual real-world demands.

4.1 Power System Fault Detection

This section highlights the effectiveness of the *i*Forest algorithm in swiftly identifying events through the usage of low-quality data measurements. The algorithm's validation is achieved using synthetic PMU data, produced by Siemens PSS/E [38], and genuine data obtained from ISO New England [39]. Figure 6 provides a visual representation of the system topology. In every simulation, a subsampling size, $\psi = 256$, is applied to train an *i*Tree, and 500 *i*Trees are utilized to build an *i*Forest. To prevent false alerts, a filtering technique is applied, which quickly adjusts scores that could be affected by measurement noise. As per IEC Std. 61000-4 [40], it is recommended that the observation interval encompasses ten cycles of the established system frequency. Additionally, Frigo et al. [41] employ a 200 ms filter for PMU noise when the system is functioning at 50 Hz frequency. Within the context of this research, the filtering method flags an anomaly only when the anomaly score falls below zero for a duration equivalent to ten cycles of the nominal system frequency.

In an effort to emulate practical PMUs, Gaussian noise, characterized by a signal-to-noise ratio (SNR) of 92 dB, is introduced to the synthetic training data. Assuming each PMU samples three buses, the training set hence integrates 19 features across all three, encompassing frequency, power flows, complex voltages, and complex currents. The duration of the training data, T_{trn}, is set at 1000 s. Moreover, we determine the augmentation coefficient (λ) as 500, subsequently incorporating 20000 artificial

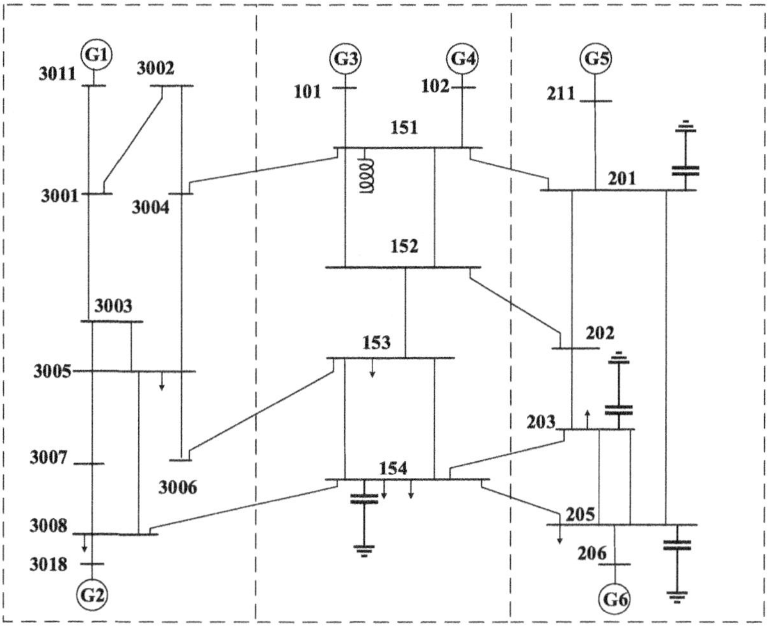

Fig. 6 Topology of PSS/E 23-bus system

instances into the training data to facilitate event detection amid low-quality data measurements. Conversely, concerning real-world data, oscillatory events within actual power systems are assembled through PMU data measurements. In particular, we select realistic instances that include the oscillatory event transpiring on July 20, 2017 [39]. A technical malfunction in a substantial generator initiated 1.13 Hz oscillations, escalating in magnitude over 80 s. PMU data from various substations is selected for both training and testing. Additionally, we define the augmentation coefficient (λ) as 10, followed by the addition of 1000 artificial instances into the training data to assist in event detection.

4.1.1 Line Tripping and Line Closure Event

In showcasing the adaptive training method, we introduce two fault events into a designated system. If the detector considers adaptive characteristics, it discerns both incidents; if not, only the first is recognized. Suppose a tripping event in Line 151–152 occurs at $t = 1.0$ s and is followed by a closure of the same line at $t = 21.0$ s. Figure 7 displays notable modifications in features, including bus voltage and bus frequency deviation, at $t = 1.0$ s and $t = 21.0$ s. Without adaptive training, the scores of iF are illustrated in Fig. 8. iF senses the line tripping event at 1.01 s, with its anomaly scores sharply descending from 0.1359 to −0.1969. Following $t = 10.0$ s, the system transitions into a new operational phase, as demonstrated in Fig. 7. At $t = 21.0$ s, the second incident—the line closure—cannot be differentiated from the first by the iF detector in Fig. 8, due to the lack of adaptive training.

In contrasting Fig. 8 with Fig. 9, the latter distinctly underscores the merits of adaptive training. The training is implemented to refine the proposed algorithm. While the detection of the first event at $t = 1.01$ s in Fig. 9, the power system transitions into a new operational state post $t = 10.0$ s. Notably, upon the occurrence of the second event at $t = 21.0$ s, the score curves between Fig. 8 and 9 diverge significantly from 20 to 50 s. The scores in Fig. 9 jump from a positive to a negative value at $t = 21.01$ s, while those in Fig. 8 persistently remain negative from 20 s to 50 s. This suggests that the second event is detected at $t = 21.01$ s by the iF detector in Fig. 9, but goes undetected in Fig. 8. This result occurs because the adaptive training process uses PMU data from the new operational state to update the iForest-based detection model. As a result, the proposed algorithm successfully detects both events because of this adaptive training.

4.1.2 Oscillation Event

To confirm how well the proposed method works, we conduct case studies using realistic PMU data. Our goal is to detect the oscillation event. Figure 10 reveals the occurrence of an oscillation event from approximately 70 to 150 s. The frequency and complex voltage profiles during this event are also displayed in Fig. 10. Within this simulation, the training data set comprises 72,000 instances, each characterized

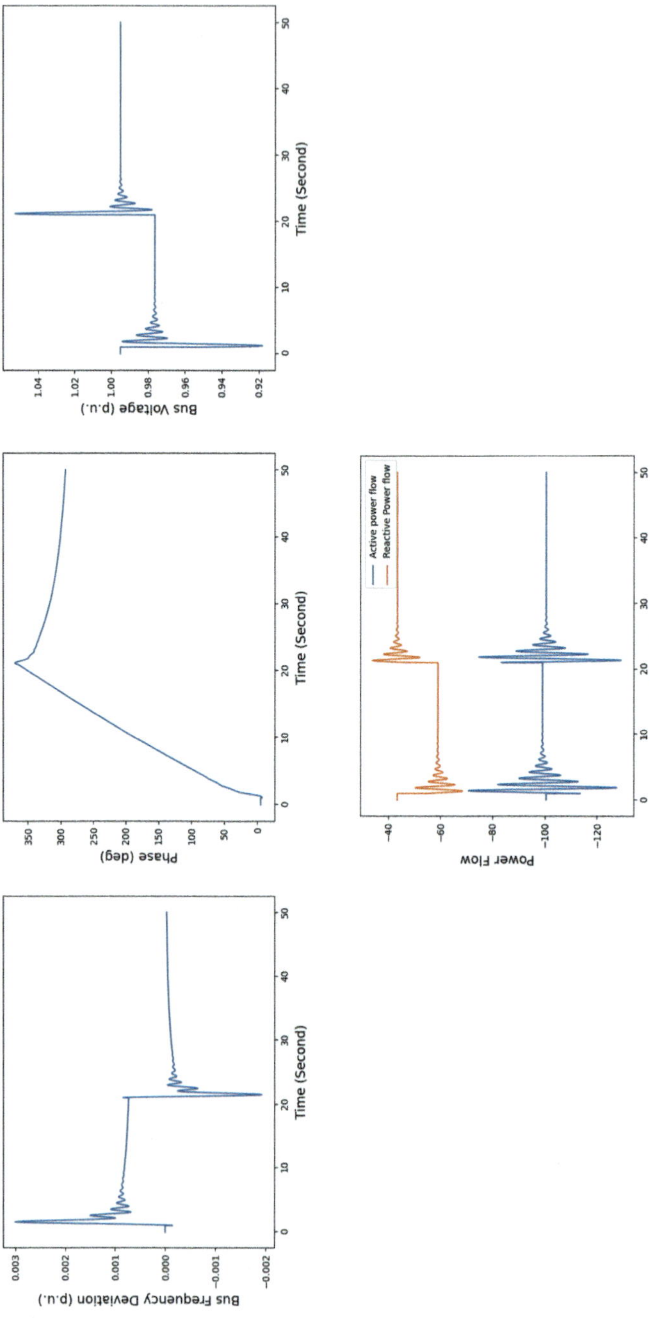

Fig. 7 The voltage magnitude, frequency, voltage phase, active and reactive powers profiles during line tripping and line closure event

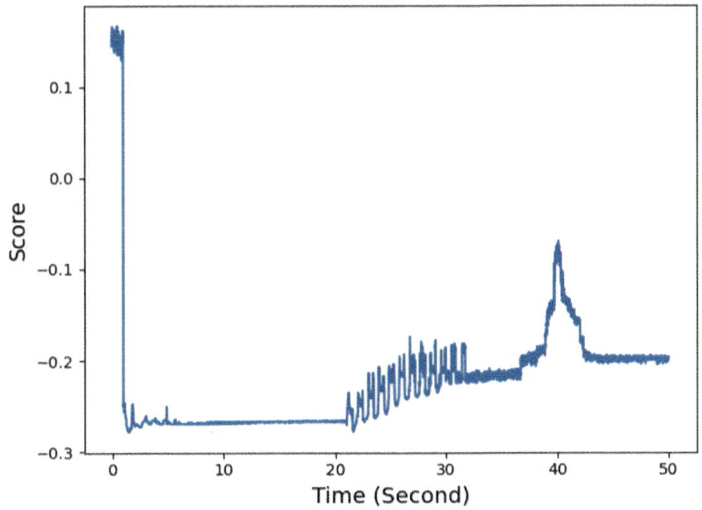

Fig. 8 The anomaly scores of iF during line tripping and line closure events

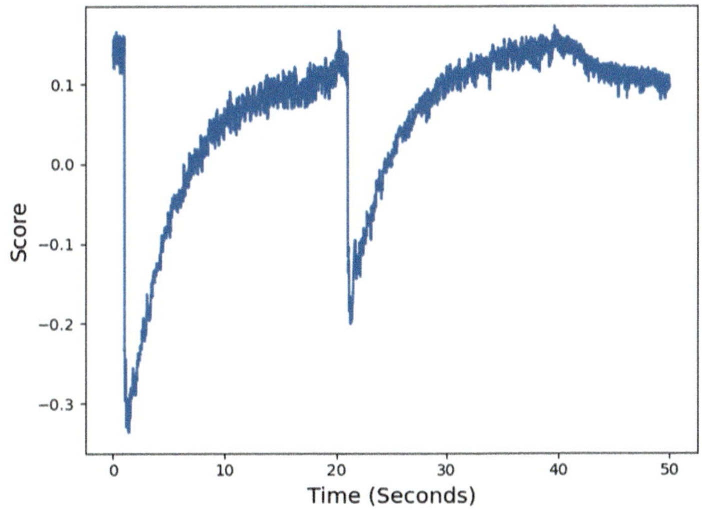

Fig. 9 The result of the iF detector iF with the adaptive training

by 170 features [39]. Figure 11 demonstrates the detection of the oscillation event at 70 s by iF. Prior to $t = 70$ s, the anomaly scores from the iF detector remain positive, indicating that the system is considered to be in a normal operational state. After 70 s, the anomaly scores from the iF detector consistently become negative, indicating the start of the oscillation event.

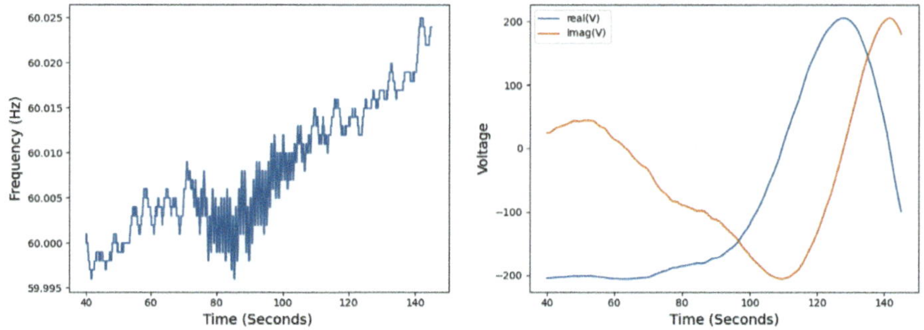

Fig. 10 The frequency and complex voltages profiles during the oscillation event

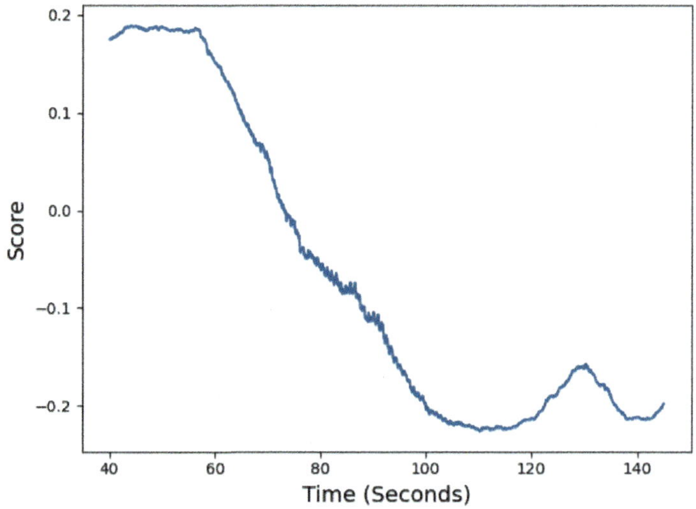

Fig. 11 The anomaly scores of iF during oscillation event

4.1.3 PMU Measurements with High Sensing Noise During the Fault Events

In this part, we showcase that low-quality measurements do not trigger false alarms using the proposed method. Figure 12a displays a bus fault event occurring at $t = 1.0$ s and resolving at $t = 1.1$ s. Following this bus fault, the voltage magnitude measurements endure Gaussian noises with a 40 dB SNR, persisting from 39.95 to 40.95 s. Figure 12b presents the iF anomaly scores s_{L1}. Initially, iF identifies this bus fault event at 1.01 s. Post $t = 25.0$ s, the system transitions into a new (normal) operating condition. From this point forward, detector iF maintains positive scores. Therefore, the presence of low-quality measurements does not trigger any false alarms from the iF detector.

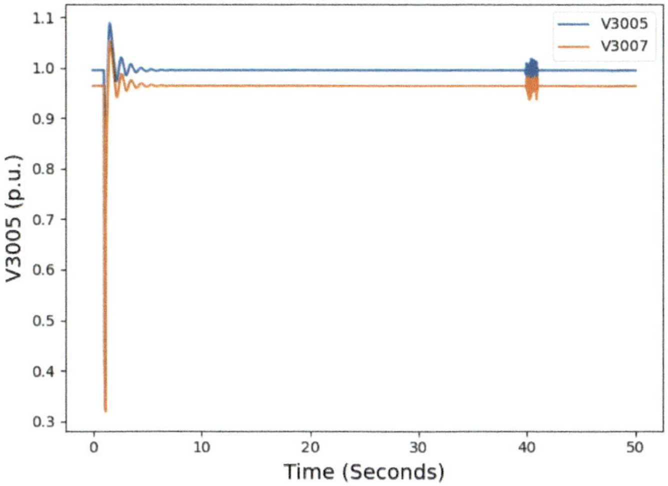

(a) V_{3005} and V_{3007} profile during High Sensing Noise.

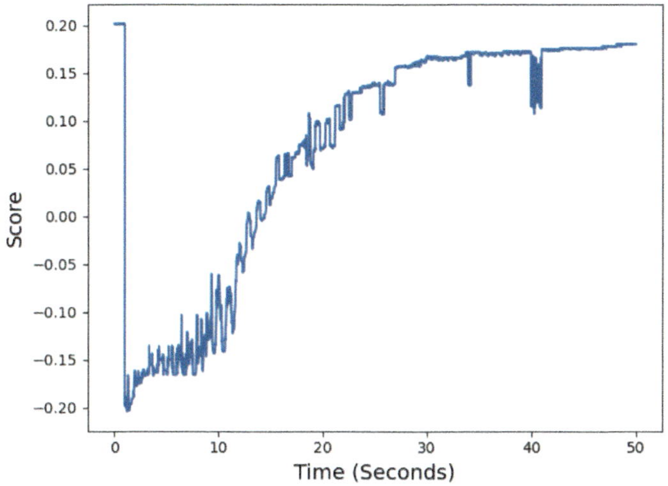

(b) Evaluation Score of *i*F detectors during High Sensing Noise.

Fig. 12 PMU measurements with high sensing noise during the events

4.1.4 PMU Measurements with One Asynchronous PMU During the Fault Events

In the ensuing simulation, we utilize PMU measurements from different substations, incorporating either 10 PMUs, within the training data set. Notably, measurements from one PMU are asynchronous, exhibiting a 0.1-s delay. As illustrated in Fig. 13,

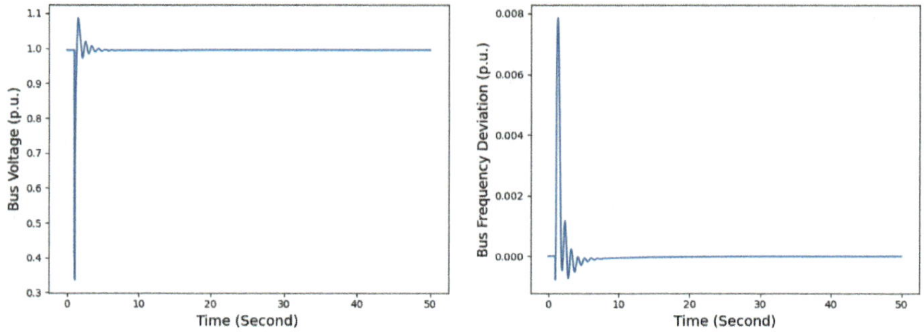

Fig. 13 The frequency and complex voltages profiles during bus fault event

we assume that a fault occurs at bus 151 at 1.0 s and is fixed after 0.1 s. Figure 13 also depicts the voltage and frequency profiles for bus 3005. Figure 14 reveals that, before the fault, the *i*F detector scores (represented by red curves) remain positive, even with one asynchronous PMU. Consequently, a single asynchronous PMU does not trigger any fault event alarm for the detector *i*F. This suggests that the *i*F detector is more robust when the training dataset includes more PMUs.

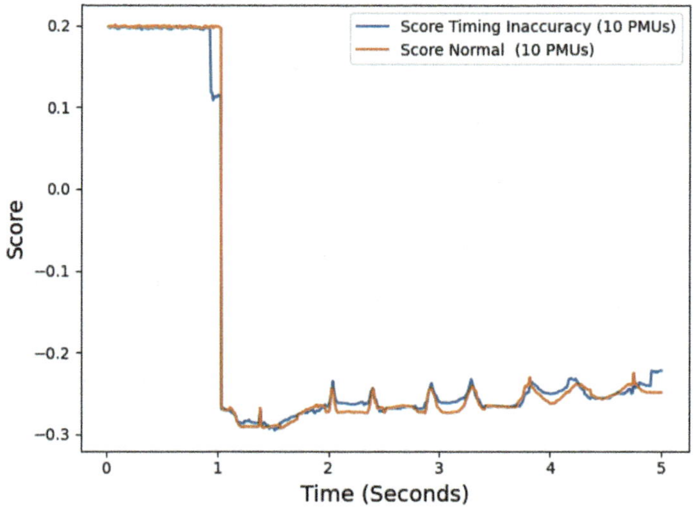

Fig. 14 The anomaly scores of *i*F during timing inaccuracy

4.2 Parameter Settings

In this study, we evaluate the performance of a proposed semi-supervised deep learning algorithm by using different amounts of labeled and unlabeled samples for training across various bus systems. Specifically, we use 300, 400, 500, and 600 labeled samples for the IEEE 9-bus, 14-bus, 30-bus, and 39-bus systems, respectively, each supplemented with 5000 unlabeled samples. For the Texas 2000-bus system, training includes 40, 60, 80, and 100 labeled samples, along with 7000 unlabeled samples. Additionally, to reduce computational complexity, 800 features are randomly selected from a total of 8000 as training samples.

Our proposed semi-supervised deep learning algorithm is compared with existing algorithms–SVC, LinearSVC, NuSVC, Random Forest (RF), AdaBoost SAMME.R (AB_R), AdaBoost SAMME.D (AB_D), and deep neural networks (DNN) [10]—using the same number of samples for training. To elucidate:

- SVC, or the support vector machine, facilitates multi-class classification [42].
- LinearSVC represents the SVC, employing a linear kernel [42].
- NuSVC utilizes SVC with a radial basis function kernel [42].
- Random Forest is an ensemble learning algorithm that uses multiple decision tree classifiers, each applied to different sub-samples of the dataset. It then uses averaging to improve predictive accuracy [9].
- AdaBoost is an ensemble learning algorithm that not only fits a classifier to the original dataset but also creates additional copies of the classifier on the same dataset. In this scenario, AB_R refers to a real boosting algorithm, whereas AB_D denotes a discrete boosting algorithm [43].

In Algorithm 2, we set the value of T to 50 and establish a threshold σ of 0.995 for all experiments. Cross-entropy between the outputs and labels is used as our loss function. Additionally, we set the training batch size to 128 and use the Adam optimizer, setting the learning rate L_0 to 0.001.

4.3 Performance of the Semi-supervised Deep Learning Method in the Stage 1

The performance is measured by accuracy, which is defined as

$$\text{Accuracy} = \frac{\text{Number of correct predictions}}{\text{Total number of predictions}}. \tag{8}$$

Here, we repeat the train-test process 10 times and calculate the average classification accuracy. The training set consists of the labeled samples and unlabeled samples, where the unlabeled samples are selected randomly from a set of $\mathcal{U}$ with 5000 candidates. Figure 15 shows simulation results for 300, 400, 500, 600 labeled samples

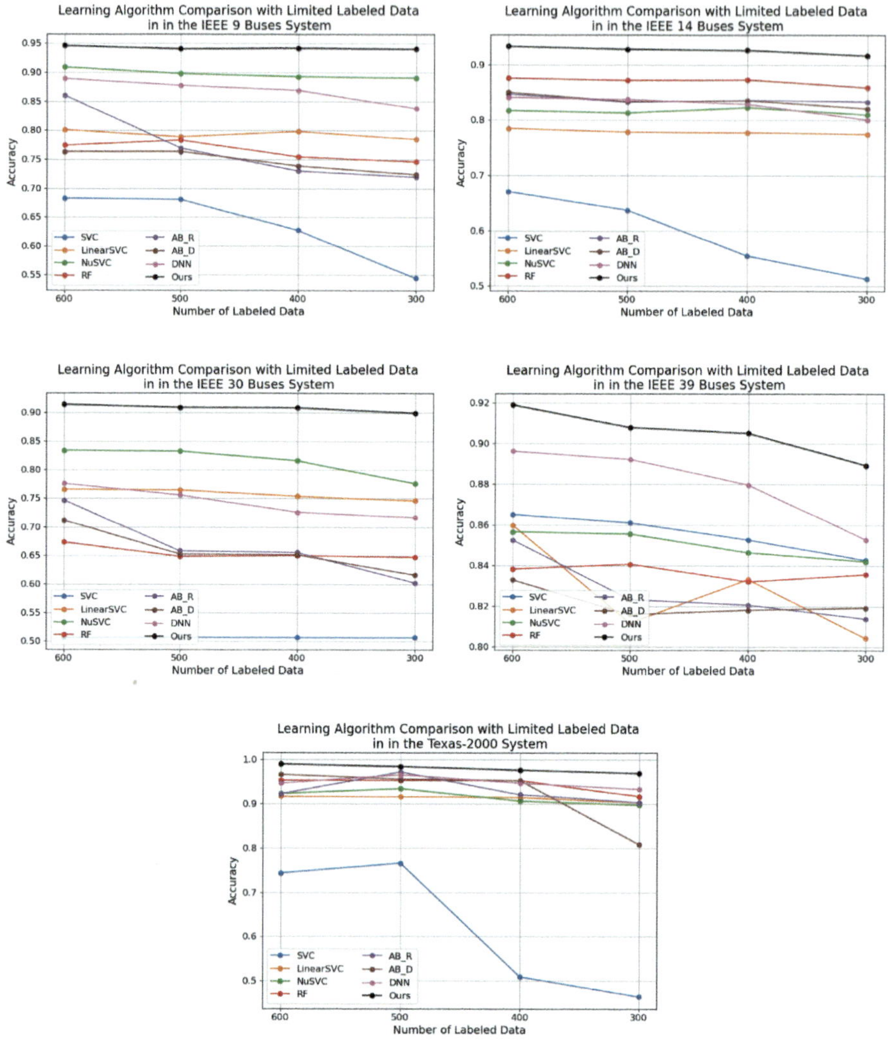

Fig. 15 Learning algorithm comparison with limited labeled data in the IEEE 9-bus, 14-bus, 30-bus, 39-bus, Texas 2000 systems

in the IEEE 9-bus, 14-bus, 30-bus, 39-bus systems and 40, 60, 80, 100 labeled samples in the Texas 2000-bus system.

To properly evaluate the performance of the proposed semi-supervised deep learning method, we make use of classification metrics for each class. Take stable OPs, for example. Let True Negatives (TN) refer to the stable OPs that are misclassified as alert or critical OPs. True Positives (TP) refer to the stable OPs correctly predicted to be stable. False Negatives (FN) refer to the alert or critical OPs that are predicted to be stable, and False Positives (FP) refer to the stable OPs that are predicted to be alert or critical. Moreover, Precision is defined as the ratio of the number of True Positives to the total number of OPs that are actually stable.

$$\text{Precision} = \frac{\text{TP}}{\text{TP} + \text{FP}}. \tag{9}$$

Likewise, recall (True Positive Ratio (TPR)) is defined as the ratio of the number of True Positives to the number of true stable OPs

$$\text{Recall} = \frac{\text{TP}}{\text{TP} + \text{FN}}. \tag{10}$$

To strike a balance between the precision and recall, we define F_1-Score as the geometrical average of the precision and recall. Equation (11) shows the F_1-Score calculation.

$$F_1\text{-Score} = 2 \times \frac{\text{Precision} \times \text{Recall}}{\text{Precision} + \text{Recall}}. \tag{11}$$

Moreover, we calculate its weighted average, Wgt-F_1, where the weights are determined by the number of true instances for each class. This metric evaluates classification problems involving multiple classes.

With a small number of labeled data for training, the proposed semi-supervised deep learning method outperforms the traditional learning methods by employing unlabeled samples. Specifically, with 600 labeled samples in the IEEE 9-bus system, the semi-supervised deep learning method achieves an accuracy of 94.62% in contrast to 68.27, 80.07, 90.93, 77.41, 86.04, 76.33, and 88.96% achieved by SVC, LinearSVC, NuSVC, Random Forest, AdaBoost SAMME.R, AdaBoost SAMME.D, and DNN, respectively. As the number of labeled samples decreases to 300 samples, the semi-supervised deep learning method achieves an accuracy of 94.03% in the IEEE 9-bus system, compared to 89.08% achieved by NuSVC, which is the highest among the traditional learning methods.

In Fig. 16, we can observe a noticeable positive correlation between the number of labeled samples and the Wgt-F_1 score, with the score progressively increasing as more labeled samples are added. For example, the Wgt-F_1 score for the IEEE 30-bus cases increases from 89.95 to 91.61% as the number of labeled samples grows from 300 to 600. Additionally, the performance of the semi-supervised deep learning method in terms of Wgt-F_1 score significantly improves as the number of buses in the system increases. For instance, in the Texas 2000-bus system, the method requires only 40 labeled samples to achieve a Wgt-F_1 score of 97.04%.

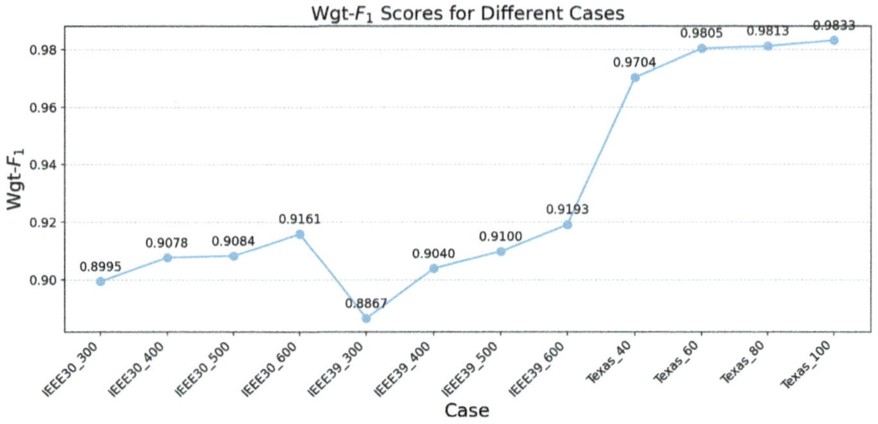

Fig. 16 The F1 Score of the semi-supervised deep learning method for the voltage stability monitoring

5 Conclusion

This chapter has presented two crucial algorithms, each designed to enhance power system monitoring and event detection by effectively utilizing PMU data. The first method, an online and adaptive approach using the *i*Forest detection algorithm, efficiently identifies events even in the presence of low-quality data without requiring a detailed system model or extensive communication between buses. The *i*Forest algorithm not only ensures swift responses and maintains low computational and storage demands but also adapts to new operating conditions very swiftly, ensuring its effectiveness even as network conditions or topologies evolve. The method's effectiveness, validated through case studies employing synthetic and real-world PMU data, underscores its proficiency in detecting events even with low-quality measurements.

Simultaneously, a semi-supervised deep learning approach for static voltage stability monitoring has been presented. This method leverages a large amount of unlabeled PMU data alongside a smaller set of labeled data, significantly reducing the time costs typically associated with data labeling. The algorithm is particularly effective, requiring only a few labeled samples from new settings to achieve high accuracy; it has been shown to achieve over 94.03% accuracy with just 300 labeled samples in IEEE benchmark system studies. After incidents such as line tripping, it continues to perform well, maintaining an accuracy of over 95.48% in new environments with as few as 5 samples per category.

References

1. Wu T, Zhang Y-JA, Tang X (2018) Isolation forest based method for low-quality synchrophasor measurements and early events detection. In: 2018 IEEE international conference on communications, control, and computing technologies for smart grids (SmartGridComm), pp 1–7
2. Wu T, Zhang Y-JA, Tang X (2020) Online detection of events with low-quality synchrophasor measurements based on *i*Forest. IEEE Trans Indus Inform 17(1):168–178
3. Wu T, Zhang Y-JA, Liu Y, Lau WC, Xu H (2020) Missing data recovery in large power systems using network embedding. IEEE Trans Smart Grid 12(1):680–691
4. Xie L, Chen Y, Kumar PR (2014) Dimensionality reduction of synchrophasor data for early event detection: linearized analysis. IEEE Trans Power Syst 29(6):2784–2794
5. Li S, Ajjarapu V, Djukanovic M (2018) Adaptive online monitoring of voltage stability margin via local regression. IEEE Trans Power Syst 33(1):701–713
6. Wu M, Xie L (2017) Online detection of low-quality synchrophasor measurements: a data-driven approach. IEEE Trans Power Syst 32(4):2817–2827
7. Zhang QF, Venkatasubramanian VM (2014) Synchrophasor time skew: formulation, detection and correction. In: North American power symposium (NAPS). September 2014, pp 1–6
8. Five year sychrophasor plan. California ISO (2011)
9. Su H, Liu T (2018) Enhanced-online-random-forest model for static voltage stability assessment using wide area measurements. IEEE Trans Power Syst 1–1
10. Zhou DQ, Annakkage UD, Rajapakse AD (2010) Online monitoring of voltage stability margin using an artificial neural network. IEEE Trans Power Syst 25(3):1566–1574
11. Glavic M, Cutsem TV (2009) Wide-area detection of voltage instability from synchronized phasor measurements. Part I: Principle. IEEE Trans Power Syst 24(3):1408–1416
12. Su H, Liu C (2016) Estimating the voltage stability margin using PMU measurements. IEEE Trans Power Syst 31(4):3221–3229
13. Lim JM, DeMarco CL (2016) SVD-based voltage stability assessment from phasor measurement unit data. IEEE Trans Power Syst 31(4):2557–2565
14. Malbasa V, Zheng C, Chen P, Popovic T, Kezunovic M (2017) Voltage stability prediction using active machine learning. IEEE Trans Smart Grid 8(6):3117–3124
15. Yadav R, Pradhan AK, Kamwa I (2019) Real-time multiple event detection and classification in power system using signal energy transformations. IEEE Trans Ind Inform 15(3):1521–1531
16. Jena MK, Panigrahi BK, Samantaray SR (2018) A new approach to power system disturbance assessment using wide-area postdisturbance records. IEEE Trans Ind Inform 14(3):1253–1261
17. North American SynchroPhasor initiative. PMU Data Quality (2017)
18. Wu T, Scaglione A, Arnold D (2023) Constrained reinforcement learning for predictive control in real-time stochastic dynamic optimal power flow. In: IEEE transactions on power systems. IEEE
19. Sunitha R, Sreerama Kumar R, Mathew AT (2013) Online static security assessment module using artificial neural networks. In: IEEE transactions on power system, vol 28, no 4, pp 4328–4335, November 2013
20. Beiraghi M, Ranjbar AM (2013) Online voltage security assessment based on wide-area measurements. IEEETrans Power Del 28(2):989–997
21. Bahmanyar A, Karami A (2014) Power system voltage stability monitoring using artificial neural networks with a reduced set of inputs. Int J Electr Power Energy Syst 58:246–256
22. Liu FT, Ting KM, Zhou Z-H (2012) Isolation-based anomaly detection. ACM Trans Knowl Discov Data 6(1):3:1–3:39
23. Wu T, Zhang Y-JA, Wang S (2021) Deep learning to optimize: security-constrained unit commitment with uncertain wind power generation and BESSs. IEEE Trans Sustain Energy 13(1):231–240
24. Wu T, Carreño IL, Scaglione A, Arnold D (2023) Spatio-temporal graph convolutional neural networks for physics-aware grid learning algorithms. IEEE transactions on smart grid. IEEE

25. Xu Y, Dong ZY, Zhao JH, Zhang P, Wong KP (2012) A reliable intelligent system for real-time dynamic security assessment of power systems. IEEE Trans Power Syst 27(3):1253–1263
26. Wu T, Scaglione A, Arnold D (2022) Reinforcement learning using physics inspired graph convolutional neural networks. In: 2022 58th annual Allerton conference on communication, control, and computing (Allerton). IEEE, pp 1–8
27. Chen D-D, Wang W, Gao W, Zhou Z-H (2018) Tri-net for semi-supervised deep learning. In: Proceedings of the twenty-seventh international joint conference on artificial intelligence, IJCAI-18. International joint conferences on artificial intelligence organization, vol 7, pp 2014–2020
28. Krishnan V, McCalley JD, Henry S, Issad S (2011) Efficient database generation for decision tree based power system security assessment. IEEE Trans Power Syst 26(4):2319–2327
29. Maaji SS, Cosma G, Taherkhani A, Alani AA, McGinnity TM (2018) On-line voltage stability monitoring using an ensemble AdaBoost classifier. In: 2018 4th international conference on information management (ICIM), May 2018, pp 253–259
30. Wu T, Zhang YJ, Tang X (2020) Online detection of events with low-quality synchrophasor measurements based on *i*Forest. IEEE Trans Ind Inform 1–1
31. Jaitly N, Hinton GE (2013) Vocal tract length perturbation (VTLP) improves speech recognition. In: Proceedings of ICML workshop on deep learning for audio, speech and language, vol 117
32. Tang Y, Eliasmith C (2010) Deep networks for robust visual recognition. In: Proceedings of the 27th international conference on machine learning (ICML-10). Citeseer, pp 1055–1062
33. LeCun Y, Bengio Y, Hinton G (2015) Deep learning. Nature 521(7553):436–444
34. Wu T, Scaglione A, Arnold D (2023) Complex-value spatio-temporal graph convolutional neural networks and its applications to electric power systems AI. In: IEEE transactions on sustainable energy
35. He H, Garcia EA (2009) Learning from imbalanced data. IEEE Trans Knowl Data Eng 21(9):1263–1284
36. Ajjarapu V, Christy C (1992) The continuation power flow: a tool for steady state voltage stability analysis. IEEE Trans Power Syst 7(1):416–423
37. Wu T, Zhang Y-JA, Wen H (2020) Voltage stability monitoring based on disagreement-based deep learning in a time-varying environment. IEEE Trans Power Syst 36(1):28–38
38. Siemens (2009) PSSE 30.2 program operational manual
39. Maslennikov S, Wang B, Zhang Q, Ma A, Luo A, Sun A, Litvinov E (2016) A test cases library for methods locating the sources of sustained oscillations. In: 2016 IEEE power and energy society general meeting (PESGM), July 2016, pp 1–5
40. IE Commission et al (2016) Electromagnetic compatibility (EMC)—part 4-1: testing and measurement techniques—overview of IEC 61000-4 series. IEC 61000-4-1, October, Technical report
41. Frigo G, Derviškadić A, Bach A, Paolone M (2019) Statistical model of measurement noise in real-world pmu-based acquisitions. In: 2019 international conference on smart grid synchronized measurements and analytics (SGSMA), May 2019, pp 1–8
42. Chang C-C, Lin C-J (2011) LIBSVM: a library for support vector machines. ACM Trans Intell Syst Technol 2(3):27:1–27:27
43. Hastie T, Rosset S, Zhu J, Zou H (2009) Multi-class AdaBoost. Stat Its Interface 2(3):349–360

Integration of Renewable Energy Sources Into Local Energy Systems

Hao Xiao, Wei Pei, Daoxin Han, and Xiaowei Pu

Abstract With the increasing demand for environmental conservation and the rapid development of new energy sources, the integration of renewable energy into local energy systems has become a highly promising direction for development. This chapter begins by summarizing the challenges faced in the integration of renewable energy in areas such as power forecasting, stability analysis and computation, digital technology in power systems, and flexible interaction techniques. The digital technologies for renewable energy forecasting and prediction method based on an improved Temporal Convolutional Network (TCN) network, as well as feature modeling, are then proposed. Finally, the concept of a virtual power plant and its role in decentralized energy generation are discussed, and an artificial intelligence-empowered optimization model for peer-to-peer energy transactions between virtual power plants, based on the Deep Deterministic Policy Gradient (DDPG) algorithm, is presented. The effectiveness of the proposed method is verified with a test case study.

1 Challenges in Integrating Renewable Energy Sources

Renewable energy integration is a novel energy system development approach in the context of the global energy transition. It aims to combine various renewable energy sources such as solar, wind, and hydro power with energy storage technologies and smart grid techniques, achieving efficient utilization and optimized distribution

H. Xiao (✉) · W. Pei · D. Han · X. Pu
Institute of Electrical Engineering, Chinese Academy of Sciences, Beijing, China
e-mail: xiaohao09@mail.iee.ac.cn

W. Pei
e-mail: peiwei@mail.iee.ac.cn

D. Han
e-mail: handaoxin@mail.iee.ac.cn

X. Pu
e-mail: puxiaowei@mail.iee.ac.cn

W. Hua et al. (eds.), *Digitalisation of Local Energy Systems*, Green Energy and Technology, https://doi.org/10.1007/978-3-031-77833-9_5

of energy [1]. As climate change and energy security issues become increasingly pronounced, the limitations and unsustainability of traditional fossil fuel energy models have become evident. Renewable energy, due to its environmentally friendly, renewable, and widely distributed characteristics, has become a key pathway for sustainable energy development [2, 3]. Renewable energy integration not only helps reduce reliance on fossil fuels and lower greenhouse gas emissions but also enhances the stability, reliability, and flexibility of energy systems, driving innovation in the energy industry.

Currently, research and practice in the field of renewable energy integration are rapidly advancing. On one hand, researchers are devoted to optimizing the integration of various renewable energy sources, exploring how to maximize the utilization of resources like solar, wind, and hydro power under different regional and climatic conditions [4, 5]. On the other hand, smart grid technology as well as digital technology are effectively promoting the complementary utilization and efficient consumption of widely distributed energy sources through information and physical integration modeling. Besides that, concepts like microgrids, virtual power plants, and energy interconnection networks are also being implemented gradually, promoting a more intelligent and efficient operation of energy systems. However, challenges persist in renewable energy integration. These include power forecasting technology, stability analysis and computation, digitalization of power systems, and other issues that necessitate further research and collaboration for resolution.

1.1 Challenges in Power Forecasting Technology to Support Renewable Energy Integration

Renewable energy prediction technology is the primary key technology supporting the integration of renewable energy, and has made significant progress after years of development. However, in order to further improve the accuracy and usefulness of power forecasting results, as well as to support large-scale renewable energy grid integration, new energy power forecasting techniques still face challenges in the following areas.

- Real-Time Data Collection Standards: The establishment of standards for real-time data collection is essential to support ultra-short-term power forecasting, ensuring high-quality and standardized data, especially in the development and construction of wind farms and photovoltaic power stations.
- Advanced Numerical Weather Forecasting: The establishment of standards for real-time data collection is essential to support ultra-short-term power forecasting, ensuring high-quality and standardized data, especially in the development and construction of wind farms and photovoltaic power stations.
- Enhancing Predictive Capabilities: The utilization of technologies such as artificial intelligence and big data mining is crucial to enhance the conditional forecasting capabilities of prediction models. This includes the application of machine learn-

ing and image processing computer vision techniques for cloud trajectory and irradiance cloud attenuation forecasting, as well as the use of machine learning pattern recognition techniques for weather and power fluctuation type identification, laying the foundation for renewable energy power forecasting and slope event prediction.

1.2 Challenges in Stability Analysis and Computation to Support Renewable Energy Integration

Power system stability analysis and computation refer to the method of studying and assessing the system stability when integrating renewable energy sources on a large scale, ensuring the safety and large-scale integration of renewable energy. As wind and solar power sources gradually replace traditional coal-fired units, power systems stability analysis and computation face three key challenges:

- Fundamental Shift in Stability Principles: Traditional synchronous generators rely on electromagnetic-mechanical balance for stability. However, the new "dual-high"' power systems with reduced rotational inertia, decreased frequency regulation capability, and lower performance standards for renewable energy devices can lead to grid disconnections and system oscillations under various disturbances, posing risks to the security, stability, and economic operation of the power system.
- Emergence of New Stability Issues in Power Electronic Systems: The stability mechanisms in power electronic-based power systems have undergone profound changes, introducing new stability problems such as wide-band oscillations. These problems include not only negative-damping oscillations but may also involve weak damping and multi-modal overlapping oscillations, for which currently there is a lack of corresponding analytical analysis and computational methods.
- Shift in Renewable Energy Penetration: The increasing penetration of renewable energy alters the role of these sources in system stability. When renewable energy and traditional synchronous generators have similar shares, research is needed to coordinate their dynamics to enhance overall stability. However, as renewable energy penetration continues to rise, attention must also be given to the impact of traditional equipment on the stability of systems dominated by power electronics, further complicating stability issues.

1.3 Challenges in Digitalization Technology of Power Systems to Support Renewable Energy Integration

The core to the new power system lies in digital twin technology, which combines electrical physics, information technology, and digital computational science to enhance the integration ability of the power system with renewable energy through digital and data-driven scientific calculations. To achieve digital new energy power systems, three core technologies need to be tackled.

- Digital Chips: Intelligent devices form the foundation of the system and need to possess features such as state awareness, prediction, adjustability, and intelligence. The key to digital chips lies in the full industry chain autonomy, including concept design, prototype design, testing, and validation, for the development of intelligent devices and transparent power systems.
- Autonomous and Controllable Simulation Software: Robust software and big data support digital twin systems, and the critical aspect is autonomous and controllable power source simulation and analysis software. This includes intelligent simulation computing engines based on deep learning and next-generation industrial Internet of Things (IoT) integrated multi-energy system simulation software.
- Microsensor Technology: Sensors are the technological basis for realizing the observability of the system's state and require micro-sized, low power consumption, self-organizing, high-precision, high-reliability microsensors. Research and development of high-performance power-specific sensors based on new materials and micro-nano structures are essential, along with the establishment of perception standards to support applications such as state awareness, data transmission, environmental monitoring, and behavior tracking.

1.4 Challenges in Flexible Interaction Technology of Power Systems to Support Renewable Energy Integration

The power system supporting the integration of renewable energy sources needs to have a high degree of flexibility and interactivity to accommodate the variable energy generation and electricity demand. To meet the demands of the new power system for real-time power balance and flexibility adjustments, three key technological challenges need to be addressed.

- Enhancing Load Management: The strengthening of load monitoring systems, load aggregation platforms, and load autonomous control terminals is necessary to enhance precise control of adjustable loads, thereby supporting electricity trading and auxiliary services.
- Decentralized Integration of Distributed Energy: In response to the characteristics of distributed new energy sources, propose a decentralized operation and control technology and market mechanism to facilitate the integration of distributed new energy sources and ensure the security of the power system. Simultaneously, research game models and strategies to drive the development of power grid companies.
- AI-Driven "Source-Grid-Load-Storage" Optimization: Leveraging artificial intelligence to explore the interactive potential of the "source-grid-load-storage"

paradigm is essential to enhance the analysis and control capabilities of new power systems. It also involves mastering power grid security analysis methods, developing data-driven collaborative optimization, and interactive control technologies to enable the flexibility and secure stability of new power systems.

2 Digital Technologies for Renewable Energy Forecasting

To effectively mitigate the impact of wind power and solar power on the power system and enhance the utilization efficiency of renewable energy electricity, pre-emptively forecasting the power generated by wind and solar sources is a proven method. Against this backdrop, digital technologies based on various methods for renewable energy forecasting are showing promising prospects for development. At present, a lot of research works have been conducted on digital technologies for renewable energy forecasting. The existing forecasting approaches can be broadly classified into three categories according to the methods involved: physical, statistical, and artificial intelligence methods [6].

Physical methods rely on mathematical models to capture the underlying physical processes of power generation [7, 8]. Physical methods require more detailed parameters such as numerical weather prediction (NWP) data and geographical location, as well as supercomputers, so they are more suitable for the prediction of power outputs in new energy power plants where data are comprehensive and accurate, and the methods are more effective for medium- or long-term wind farm forecasting [9]. However, the computational complexity of the physical methods is high and not applicable to most scenarios. Statistical methods are primarily used to predict future power generation sequences by learning the laws of past power generation sequences. The main methods include autoregressive moving average [10], Gaussian process regression [11], grey prediction, and Kalman filter [12]. And are suitable for the output power prediction of new energy power plants lacking meteorological feature data. However, the prediction accuracy of such methods is hard to improve and difficult to fit complex scenarios. With the development of big data, artificial intelligence methods demonstrate the advantage of high prediction accuracy. Consequently, new energy generation forecasting based on artificial intelligence neural networks has attracted considerable attention in recent years, with the primary artificial intelligence methods including multilayer perceptron (MLP), support vector regression (SVM), random forest (RF), LSTM, CNN, GRU, and so on.

Besides that, for digital technologies of renewable energy forecasting, data feature selection and expansion are also key issues affecting prediction accuracy. Most of the existing studies directly employ available weather forecast data (NWP) as input features, which may ignore certain key weather factors [13]. For PV power generation prediction, astronomical data, such as the solar time angle, are extremely important for power generation prediction, and they can be used as extended features [14]. However, in practice, some extended factors, such as astronomical features, also greatly affect the accuracy of the prediction, and further theoretical analysis of astronomical features is necessary.

To this content, we proposed an improved prediction method based on enhanced Temporal Convolutional Networks (TCN) and feature modeling to better support renewable energy prediction. The TCN network was first proposed in 2018 with a convolutional structure and residual blocks to extract depth features and long-term temporal relationships, allowing extremely long input sequences to be processed as a whole. The network has a flexible perceptual field, which denotes its ability to look deeply into the past and make predictions; additionally, unlike recurrent neural networks, its gradient stability does not suffer from gradient explosion. The evident advantages of the TCN network have led to its extensive use in time-series prediction.

To enable the network to learn more historical data, which widens the received data perception field exponentially without increasing the complexity of the model, the TCN network uses dilated convolution. Let us assume that the dilated convolution is denoted by $*$, the input sequence is denoted by x, and the single filter is denoted by $f \in \{0, ..., k-1\}$. Equation (1) can be obtained based on a dilated convolution operation on element s.

$$F(s) = \sum_{i=0}^{k-1} f(i) \cdot I_{s-d \cdot i} \quad (1)$$

The receptive field calculation formula can then be obtained, as in (2).

$$r = 1 + 2 \times (k-1) \times \frac{b^n - 1}{b - 1} \quad (2)$$

where, d denotes the dilation factor, and k refers to the filter size. Further, if we set the dilation base to b, the dilation factor d of each layer becomes b^{n-1}. We now let the size of the TCN network receptive field be r.

Considering that the TCN model suffers from underfitting of nonlinear capability in renewable energy forecasting and accumulation of errors in the rolling forecasting mechanism, an improved TCN network with two main improvements was proposed as shown in Fig. 1 to improve the accuracy of the prediction.

The first improvement for the proposed improved TCN network is adding a projection head layer to utilize more useful information and improve the quality of learning, and the improved structure is labeled as TCN-P. The projection head layer is shown in (3).

$$y_i = W^{(2)} \sigma(W^{(1)} h_i) \quad (3)$$

where h_i is the output of the TCN network, σ is the expression of the nonlinear activation function, and $W^{(1)}$, $W^{(2)}$ denote the weight parameters connecting the front and back layers of the nonlinear activation function layer, respectively.

In the design structure, σ denotes the sigmoid activation function, and the function expression is shown in (4):

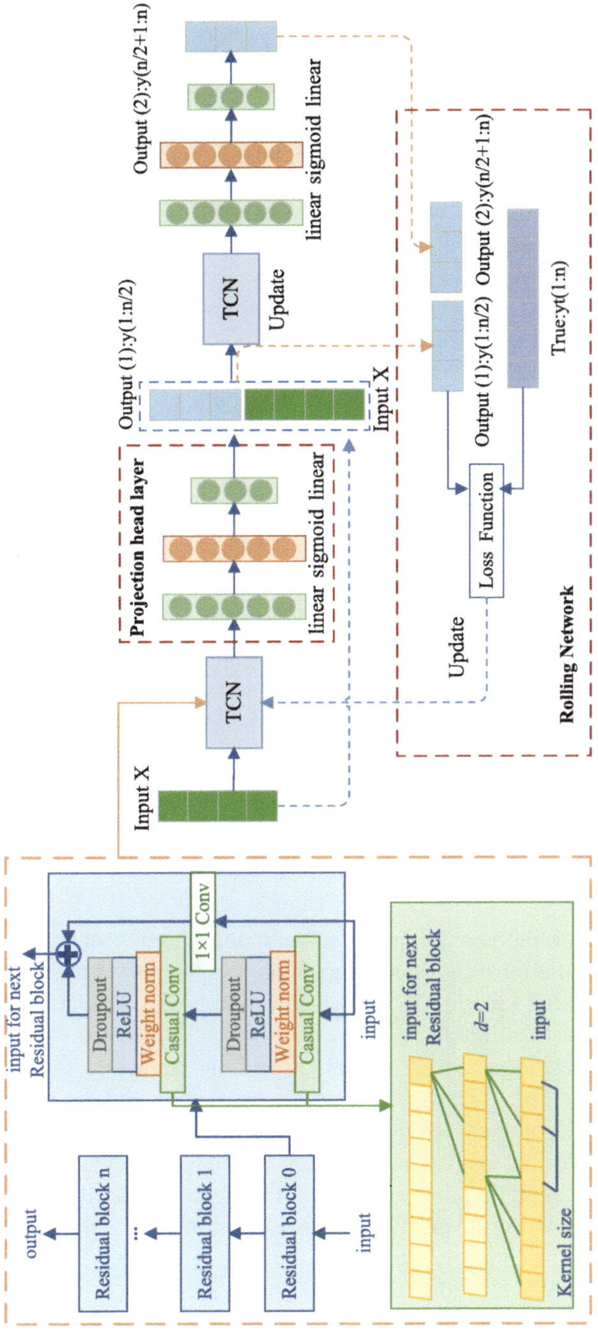

Fig. 1 The proposed improved TCN network structure

$$\sigma(x) = \frac{1}{1 + e^{-x}} \tag{4}$$

The second improvement for the proposed improved TCN network is the development of a rolling prediction network. The improved network structure is labeled as TCN-PG. Unlike the existing common rolling prediction mechanisms, which usually construct a network for the first part of the data and then use the same network or construct a new network to input the first part of the predicted data for the second half of the prediction, a rolling network is designed to reducing the complexity of the scrolling mechanism while improving the accuracy of the predictions, the loss generated from the second half of the prediction results will update the whole network and dynamically adjust when predicting the second half data. The parameters of the first half of the network are dynamically adjusted when predicting the second half, thus ensuring the accuracy and validity of the prediction.

In TCN-PG, the first several predicted values are added as the inputs of the last several predicted values for training; the predicted value is calculated in (5).

$$y_{nT+1}^{(n+1)T} = f\left(\left[X, y_{(n-1)T+1}^{nT}\right] \right) \tag{5}$$

where T is the total prediction scale, T_1 is the iteration scale, N is the total number of iterations, $n \in \{1, 2, \ldots, N\}$ is the number of intermediate iterations. Then, $T = N \times T_1$. Further, the TCN network function expression is represented by f. Then, the input is characterized as X, and Y is the output, $Y = [y_1, y_2, \ldots, y_T]$.

Taking the error accumulation in the entire network as the objective function, which can address the problem of error accumulation. The loss function used in the network is designed as. The network model gradient is calculated and updated in the entire network.

$$\text{Loss} = \sum \left(y^n - Lf_1(X^m), f_2(X^m + f_1(X^m))\right)^2 \tag{6}$$

where X^m refers to the leveled input features, and y^n represents the true output, n represents the prediction of n future data points, the first part of the TCN-PG network is represented by the function f_1, and the second part of the TCN-PG network is represented by the function f_2.

The overall prediction process combined with the improved TCN network structure is as shown in Fig. 2. First, data anomaly processing and normalization are performed, and meteorological features which are highly relevant to renewable energy generation are screened using the correlation coefficient calculation. Meanwhile, the historical power generation data are used as input features. Following this, the TCN network structure is improved by adding projection layers and a rolling network structure, and the training set data are input into the improved TCN model for training. Simultaneously, the validation set determines the prediction performance of the model and changes the network parameter selection. The final model structure with the best results is used as the final model, and the test set is input into the final model to obtain the prediction results.

3 Virtual Power Plants and Their Role in Decentralized Energy Generation

3.1 The Concept of a Virtual Power Plant

The term "Virtual Power Plant" (VPP) originated from Dr. Shimon Awerbuch's 1997 book titled "Virtual Utility: Emerging Technologies and the Competitive Power Industry." In this book, a virtual utility as a flexible collaboration among independent and market-driven entities that can provide efficient electricity services to consumers without the need to own the corresponding assets is defined. Similar to how virtual utilities leverage emerging technologies to offer consumer-oriented electricity services, a VPP does not change the way each distributed generator (DG) is connected to the grid. Instead, it employs advanced control, metering, communication technologies, and more to aggregate various types of distributed energy resources (DERs) such as DGs, energy storage systems, controllable loads, electric vehicles, etc. The higher-level software architecture optimizes the coordinated operation of multiple DERs, promoting rational resource optimization and utilization.

The concept of a VPP emphasizes its outward functions and effects, bringing about updated operational concepts and generating socio-economic benefits. Its primary application scenario is in the electricity market [15]. This approach aggregates DERs to ensure stable transmission on the public grid and provides fast-response ancillary services without requiring significant grid modifications. It effectively integrates DERs into the electricity market, mitigating the risk of isolated operation and enabling economies of scale[16]. Simultaneously, the visualization of DERs and the coordinated control optimization of the VPP significantly reduce the impact of DER integration on the grid, alleviate scheduling difficulties caused by DG expansion, rationalize distribution management, and enhance system stability [17].

In summary, the core of the VPP concept can be summarized as "communication" and "aggregation." A VPP can be understood as a specialized power coordination and management system that aggregates and optimizes DERs such as DGs, energy storage systems, controllable loads, electric vehicles, through advanced information communication technology and software systems, enabling their participation in the electricity market and grid operations as a distinct power plant, thus enhancing renewable energy integration capabilities.

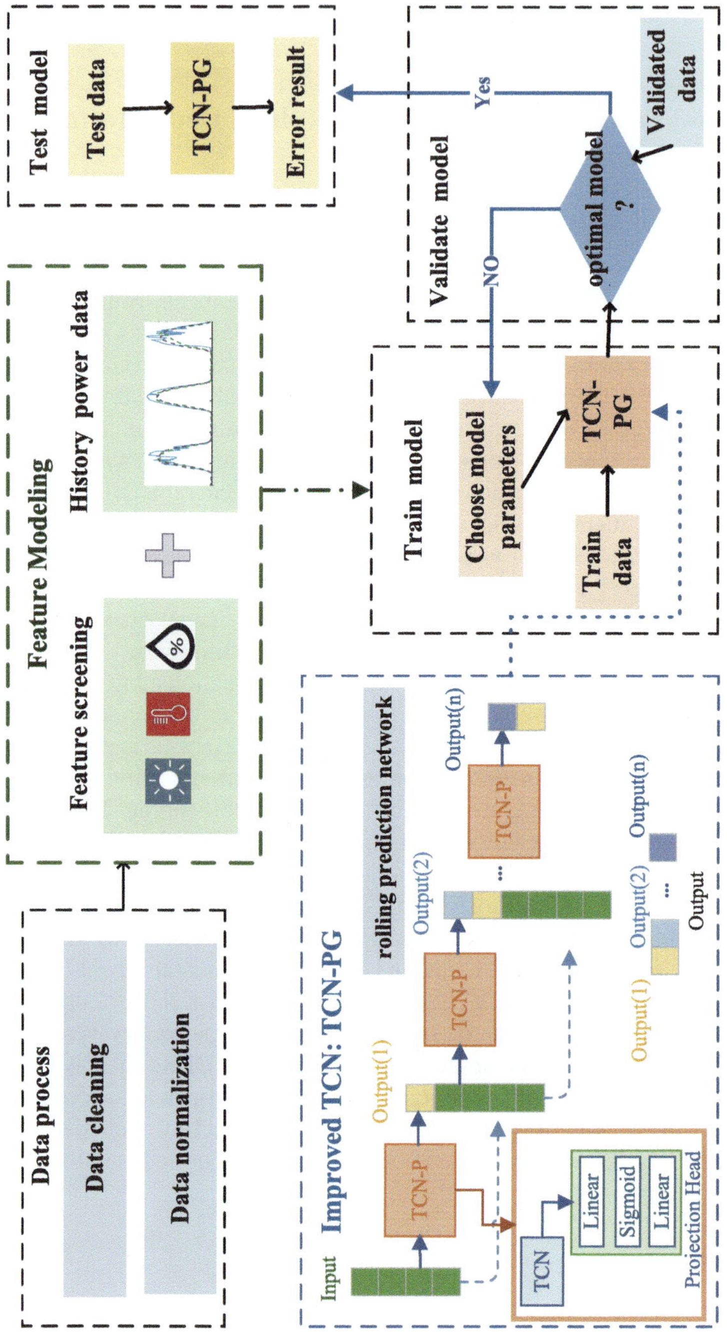

Fig. 2 Overall power predication flowchart based on feature modeling and improved TCN

3.2 AI-Empowered VPP Interactive Operation and Renewable Energy Integration Model

In the realm of modern energy systems, the integration of renewable energy sources and the optimization of their utilization pose significant challenges. The emergence of AI-empowered VPPs presents a transformative solution to these challenges. This innovative approach combines the capabilities of artificial intelligence with the concept of VPPs, revolutionizing the way energy is generated, consumed, and managed.

The AI-empowered VPP entails a sophisticated framework that leverages advanced communication technologies, smart algorithms, and real-time data analytics. This framework enables interactive optimization of the power plant's operations, facilitating seamless coordination between various distributed energy resources. By aggregating sources like distributed generators, energy storage systems, controllable loads, and even electric vehicles, the VPP optimally balances energy supply and demand. A cornerstone of this paradigm is the efficient integration of renewable energy sources, addressing the intermittent nature of resources like solar and wind power. The model harnesses AI to predict energy generation patterns, allowing for proactive load management and resource allocation. Consequently, renewable energy sources become more dependable, contributing to grid stability and reducing reliance on conventional, non-renewable sources.

This integration framework extends its benefits to the larger energy landscape by offering valuable ancillary services, fast-response capabilities, and enhanced grid resilience. The AI-empowered VPP's role in revolutionizing energy consumption and interaction not only amplifies the potential for renewable energy integration but also propels the energy sector toward a more sustainable and intelligent future.

In this context, we proposed an AI-empowered VPP Peer-to-Peer (P2P) energy trading method, which is shown in Fig. 3. The following sections will sequentially describe the external environment modeling of this method, modeling of the Deep Deterministic Policy Gradient (DDPG) [18] intelligent agent and algorithm, as well as a specific case analysis.

3.2.1 External Environment Modeling for P2P Energy Trading of VPPs

The external environment is mainly related to the interactive transaction's behavior of other VPPs. The non-privacy-related state data disclosed by other VPPs can be obtained through historical data, such as the historical P2P transaction volume and the historical P2P transaction price of other VPPs. However, this part of the state involves the decision-making strategy and internal parameters of other VPPs. When it comes to user privacy and security, it is impossible to obtain parameters to establish a physical model. For this part of the virtual environment, using historical data information to learn a deep neural network that does not involve privacy and characterizes the interaction characteristics of VPPs, this part of the virtual environment can be well simulated.

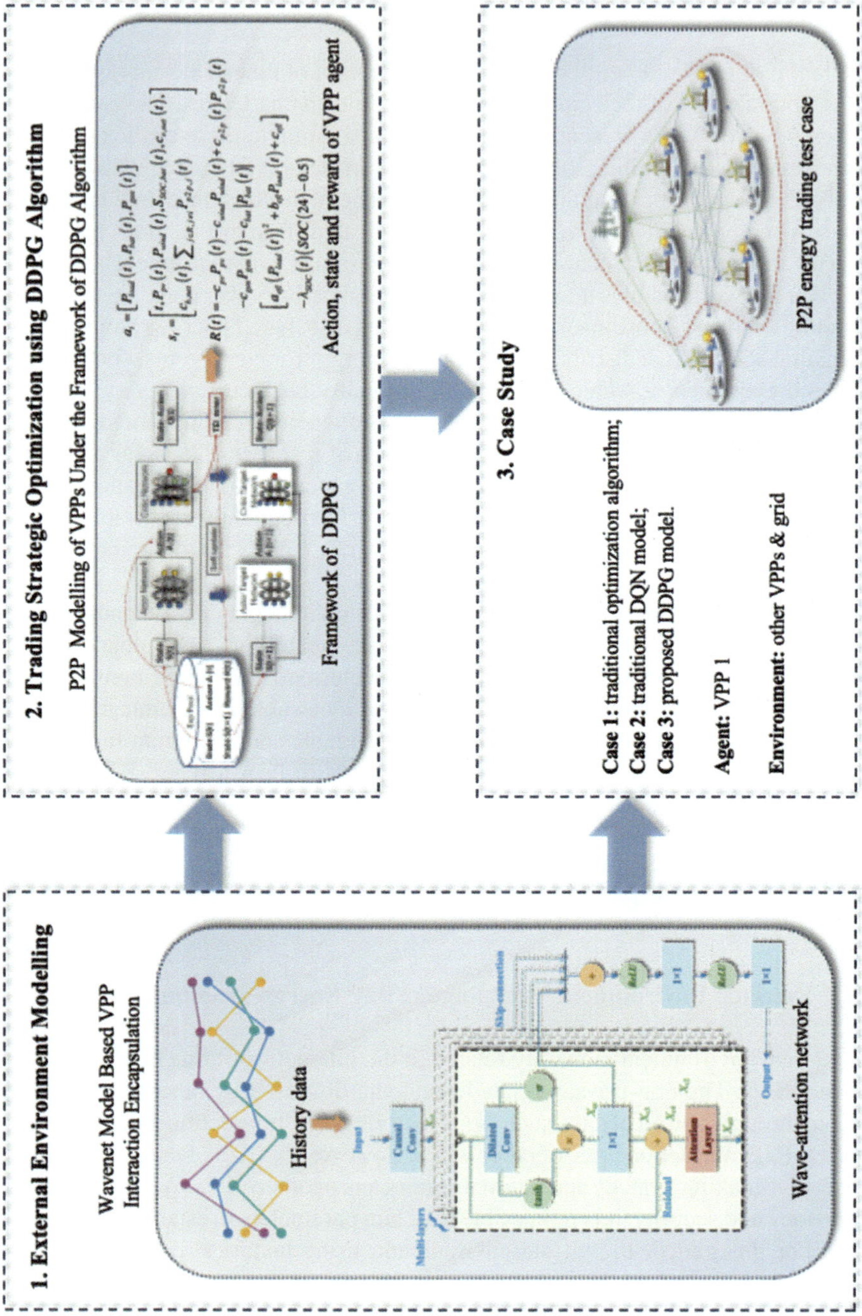

Fig. 3 Structure of AI-empowered VPP P2P energy trading

Considering that the interactive data of VPP has the characteristics of high privacy and strong coupling, it is difficult for traditional deep learning methods to reveal the intrinsic characteristics of the interactive behavior of VPP. In order to solve this problem, a novel wave-attention network is proposed for VPP interaction based on the Wavenet model. Wavenet is a deep neural network model used to generate high-quality audio waveforms, which has the advantage of capturing long-term dependence relationships. In order to make Wavenet more suitable for VPP interaction power prediction, an attention mechanism is introduced into the model. The wave-attention network diagram is shown in Fig. 4.

The transmission of network parameters satisfies the relationship shown below:

$$X_{ac} = \tanh(W_1 * X_{c2}) \odot \sigma(W_2 * X_{c2}) \tag{7}$$

$$X_{C5} = X_{c3} + X_{c1} \tag{8}$$

where X_{c2} is the output of dilated causal convolution layer, X_{ac} is the output of activation function layer, tanh() is the tanh activation function, σ is the sigmoid activation function, W_1 and W_2 are the corresponding weights, $*$ is the convolution operator, $\circ$ is the multiplication operator. X_{C5} is the residual connection output and X_{c3} is the 1×1 convolutional layer output.

Next, the self-attention mechanism is introduced to enhance the extraction of internal coupling features of VPP operational data. The calculation of the self-attention layer is as follows.

$$X_{\mathrm{LN}} = \mathrm{LayerNorm}(X_{C5}) \tag{9}$$

$$X_{\mathrm{att}} = \mathrm{softmax}\left(\frac{QK^T}{\sqrt{d_k}}\right)V \tag{10}$$

$$\begin{aligned} Q &= W^Q \cdot X_{\mathrm{LN}} \\ K &= W^K \cdot X_{\mathrm{LN}} \\ V &= W^V \cdot X_{\mathrm{LN}} \end{aligned} \tag{11}$$

where LayerNorm() represents the layer norm operator, X_{LN} is the output of layer norm operator, W^Q, W^K and W^V are the weight of query, key and value vector, softmax() represents the softmax activation function, d_k is the variance of input X_{LN}.

On the one hand, the output of the self-attention layer will be the input data of the next layer. On the other hand, the output of each 1×1 convolution layer forms the skip-connection together. After that, the highly extracted information will pass another two 1×1 convolution layer with ReLU activation function. Finally, the prediction of VPP interactive power will be output.

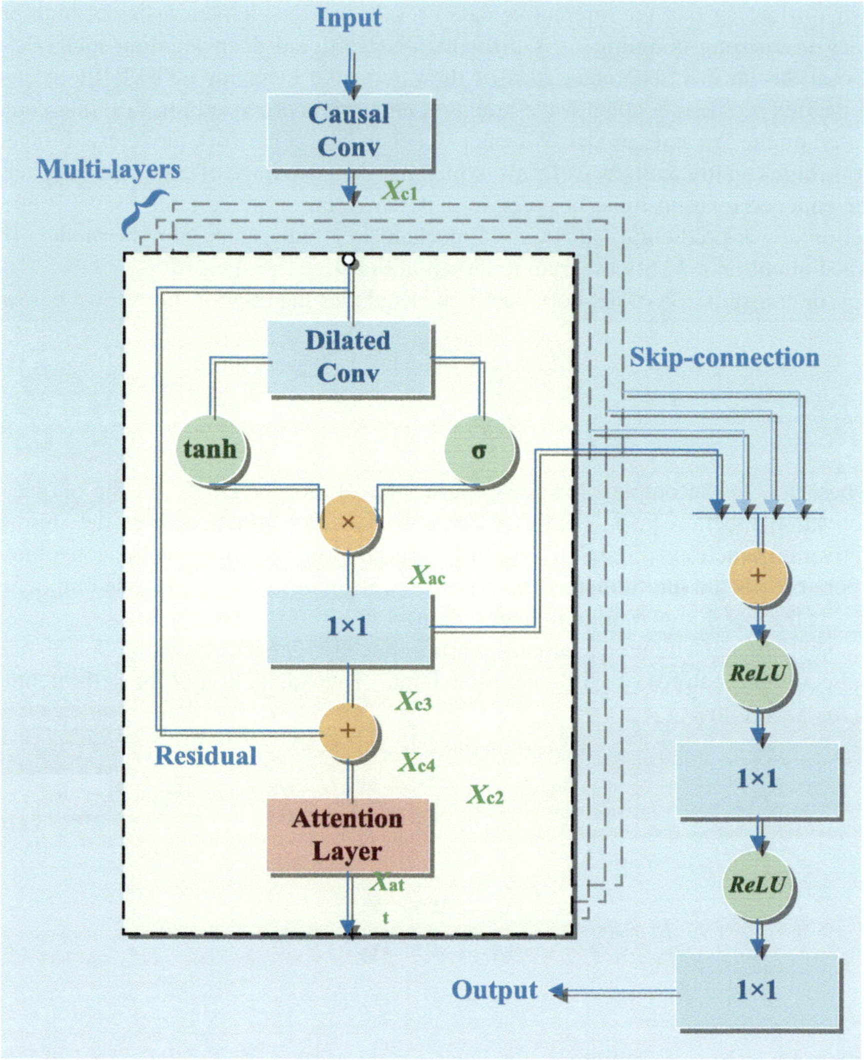

Fig. 4 The structure of wave-attention network

3.2.2 P2P Energy Trading Strategic Optimization Using DDPG Algorithm

After virtual environment modeling, the optimization is carried out using the DDPG algorithm. The DDPG algorithm is proposed *r* to solve the problem of poor convergence of the DPG algorithm by combining the advantages of the Actor-Critic algorithm, Deep Q-Learning (DQN) algorithm, and DPG algorithm. By using the target network and experience replay memory of the DQN algorithm to ensure stable convergence of the algorithm [19], the dilemma of poor convergence of DPG is solved. The framework based on the DDPG algorithm is shown in Fig. 5.

In Fig. 5, the two added target networks, along with the Actor network and the Critic network, are used to calculate TD errors for network updates. The Actor network and Critic network are updated in real time during learning, and the target network corresponding to the two networks is soft-updated regularly based on these two networks. The experience pool stores the training data and samples the data from the stored data to accelerate the network training process. The data playback of the experience pool improves the data utilization efficiency, reduces the correlation of the data in the data sampling, and improves the training stability of the system.

The update of the Actor network in the system is based on the strategy gradient obtained by the DDPG algorithm, and the expression is shown in (12); the update of the Critic network is obtained by the loss minimization function, and the update expression is shown in (13).

$$\begin{aligned} &\nabla_{\theta_\pi}\left(E[R(0)]+\gamma R(1)+\gamma^2 R(2)+\cdots\right) \\ &=\frac{1}{N}\sum_{i\in U(D)} \nabla_{\theta_\pi}\pi(s\mid\theta^\pi)\Big|_{s=s_i}\cdot\nabla_a Q(s,a\mid\theta^Q)\Big|_{\substack{s=s_i\\ a=\pi(s_i)}} \end{aligned} \tag{12}$$

$$L(\theta^Q)=\mathbb{E}^{U(D)}_{s,a,R,s'}\left[\left(R(t)+\gamma Q'(s(t+1),\pi(s(t+1)|\theta^{\pi'})|\theta^{Q'})-Q(s,a|\theta^Q)\right)^2\right] \tag{13}$$

where ∇ is to obtain the gradient of the corresponding parameter, $L(\theta^Q)$ is the calculation of the loss function of the Critic network, $E(\cdot)$ is the calculation of the expected value, $U(\cdot)$ is to extract random data from the experience pool, D is the data set of the experience pool, N is the number of data extracted each time, i is the serial number of each data extracted, and each extracted data includes content (s, a, R, s'). $\theta^{\pi'}$ is the parameter of the Actor target network, $\theta^{Q'}$ is the parameter of the Critic target network.

The target network is a network whose update frequency is lower than that of the Actor and Critic in order to improve the stability of the system. The parameter update of the target network is updated after training a minimum training round. The parameter update method uses the moving average algorithm, and its update mode is shown in (14) and (15).

$$\theta^{\pi'} \leftarrow \sigma\theta^\pi + (1-\sigma)\theta^{\pi'} \tag{14}$$

$$\theta^{Q'} \leftarrow \sigma\theta^Q + (1-\sigma)\theta^{Q'} \tag{15}$$

where τ is a hyperparameter that disperses the timing of the soft update of the target network to each step of the update, and its parameter characteristics meet. In theory, in order to ensure the stability of the system training, the update frequency of the two target networks is different from that of the corresponding main network. By

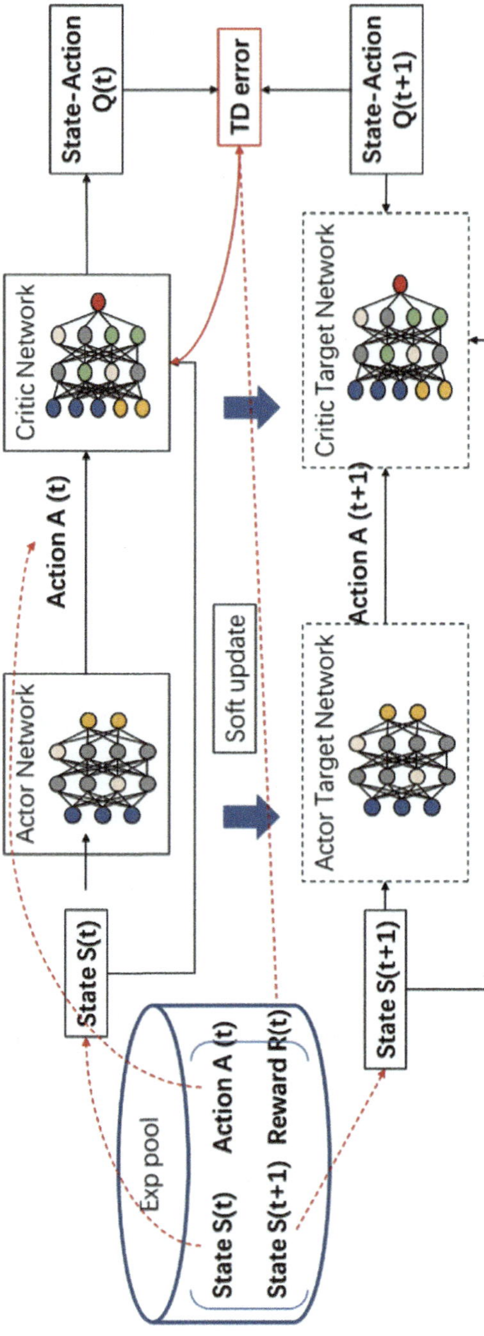

Fig. 5 Framework of deep deterministic policy gradient algorithm

using hyperparameters τ, the update quantity can be distributed more evenly on the time axis, realizing the same frequency as the update of the main network. Moreover, extending the time axis will make the soft update of the target network more stable.

Finally, for the action exploration of Actor network, since it is a deterministic probabilistic strategy model, the exploration noise is added to increase the exploration space of the network. The output of the action strategy model after adding noise is shown in (16):

$$\pi'(s(t)) = \pi(s(t))\theta' + N(t) \tag{16}$$

$N(t)$ is the noise interference model, and the Gaussian model is generally selected, whose variance decreases with the training time, and the exploration space is dynamically adjusted.

For the VPP_i participating in the interactive game theory learning problem in the P2P electricity market, VPP_i's control actions include its own flexible load power $P_{\text{load}}(t)$, energy storage action power $P_{\text{bat}}(t)$, micro-turbine power output $P_{\text{aen}}(t)$, P2P power traded by VPP_i in the electricity market $P_{p2p}(t)$, and the bidding strategy for participating in P2P electricity market transactions $c_{p2p}(t)$. The environmental state variables observed by VPP_i include the current time point t, its own renewable energy photovoltaic output $P_{\text{pv}}(t)$, its own renewable energy wind power output $P_{\text{wind}}(t)$, the SOC of its energy storage devices $S_{\text{soc, bat}}(t)$, photovoltaic generation cost c_{pv}, wind power generation cost c_{wind}, micro-turbine generation cost c_{gen}, energy storage charging and discharging cost c_{bat}, selling electricity price $c_{s,\text{net}}(t)$, grid buying-back electricity price $c_{b,\text{net}}(t)$, and the transaction prices $c_{\text{p2p},i}(t)$ and transaction powers $P_{\text{p2p},i}(t)$ submitted by all other VPP_j participants in the P2P electricity market, where the condition is satisfied $\forall j \in R, j \neq i$.

Under the Supply and Demand Ratio (SDR) market mechanism [20], VPPs initially submit their supply or demand for electricity within the regions, with any surplus or shortfall being supplemented by the power grid. At this point, the action dimension for deep reinforcement learning is:

$$a_t = \left[P_{\text{load}}(t), P_{\text{bat}}(t), P_{\text{gen}}(t)\right] \tag{17}$$

The state space for deep reinforcement learning is

$$s_t = \left[t, P_{\text{pv}}(t), P_{\text{wind}}(t), S_{\text{SOC,bat}}(t), c_{s,\text{net}}(t), c_{b,\text{net}}(t), \sum_{j\in R, j\neq i} P_{2p,j}(t)\right] \tag{18}$$

The reward obtained by VPPs is economic gain, as shown in (19).

$$\begin{aligned} R(t) = &- c_{\text{pv}} P_{\text{pv}}(t) - c_{\text{wind}} P_{\text{wind}}(t) + c_{2p} P_{2p}(t) - c_{\text{gen}} P_{\text{gen}}(t) - c_{\text{bat}} P_{\text{bat}}(t) \\ &+ \left[a_{\text{efl}}\left(P_{\text{load}}(t)\right)^2 + b_{\text{efl}} P_{\text{load}}(t) + c_{\text{efl}}\right] \\ &- \lambda_{\text{soc}}(t)\left(\text{SOC}(24) - \text{SOC}(1)\right) \end{aligned} \tag{19}$$

where a_{efl},b_{efl}, and c_{efl} represent the quadratic coefficient, linear coefficient, and constant coefficient of the economic benefit function, respectively.

In this case, the VPP P2P transaction electricity price is calculated uniformly in the electricity market. It represents the penalty term parameter for the SOC constraint of the VPP energy storage devices, and its parameter calculation method is as shown in (20).

$$\lambda_{\text{soc}}(t) = \begin{cases} 0, & \text{when } 0 < t < T \\ \lambda_{\text{soc}}, & \text{when } t = T \end{cases} \tag{20}$$

The VPP power constraint at this moment is as follows:

$$P_{\text{bat}}(t) + P_{\text{gen}}(t) + P_{\text{pv}}(t) + P_{\text{wind}}(t) + P_{2p}(t) = P_{\text{load}}(t) \tag{21}$$

Using the modeling approach described above, the VPP agent iteratively interacts with the environment using the DDPG algorithm encapsulated with an external wave-attention network until convergence, aiming to obtain the optimal VPP P2P trading strategy.

3.3 *Case Study*

3.3.1 Test System

In order to assess the effectiveness of the proposed P2P energy trading algorithm between VPPs, a test case comprising six VPPs, as illustrated in Fig. 6, is utilized for validation. In the test case, a total of 6 VPPs participate in the P2P power market transaction, and the P2P transaction under the SDR market mechanism is used for method verification. VPP 1 is the main agent and is trained to learn its operation and interaction strategy. Other VPP, power grid, and power market mechanisms are treated as the external environment of the agent of VPP 1. The training parameters of the deep reinforcement learning model are shown in Table 1, including the maximum number of training rounds, the size of memory bank, the size of sampling, reward attenuation, and the update hyperparameter of the target network. Three cases are designed to validate the proposed solution's effectiveness.

3.3.2 Comparative Analysis of Different Cases

- Case 1: P2P energy trading under SDR market based on traditional model-based optimization algorithm;
- Case 2: P2P energy trading under SDR market based on traditional DQN model;
- Case 3: P2P energy trading under SDR market based on proposed DDPG model.

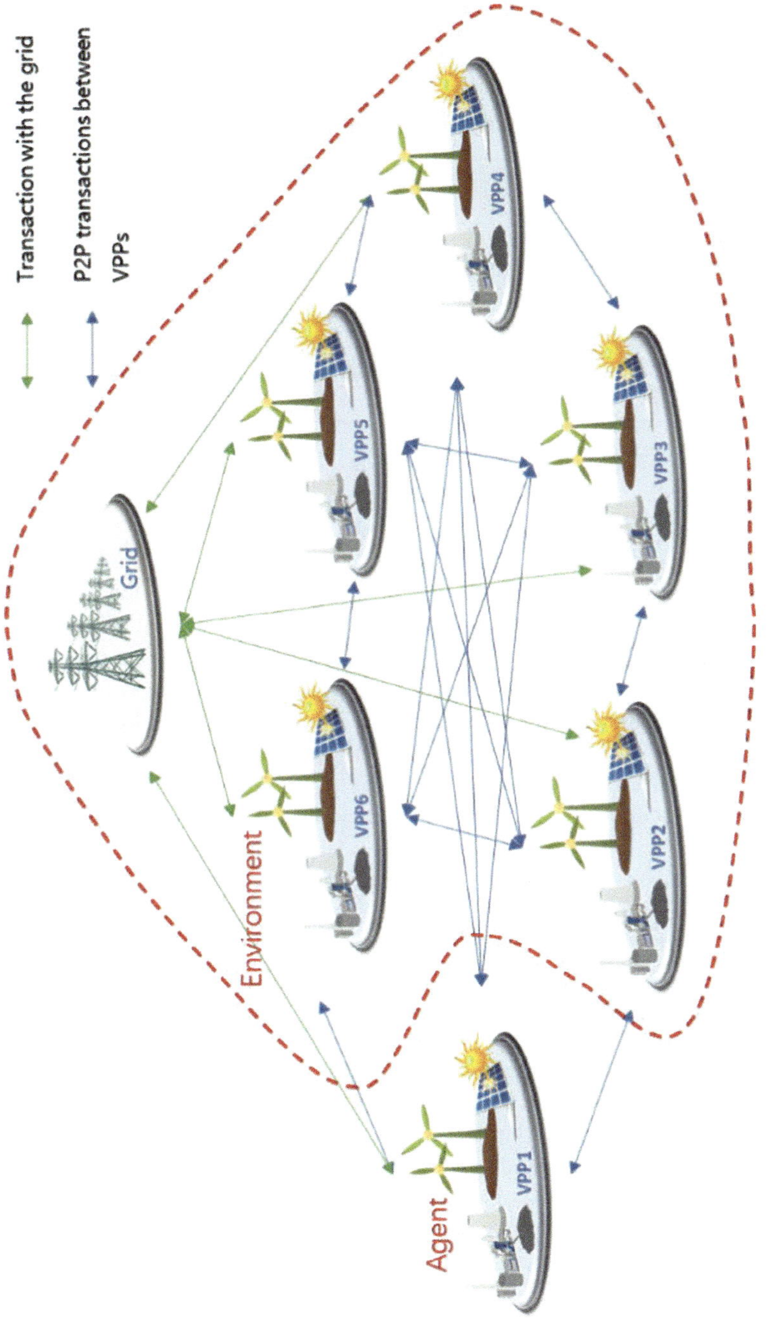

Fig. 6 P2P energy trading test case for VPPs

Table 1 Training parameters of deep reinforcement learning

Parameter name	Parameter value
Maximum training rounds	3000
Number of data in memory	500
Max number of withdrawals per reinforcement training	64
Reward attenuation coefficient γ	1
The target network updates the hyperparameter	0.15

Table 2 Parameters of actor and critic network/target network

	Actor network	Critic network
Input dimension	8	11
Output dimension	3	1
Network structure	[32, 16]	[32, 16]
Learning rate	1e-5	1e-3

The DDPG used in the test case includes four neural networks, namely, the Actor network, the Critic network, and the target networks of the two networks. The parameters of this part of the network are shown in Table 2. The input dimension of the Actor network and the target network is the state dimension, and the output dimension is the action dimension. The input dimension of the Critic network and the Critic target network is the sum of the state dimension and the action dimension, and the output evaluation value dimension is 1. The learning rate is set to ensure that the learning rate of Actor and its target network is smaller than that of Critic and its target network, which ensures the convergence stability of agent strategy learning.

The cumulative reward of the action of the VPP's agent changes with the number of training rounds by using DDPG and traditional DQN is shown in Fig. 7. It can be seen that under the supply-demand ratio pricing mechanism, the result of the DDPG algorithm has been stable convergence with about 1500 epoch of iterations. However, traditional DQN still cannot converge well under finite iterations, and the sensitivity of training convergence of the DDPG algorithm is greater due to the discretization of the action space.

Furthermore, the system operation was optimized for three different scenarios, and the energy trading and renewable energy consumption in different scenarios were analyzed. For case 1, the traditional meta-heuristics optimization method is used to solve the operation and trading strategy of VPP1 under the supply-demand ratio market mechanism. The optimal operation strategies of VPP1 under case 1 are shown in Fig. 8, and the SOC of the energy storage device inside VPP1 is shown in Fig. 9. For case 2, based on the model trained by DQN, the optimal operation strategies of VPP1 under the SDR market mechanism are shown in Fig. 10, and the SOC of the energy storage device inside VPP1 is shown in Fig. 11. For case 3, based on the DDPG deep reinforcement learning, the optimal operation strategies of the

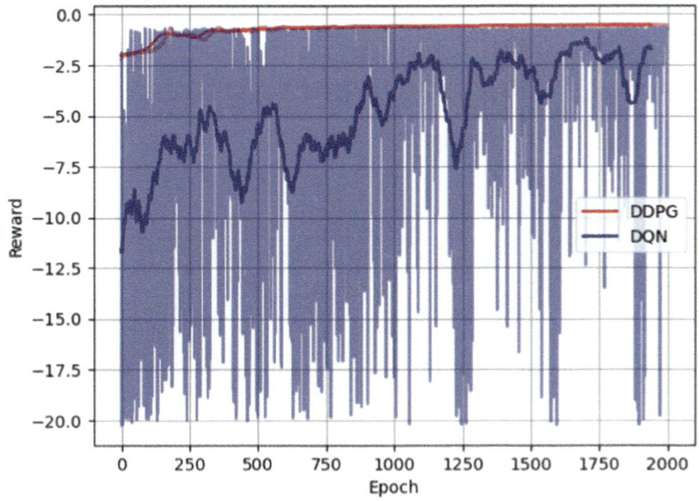

Fig. 7 Convergence process of cumulative reward by DDPG and DQN

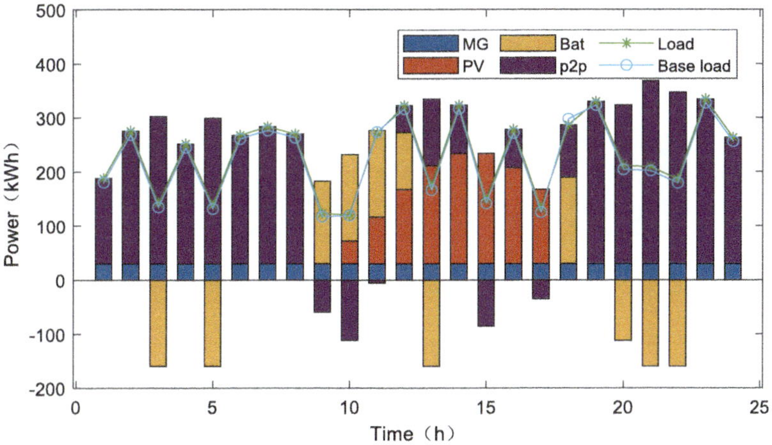

Fig. 8 Optimal operation strategies of VPP1 based on traditional meta-heuristics optimization algorithm

VPP1 under the SDR pricing strategy are shown in Fig. 12, and the SOC of the energy storage device inside VPP1 is shown in Fig. 13.

From Figs. 8, 10, and 12, we can find that for all three cases, the trading and operation strategy can satisfy the power balance of the VPP, making full use of the renewable energy of the VPP. The cost of the micro-gas turbine of the VPP is low, and the strategy allows the micro-gas turbine to fully exert its power to obtain higher returns. The economic benefit of the load response of the VPP is the secondary load curve, which fluctuates with the transaction price. While the energy storage is

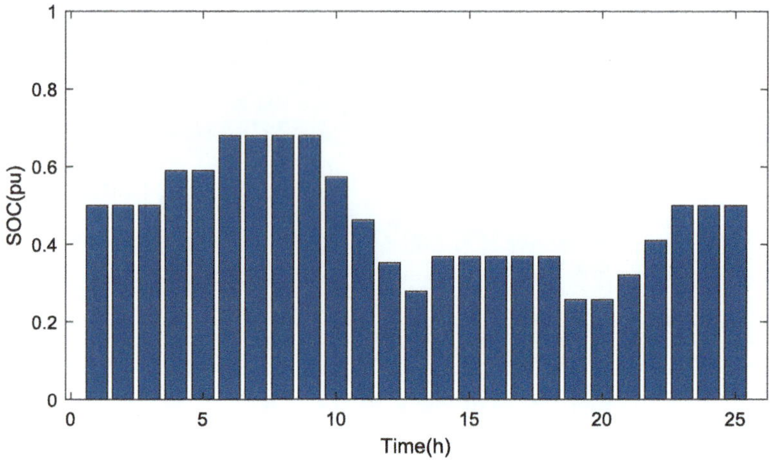

Fig. 9 SOC of energy storage system based on traditional model-based optimization algorithm

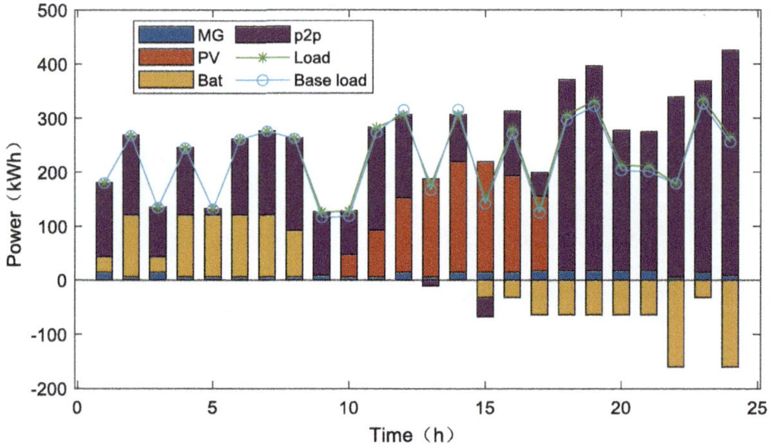

Fig. 10 Optimal operation strategies of VPP based on DQN

discharged under the pursuit of income, the charge returns to 0.5 at the end of the day, satisfying the equation constraint.

Further, the comparison of operating cost and calculation time for different cases is conducted, the results are shown in Table 3.

From Table 3, it can be found that the operating cost of the VPP1 is 4.0943e4 dollars without optimization, while the cost based on the DQN reinforcement learning algorithm in case 2 is $1.1824e^4$ dollars, which reduces the cost by 71.1% compared with no optimization. In contrast, the cost based on DDPG-based reinforcement learning in case 3 is $7.445e^3$ dollars, the cost has reduced by 81.8% compared with the non-optimization method. The cost based on centralized optimization is the lowest;

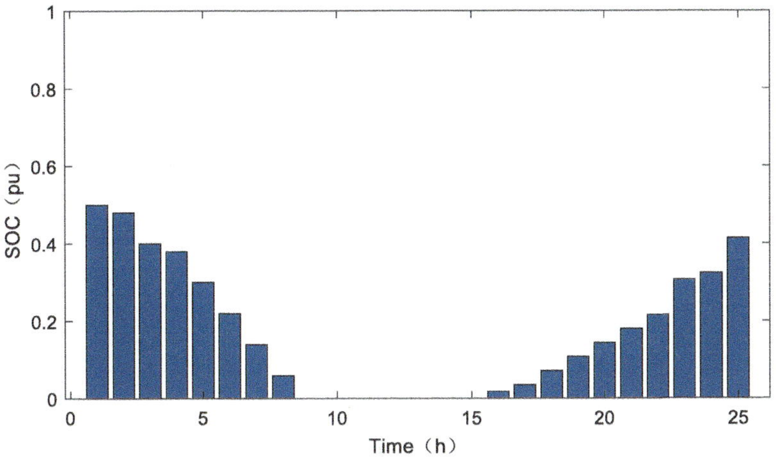

Fig. 11 SOC of energy storage system based on DQN strategies

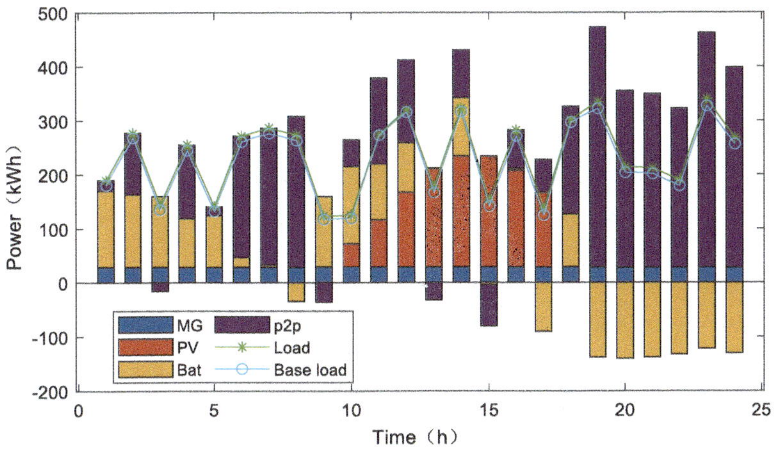

Fig. 12 Trading and operation strategies of VPP based on proposed DDPG algorithm

the cost has reduced by 82.5% compared with the cost of no optimization method. The difference between traditional optimization and reinforcement learning is that deep reinforcement learning has different effects on energy storage, action control. Compared with traditional optimization algorithms, the reinforcement learning strategy is rougher in the exploration of energy storage SOC, and the optimization exploration algorithm has more improvements in the exploration of solutions, while the exploration of the deep reinforcement learning algorithm here is random.

In terms of the calculation time, the traditional optimization algorithm takes 3.11s to obtain the strategy, where the DQN algorithm and DDPG algorithm take only 0.51 and 0.49s, respectively. The calculation time of DDPG is 84.24% less than that of the

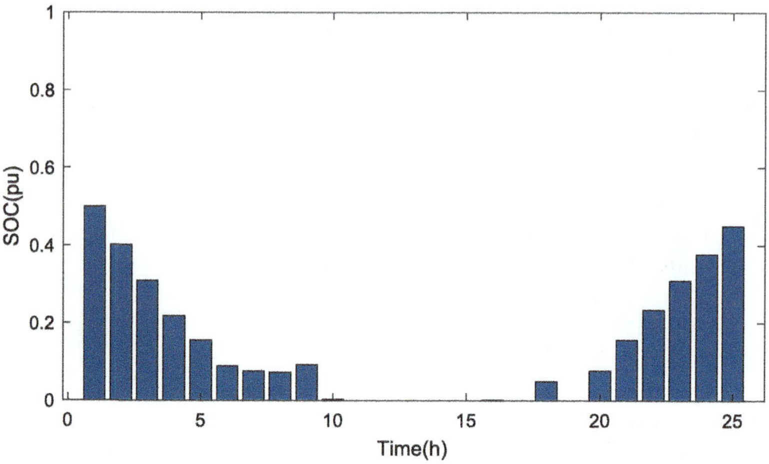

Fig. 13 SOC of energy storage system based on proposed DDPG algorithm

Table 3 Comparison of electricity cost and calculation time of different cases

	Non-opt	Case 1	Case 2	Case 3
Cost (dollar)	4.0943e4	7.163e3	1.1824e4	7.445e3
Cost reduction	/	82.5%	71.1%	81.8%
Computing time (s)	/	3.11	0.51	0.49
Time reduction	/	/	83.60%	84.24%

optimization-based method. This indicates that the DDPG algorithm can significantly improve the efficiency of decision calculation while ensuring comparable accuracy.

3.3.3 Sensitivity Analysis

Further analysis was conducted on the sensitivity of the DDPG algorithm to parameters, considering that the constraint scaling and qualification method is a common method to solve the problem that deep reinforcement learning cannot satisfy the equality constraint and inequality constraint of VPP and power market by using the scaling constraint as a penalty term. Taking the original constraint (22) as an example, it is scaled and added to the reward function to get a new Formula (23) for calculating the reward function, where $\lambda_1 > 0$, $\lambda_2 > 0$ is the penalty coefficient, and the relation between its value and the reward value is about the order of 10^{-1}–10^0 magnitude.

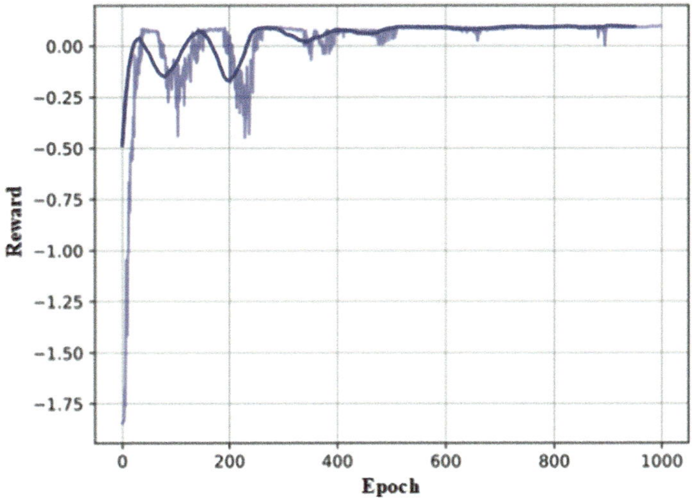

Fig. 14 SOC penalty term is distributed in nearby intervals

The larger the value, the better the condition of constraint satisfaction, but the greater the overall convergence interference of the model.

$$\begin{cases} A_1 x = b_1 \\ A_2 x < b_2 \end{cases} \tag{22}$$

$$R' = R + \lambda_1 |A_1 x - b_1| + \lambda_2 (A_2 x - b_2) \tag{23}$$

In addition, the equation constraint can also be scaled on the timeline. Take the SOC constraint of energy storage $\text{SOC}(25) = \text{SOC}(0)$ as an example, it can be rewritten as a formal penalty term $-\lambda_{\text{SOC}}|\text{SOC}(25) - \text{SOC}(0)|$. However, the distribution of this scaling on the timeline is concentrated at one point and inconsistent with the time distribution of the economic rewards of the power market uniformly distributed on the timeline. Penalty term can be distributed over several timelines, converting to $-\frac{\lambda_{\text{SOC}}}{n}\sum_{i=25-n}^{25}|\text{SOC}(25) - \text{SOC}(0)|$. This is also a better explanation in the physical sense, that is, to allow time for forced energy storage charging and discharging to restore the initial SOC. Their training differences are as Fig. 14.

As shown in Figs. 14 and 15, the two graphs are deep reinforcement learning environments where only energy storage actions are trained. The penalty terms in Fig. 14 are distributed evenly over multiple time intervals, while Fig. 15 is distributed only at the last end time point. It can be seen that the SOC penalty term distributed in nearby intervals is easier to explore better actions than the SOC penalty term only distributed at the end time point. Under feasible conditions, the distribution interval of the constraint penalty term is reasonably enlarged, which is helpful to accelerate the training convergence.

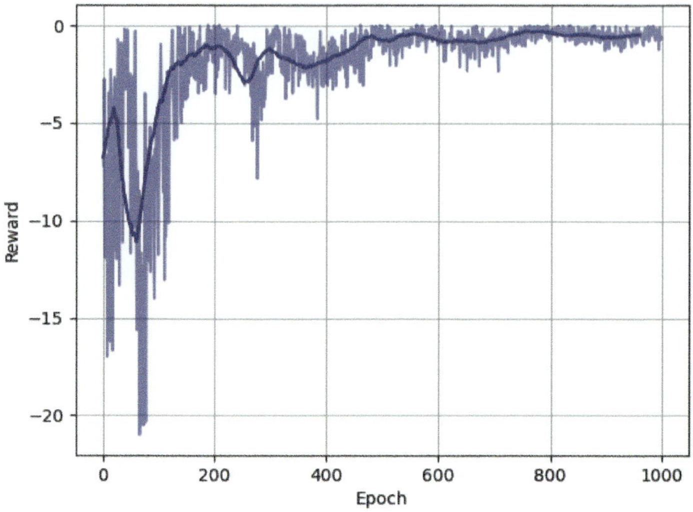

Fig. 15 SOC penalty term is only distributed at the end time point

3.4 Conclusions

This chapter discusses the integration of renewable energy sources into local energy systems. Firstly, it elucidates the various challenges faced in the integration of renewable energy sources in power forecasting technology, stability analysis and computation, and digitalization technology of power systems. Then, it summarizes existing digital technologies for renewable energy forecasting and proposes an improved prediction method based on enhanced TCN and feature modeling to better support renewable energy prediction. Finally, it introduces the concept of VPP and presents an AI-empowered VPP interactive optimization operation and renewable energy consumption model, demonstrating the effectiveness of the proposed method through practical case studies.

Acknowledgements This work was supported by National Natural Science Foundation of China (52177124, U2066211), the Institute of Electrical Engineering, CAS (E155610101), and in part by the DNL Cooperation Fund, CAS (DNL202023).

References

1. Hu H, Xie N, Fang D, Zhang X (2018) The role of renewable energy consumption and commercial services trade in carbon dioxide reduction: evidence from 25 developing countries. Appl Energy 211:1229–1244
2. Handayani K, Anugrah P, Goembira F, Overland I, Suryadi B, Swandaru A (2022) Moving beyond the NDCs: ASEAN pathways to a net-zero emissions power sector in 2050. Appl Energy 311:118580
3. Lin B, Liu X (2013) Electricity tariff reform and rebound effect of residential electricity consumption in China. Energy 59:240–247
4. Zhu F, Zhong P-A, Xu B, Liu W, Wang W, Sun Y, Chen J, Li J (2020) Short-term stochastic optimization of a hydro-wind-photovoltaic hybrid system under multiple uncertainties. Energy Convers Manag 214:112902
5. Liu B, Lund JR, Liao S, Jin X, Liu L, Cheng C (2020) Optimal power peak shaving using hydropower to complement wind and solar power uncertainty. Energy Convers Manag 209:112628
6. Ma J, Yang M, Han X, Li Z (2018) Ultra-short-term wind generation forecast based on multivariate empirical dynamic modelling. IEEE Trans Indus Appl 54(2):1029–1038. https://doi.org/10.1109/TIA.2017.2782207
7. Ma L, Luan SY, Jiang CW, Liu HL, Zhang Y (2009) A review on the forecasting of wind speed and generated power. Renew Sustain Energy Rev 13(4):915–920 May
8. Tian H, Mancilla-David F, Ellis K, Muljadi E, Jenkins P (2012) A cell-to-module-to-array detailed model for photovoltaic panels. Solar Energy 86(9):2695–2706
9. Lorenz E, Scheidsteger T, Hurka J, Heinemann D, Kurz C (2011) Regional PV power prediction for improved grid integration. Progr Photovoltaics 19(7):757–771
10. Erdem E, Shi J (2011) ARMA based approaches for forecasting the tuple of wind speed and direction. Appl Energy 88(4):1405–1414
11. Chen N, Qian Z, Nabney IT, Meng X (2014) Wind power forecasts using Gaussian processes and numerical weather prediction. IEEE Trans Power Syst 29(2):656–665
12. Liu H, Tian HQ, Li YF (2012) Comparison of two new ARIMA-ANN and ARIMA-Kalman hybrid methods for wind speed prediction. Appl Energy 98(1):415–424
13. Wang J, Zhong H, Lai X, Xia Q, Wang Y, Kang C (2019) Exploring key weather factors from analytical modelling toward improved solar power forecasting. IEEE Trans Smart Grid 10(2):1417–1427. https://doi.org/10.1109/TSG.2017.2766022
14. Alessandrini S, Delle Monache L, Sperati S, Cervone G (2015) An analog ensemble for short-term probabilistic solar power forecast. Appl Energy 157:95–110
15. Asmus P (2010) Microgrids, virtual power plants and our distributed energy future. Electr J 23(10):72–82
16. You S (2010) Developing virtual power plant for optimized DG operation and integration. Technical University of Denmark, Lyngby, Denmark
17. You S, Traeholt C, Poulsen B (2009) Generic virtual power plants: management of distributed energy resources under liberalized electricity market. In: 8th international conference on advances in power system control, operation and management (APSCOM 2009), IET, pp 1–6
18. Lillicrap TP, Hunt JJ, Pritzel A, Heess NMO, Erez T, Tassa Y, Silver D, Wierstra D (2015) Continuous control with deep reinforcement learning. CoRR arXiv:1509.02971 [Online]. https://api.semanticscholar.org/Corp
19. Mnih V, Kavukcuoglu K, Silver D, Rusu AA, Veness J, Bellemare MG, Graves A, Riedmiller M, Fidjeland AK, Ostrovski G, Petersen S, Beattie C, Sadik A, Antonoglou I, King H, Kumaran D, Wierstra D, Legg S, Hassabis D (2015) Human-level control through deep reinforcement learning. Nature 518:529–533
20. Liu N, Yu X, Wang C, Li C, Ma L, Lei J (2017) Energy-sharing model with price-based demand response for microgrids of peer-to-peer prosumers. IEEE Trans Power Syst 32(5):3569–3583. https://doi.org/10.1109/TPWRS.2017.2649558

Energy Flexible Buildings in Local Energy Systems

Maomao Hu and Rui Tang

Abstract Energy flexibility in buildings has been increasingly viewed as a key solution to address the power imbalance between supply and demand sides in electric grids, which is caused by the growing share of intermittent renewable energy generation. This chapter investigates the energy flexibility of buildings at individual and cluster levels. At the individual building level, we examine the energy flexibility of two types of space cooling and heating systems in residential buildings, including variable-speed heat pumps in Hong Kong and floor heating systems in Denmark, by using the advanced model predictive control technique. The proposed model predictive controller helps to improve thermal comfort at the beginning of occupancy, reduce power consumption during peak demand periods, and reduce daily electricity costs. We then introduce a novel supply-based feedback control to enable commercial buildings to provide fast power reduction in response to short-term or even immediate requests from smart grids. At the building cluster level, we review the state-of-the-art classification, architectures, and techniques of neighborhood-level coordination and negotiation in residential microgrids. Moreover, we explain the necessity and a method to realize the cluster-level building demand management in a distributed mode using game theory-based approaches.

1 Introduction

Renewable energy sources (RESs) have increasingly penetrated electric grids and energy markets in recent years to displace the use of fossil fuels and reduce greenhouse gas emissions. The wide integration of RESs, however, influences the reliability of power systems because RESs are intrinsically intermittent and difficult to

M. Hu (✉)
Department of the Built Environment, National University of Singapore, Singapore, Singapore
e-mail: maomaohu@nus.edu.sg

R. Tang
Institute for Environmental Design and Engineering, The Bartlett, University College London, London, UK
e-mail: rui.tang@ucl.ac.uk

W. Hua et al. (eds.), *Digitalisation of Local Energy Systems*, Green Energy and Technology, https://doi.org/10.1007/978-3-031-77833-9_6

forecast [1–3]. The peak power consumption at the demand side under severe hot or cold weather conditions sometimes causes a power imbalance, which also adversely affects the reliability of power systems [4]. Buildings are responsible for around 35% of the total energy consumption worldwide [5]. Due to the elastic nature, building energy use could be adjusted toward the needs of power grids through the ways such as peak demand charges and time-of-use rates. Heating, ventilation, and air-conditioning (HVAC) systems, accounting for more than 50% of energy demand in buildings, are excellent energy flexibility resources to reduce or shift the electricity demand during peak periods [6]. In this context, modern buildings and their HVAC systems are expected to be not only energy efficient but also energy flexible and grid-responsive [7].

As defined in IEA EBC Annex 67, the energy flexibility of a building refers to "the ability to manage its demand and generation according to local climate conditions, user needs, and energy network requirements" [8]. Over the last few decades, many studies have focused on the investigation of the energy flexibility of energy systems in individual buildings, including residential and commercial buildings. For individual residential buildings, Dréau and Heiselberg [9] investigated the flexibility potential of two residential houses with different levels of insulation and air-tightness, considering both energy performance and occupants' thermal comfort. The passive house, with its higher heating inertia, was found to have a larger influence on the control strategies for energy flexibility. Shafie-khah et al. [10] applied hybrid phase-change materials (PCMs) in a residential building to provide demand flexibility. Their results showed that the integration of PCMs affected the indoor thermal dynamics and the electricity cost savings under various demand response programs. Three different thermal energy storage tanks, i.e., water, PCMs, and thermochemical material tanks, were integrated with the heating system in an office building to provide demand flexibility [11]. Some studies have put emphasis on the development of energy flexibility control strategies for heating, ventilation, and air-conditioning (HVAC) systems in individual buildings. Temperature set-point reset [12, 13] and precooling/heating [14, 15] are two common control strategies for providing load reduction and shifting. Yin and Kara et al. [13] applied a set-point change strategy to cooling systems in both residential and commercial buildings to reduce load during on-peak hours. The hourly power reduction potential of HVAC systems for each type of building was assessed under various thermostat adjustment strategies. Hu and Xiao et al. [12, 15] combined temperature set-point reset and precooling strategies for residential air conditioners to shift the energy consumption from high-price to low-price time periods using a genetic algorithm. An advanced model predictive control (MPC) method was developed by Hu and Xiao et al. [16] for a floor heating system to provide energy flexibility during on-peak hours. Their model simultaneously considered weather conditions, occupancy, and dynamic electricity prices from electric utilities.

In commercial buildings, the control methods are normally more complicated than in residential buildings. Basically, load shifting and load shedding are the two major means of peak load management in commercial buildings. Load shedding control reduces peak electric load in a building via turning off non-essential electrical load [17]. Compared with load shedding, load shifting, which is the process of shifting on-peak load to off-peak hours so as to take advantage of the electricity rate difference in different periods, is more commonly used for demand-side management in commercial buildings. Four typical categories of facilities are widely used for peak loading shifting, including: building thermal mass (BTM) [18, 19], thermal energy storage system (TES) [20, 21], combined use of BTM with TES [22–24], and phase-change materials (PCM) [25, 26].

Although a considerable amount of literature has investigated the energy flexibility in individual buildings [9–15], to have a sufficient impact on electric grids, the energy flexibility of buildings must be managed at an aggregated level [27]. Investigating the aggregate energy flexibility can help service providers assess the environmental and financial benefits of energy flexibility services and determine feasible electricity prices for electricity consumers [28, 29]. More attention, therefore, has focused on energy flexibility at a neighborhood level [27, 30]. Taniguchi et al. [31] used a bottom-up energy modeling method to investigate the energy flexibility of 5,000 Japanese households. Various influential factors were considered to capture the diversity of energy consumption patterns, including floor area, household composition type, building insulation level, etc. In [32], a demand-side load management strategy was proposed for 10,000 residential air conditioners using a developed mathematical model of aggregate dynamics. Hu and Xiao [28] evaluated the uncertainty in the aggregate energy flexibility of a residential community based on a stochastic Markov-chain occupancy model. Simulation results indicated that the uncertainty of the aggregated energy flexibility decreased when the building clusters scaled up.

In this chapter, we aim to investigate energy flexibility in buildings at both individual and cluster levels. In Sect. 2.1, the energy flexibility of two types of space heating and cooling systems in residential buildings, including variable-speed heat pumps in Hong Kong and floor heating systems in Denmark, is explored by using the advanced model predictive control technique. Section 2.2 presents a direct load control scheme in commercial buildings, which controls the building demand by limiting the supply side of thermal systems. Section 3.1 reviews the state-of-the-art classification, architectures, and techniques of neighborhood-level coordination and negotiation in residential microgrids. In Sect. 3.2, we present the necessity of cluster-level building demand management based on a case study in a Hong Kong campus and introduce a decentralized method of optimizing the operation of buildings integrated with passive and active thermal storages.

2 Energy Flexibility in Individual Buildings

2.1 Individual Residential Buildings

Residential buildings represent around 55% of the total energy use in the building sector in the United States [33]. With the assistance of smart in-home sensors and meters, smart electric appliances, and smart home energy management systems (HEMSs), today's residential buildings have demonstrated the ability to change their normal energy consumption patterns in response to changes in the price of electricity over time, or to the incentive payments at times of high wholesale market prices or when system reliability is jeopardized. According to the U.S. Energy Information Administration, demand response programs provided 13,036 MW actual peak reductions in 2015 in total, and the residential sector provided 26% of the total peak reductions [34]. As indicated by the Federal Energy Regulation Committee, it is the residential class that represents the most untapped potential for demand response, and the residential customers were projected to provide over 45% of the potential impacts in 2019 [35]. Globally, residential distributed energy resources (DERs), including electric vehicles and on-site solar generations, were estimated with an annual growth of 60 GW in 2021, which will continue to increase in the next decade [36]. The increasing residential DESs enable residential buildings to further provide energy flexibility and improve the reliability of electric grids.

In this section, we aim to investigate the energy flexibility of two types of space heating and cooling systems in residential buildings: (i) variable-speed heat pumps in Hong Kong and (ii) floor heating systems in Denmark. The energy flexibility is achieved by adopting the advanced model-based optimal control strategy, i.e., model predictive control, which can simultaneously consider the building thermal characteristics and all the influential variables, including weather conditions, occupancy profile, and dynamic electricity pricing.

2.1.1 Residential Buildings with Variable-speed Heat Pumps

Overview

This section presents a novel MPC-based frequency control method for variable-speed heat pumps, which directly controls the operating frequency of the compressor in response to 5-minute real-time electricity prices (RTPs). The implementation structure of MPC for variable-speed heat pumps includes two main parts: offline data-driven model development and online MPC. The system of interest in this study is an integrated building energy system that includes an air-conditioned room and a variable-speed heat pump. Considering the computational efficiency for online MPC-based DR control, control-oriented models of both the room and the heat pump are required. In this study, we develop a stochastic state-space model of room thermal dynamics and 3-dimensional performance maps of variable-speed heat pump, based on the detailed physical models developed in our previous work [15], for online prediction of the system's evolution under the influence of predicted future exogenous variables such as weather conditions, occupancy, and RTP. The output of the MPC controller is the control signal of the operating frequency of the heat pump, which is used to control the compressor speed directly.

Building thermal model

The building models in commonly used simulation software such as TRNSYS, EnergyPlus, and eQUEST cannot be readily used for model-based control meth-

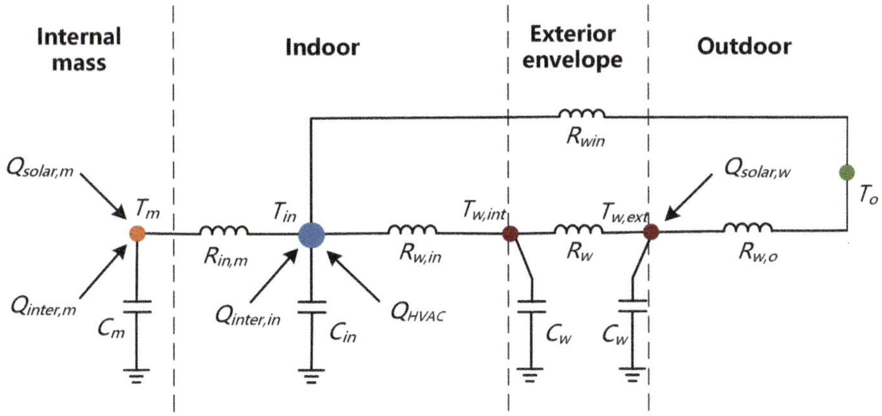

Fig. 1 Schematic of the gray-box RC room thermal model

ods because they are complex and usually time-consuming. In previous studies [12], we developed, identified, and validated a gray-box RC room thermal model. The structure of the gray-box model is illustrated in Fig. 1. The gray-box room thermal model can learn the thermal characteristics of the to-be-controlled room by analyzing the data collected from smart in-home sensors. Due to the simplicity of the model structure, the developed room thermal model can be easily integrated into the model-based controller to effectively predict the room's thermal dynamics. The energy balance for the external and internal walls, indoor air, and internal thermal mass is given by Eqs. (1)–(4).

$$C_w \frac{dT_{w,ext}}{dt} = \frac{T_o - T_{w,ext}}{R_{w,o}} + \frac{T_{w,int} - T_{w,ext}}{R_w} + f_{solar,w} A_w I_{solar} \quad (1)$$

$$C_w \frac{dT_{w,int}}{dt} = \frac{T_{w,ext} - T_{w,int}}{R_w} + \frac{T_{in} - T_{w,int}}{R_{w,in}} \quad (2)$$

$$C_{in} \frac{dT_{in}}{dt} = \frac{T_m - T_{in}}{R_{in,m}} + \frac{T_{w,int} - T_{in}}{R_{w,in}} + \frac{T_o - T_{in}}{R_{win}} + f_{inter,in} Q_{inter} + Q_{HVAC} \quad (3)$$

$$C_m \frac{dT_m}{dt} = \frac{T_{in} - T_m}{R_{in,m}} + f_{solar,m} A_{win} I_{solar} + f_{inter,m} Q_{inter} \quad (4)$$

where R and C represent the overall heat resistance and capacitance; T denotes temperature; the subscripts in, o, w, int, ext, win, and m indicate indoor air, outdoor air, exterior wall, internal wall surface, external wall surface, window, and internal mass, respectively; Q_{inter} denotes the internal heat gain, I_{solar} denotes the global solar radiation, A denotes the geometric area, and f denotes the conversion coefficients for the heat gains, which are also identified together with R and C. The detailed model identification methods can be found in our previous publication [12].

The dynamic room thermal model, as described by differential Equations (1)–(4), is an MIMO system. State space models are commonly used to model MIMO systems because of their advantage in the explicit expression of the relationships between the outputs and inputs. They can also be used to formulate convex optimization problems that can generally be solved conveniently with state-of-the-art optimization techniques. To make the model more realistic, white Gaussian noise is added to the system model. Equations (1)–(4) can be converted into a continuous-time state-space model with stochastic noise, as shown in Eq. (5).

$$dT = (AT + Bu + Ed)dt + dw(t) \tag{5}$$

where the system state $T = \left[T_{w,ext}\ T_{w,int}\ T_{in}\ T_m\right]^T$; the input vector $u = Q_{HVAC}$; the disturbance vector $d = \left[T_o\ I_{solar}\ Q_{inter}\right]^T$; the system matrix $A = \begin{bmatrix} \frac{-1}{C_w R_{w,o}} + \frac{-1}{C_w R_w} & \frac{1}{C_w R_w} & 0 & 0 \\ \frac{1}{C_w R_w} & \frac{-1}{C_w R_w} + \frac{-1}{C_w R_{w,in}} & \frac{1}{C_w R_{w,in}} & 0 \\ 0 & \frac{1}{C_{in} R_{w,in}} & \frac{-1}{C_{in} R_{w,in}} + \frac{-1}{C_{in} R_{in,m}} + \frac{-1}{C_{in} R_{win}} & \frac{1}{C_{in} R_{in,m}} \\ 0 & 0 & \frac{1}{C_m R_{in,m}} & \frac{-1}{C_m R_{in,m}} \end{bmatrix}_{4\times4}$; the input matrix $B = \left[0\ 0\ \frac{1}{C_{in}}\ 0\right]^T$; the disturbance matrix $E = \begin{bmatrix} \frac{1}{C_w R_{w,o}} & \frac{f_{solar,w} \times A_w}{C_w} & 0 \\ 0 & 0 & 0 \\ \frac{1}{C_{in} R_{win}} & 0 & \frac{f_{inter,in}}{C_{in}} \\ 0 & \frac{f_{solar,m} \times A_{win}}{C_m} & \frac{f_{inter,m}}{C_m} \end{bmatrix}_{4\times3}$; $w(t)$ is a Wiener process, which is a stochastic process with independent normal distributed increments.

In practical applications, the stochastic continuous-time state-space model needs to be discretized, and the stochastic discrete-time state-space model is given by Eqs. (6) and (7).

$$T_{k+1} = A_d T_k + B_d u_k + E_d d_k + w_k \tag{6}$$

$$y_k = C_d T_k + v_k \tag{7}$$

where A_d, B_d, and E_d are the corresponding matrices of the discrete-time state-space model, which depend on the sampling time. The observed output vector $y_k = T_{in}$; and the output matrix $C_d = \left[0\ 0\ 1\ 0\right]$. The random variables w_k and v_k represent the process and measurement noises, respectively, which are assumed to be independent and white and to have normal distribution probabilities (i.e., $w_k \sim N(0, Q)$ and $v_k \sim N(0, R)$). Q and R are the covariance matrices of the process noise and the measurement noise, respectively. Equations (6) and (7) are used in the MPC controller to online predict the system evolution.

Performance model of variable-speed heat pumps

We develop 3-dimensional performance maps of inverter ACs based on the steady-state inverter AC models developed in our previous work [15]. The model of inverter

ACs was validated using the experimental data of a Mitsubishi split-type inverter AC. The rated cooling capacity of the tested inverter AC is 2.5 kW. A performance map is a kind of graphical representation designed to show the performance of a system at different levels of disturbance. Since the performance of inverter ACs is quantified by both their cooling capacity and COP, the performance maps of inverter-driven ACs include the cooling capacity map and the COP map. The typical operating conditions are the combinations of different compressor rotational speeds ($N_{comp} = [20, 30, \cdots, 100]$), indoor air temperature ($T_{in} = [21, 24, 27, 30, 33, 36]$) and outdoor air temperature ($T_{out} = [23, 26, 29, 32, 35, 38]$). The performance data of the inverter AC under typical operating conditions are obtained from simulation tests conducted using the steady-state inverter AC model developed in our previous work. With the performance maps, the cooling capacities and COP of the inverter AC at given indoor and outdoor conditions and compressor frequencies can be obtained by 3-dimensional interpolations.

Figure 2 shows the performance maps of the variable-speed heat pump at a range of compressor speeds, i.e., 20–100 Hz. It can be found that the cooling capacity increases as the compressor speed increases, but the COP decreases as the compressor speed increases. The outdoor air temperature has a greater influence on the COP than the indoor air temperature. The performance maps of the heat pump under the full range of operating conditions are used for online MPC in this study.

Formulation and solving of optimization problem

For purposes of comparison, a DR-enabled MPC controller and an ordinary MPC controller without a DR function are developed in this study to fulfill different tasks. A general MPC framework includes four parts: cost function, constraints, system

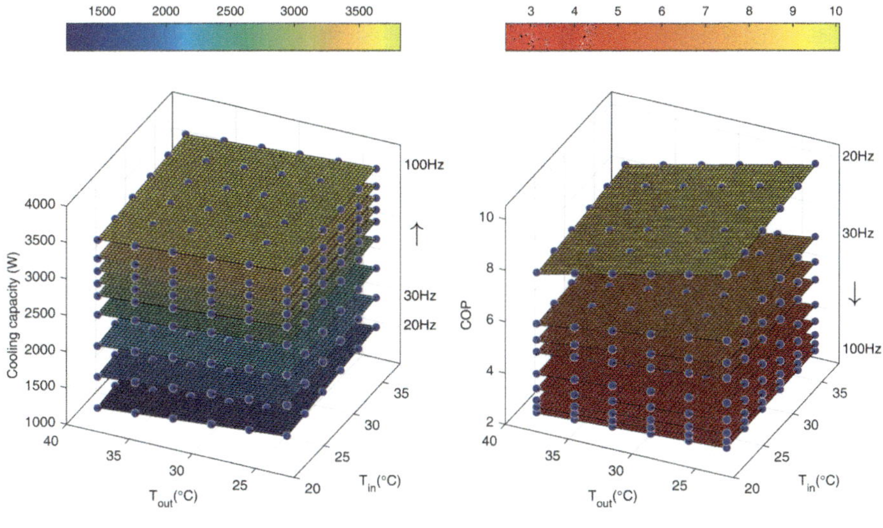

Fig. 2 Performance maps (cooling capacity and COP map) of the variable-speed pump at a range of compressor speeds

dynamics, and current state [37]. The difference between the two MPC controllers is the construction of the cost function and constraints.

The objective of the DR-enabled MPC controller is to minimize the operating costs of the variable-speed heat pump over the prediction horizon N while keeping the indoor air temperature within a pre-specified range. The optimization problem is formulated as Eqs. (8)–(13).

$$\min_{P_1, P_2, \cdots, P_{N-1}} \sum_{k=1}^{N-1} P_k \Delta t \cdot RTP_k + \rho_e e_k \tag{8}$$

$$\text{s.t.} \quad T_{k+1} = A_d T_k + B_d u_k + E_d d_k + w_k \tag{9}$$

$$y_k = C_d T_k + v_k \tag{10}$$

$$y_{lb,k} - e_k \leq y_k \leq y_{ub,k} + e_k \tag{11}$$

$$e_k \geq 0 \tag{12}$$

$$P_k = 0 \quad or \quad P_{min} \leq P_k \leq P_{thsh,k} \tag{13}$$

where N is the prediction horizon; Δt is the prediction interval; P_k and COP_k are the AC power consumption and COP at time step k, respectively; P_{thsh},k is the dynamic power threshold; T_k, y_k, and d_k are the state vector, output vector, and disturbance vector in the state-space system model, respectively; w_k and v_k are the process and measurement noise, respectively, which are assumed to be independent and white and to have a normal distribution. A_d, B_d, C_d, and E_d are the state-space matrices of the discrete-time state-space model.

The objective function, Eq. (8), aims to minimize the operating costs over the prediction horizon. If the constraints are not violated, the optimal power consumption remains at 0 kW, i.e., the AC is turned off. Equations (9) and (10) depict the system dynamics in a state-space form. The input variable, the cooling capacity ($Q_{HVAC,k}$), is obtained by the product of COP_k and power consumption (P_k). Equation (11) is a dynamic soft constraint that maintains the indoor air temperature within a certain range. The temperature constraint may not be satisfied in practice even if the AC runs at its full cooling capacity, such as on an extremely hot day. This situation may lead to a failure to solve the optimization problem. To address this issue, we can relax the MPC problem by introducing a slack variable e_k. The hard constraint is then converted to a soft constraint, which has a similar function as the dead-band plays in the on-off control. To ensure that the preset temperature range is met most of the time (i.e., $e_k = 0$), an especially large penalty ρ_e can be imposed on the slack variable e_k in the objective function. Equation (13) is a time-varying hard constraint to limit the maximum power consumption of AC based on the RTP from electric utilities. An inverter AC has a minimum operating frequency (e.g., 20 Hz) and corresponding minimum power consumption P_{min}. Therefore, the power consumption should range between P_{min} and $P_{thsh,k}$, which can be determined by pre-defined rules [38] or equal zero when it is turned off.

The prediction horizon N and prediction interval Δt play significant roles in MPC controllers. The optimal prediction horizon is a trade-off between the accuracy of the predictions of exogenous inputs and a sufficient length of the prediction horizon. A long horizon also increases the computation time for each optimization step. The room concerned in this study is located in a cooling-dominated district and has a relatively small thermal mass. In addition, the inverter AC delivers cooling to the space via heat convection, which is a direct and quicker mode of heat transfer than heat radiation in a radiant heating system. Considering the short thermal response time of the target room to the cooling system, the prediction horizon N in this study is set to 3 h, which is sufficient for the inverter AC to respond in advance to upcoming disturbances.

Compared with the DR-enabled MPC controller, the ordinary MPC controller without the DR function does not consider the dynamic RTP in formulating the optimization problem. The optimization problem of the ordinary MPC scheme, which aims to save energy use, can be described by the objective function Eq. (14), and the constraints including Eqs. (9)–(12). In contrast to the time-varying constraint (Eq. 13), in the DR-enabled MPC controller, Eq. (15) is a static constraint that limits power consumption in a constant power range.

$$\min_{P_1, P_2, \cdots, P_{N-1}} \sum_{k=1}^{N-1} P_k \Delta t + \rho_e e_k \tag{14}$$

$$\text{s.t.} \quad P_k = 0 \quad or \quad P_{min} \leq P_k \leq P_{max} \tag{15}$$

Current state estimation using the Kalman filter

At the beginning of each optimization step, the current state is needed as the starting point for optimization. In our case, the system state includes four variables: $T = \left[T_{w,ext}\ T_{w,int}\ T_{in}\ T_m\right]^T$. However, only the indoor air temperature T_{in} is measured in real applications. To address this issue, the Kalman filter is commonly used to estimate the immeasurable variables and to filter the noise in measurements [39, 40]. The Kalman filter is an algorithm that provides a computationally efficient solution to estimate the state of a process in a manner that minimizes the prediction mean-squared error. The solution is recursive in such a way that each updated estimate of the state is computed from the previous estimate and the new measured data. As shown in Fig. 3, each Kalman filter cycle includes two steps: time update (priori estimate) and measurement update (posteriori estimate). P, Q, and R are the covariance matrices of the state estimate error, the process noise, and the measurement noise, respectively. $\hat{T}_k'$ denotes a priori state estimate at time step k. $\hat{T}_k$ denotes the posteriori state estimate given measurement y_k (i.e., indoor air temperature T_{in}) at time step k. K_k is the Kalman gain determined by solving an algebraic Riccati equation. I is an identity matrix.

Case study

A prototype room in a residential flat in Hong Kong is modeled in TRNSYS. The residential room (L × W × H: 4.8 × 3.6 × 3 m) has one south-facing exterior wall

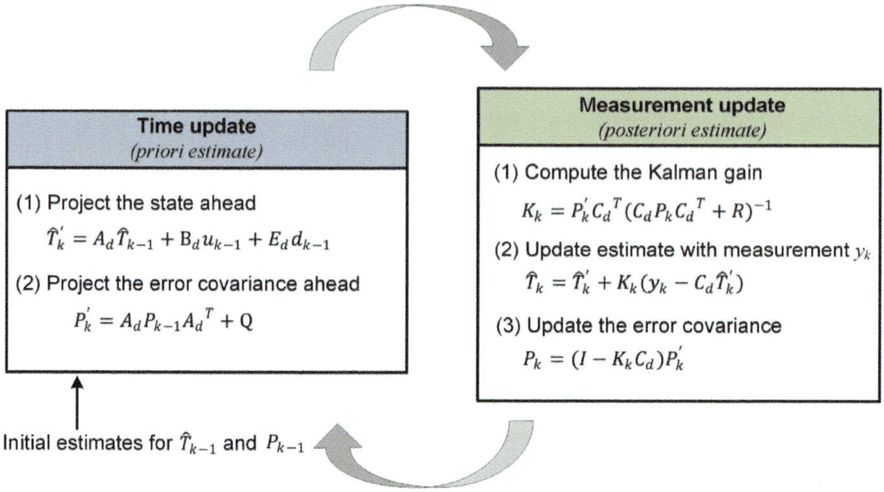

Fig. 3 Kalman filter cycle for current state estimate in MPC

(3.6 × 3 m) and one east-facing exterior wall (4.8 × 3 m). Both exterior walls have single-glazed windows, and the window-wall ratio of each wall is 0.2. The overall heat transfer coefficients of the exterior wall and the single-glazed window are 2.57 and 5.69 $W/(m^2 \cdot K)$, respectively. Instead of using the default TMY weather data, real historical weather data at 1-minute intervals from the Hong Kong Observatory are used in simulation tests.

TRNSYS and MATLAB are combined in this study for Computer-based dynamic simulation. In this study, the component of Type 155 is used to establish the communication between the models in TRNSYS 18 (32-bit) and MATLAB 2014a (32-bit) in the Windows 10 64-bit operating system. The MPC controller is implemented in MATLAB using the YALMIP optimization toolbox with the Gurobi optimization solver. The simulation tests with the two types of MPC controllers are conducted separately under conditions identical to those used in the reference case (PID case). The simulation time step in TRNSYS is set as 5 min, and at each time step, the MPC controller in MATLAB is called by TRNSYS, and the frequency control signal is sent to the inverter AC model.

Figure 4 shows the system performances under the PID control, the ordinary MPC without DR function, and the DR-enabled MPC. Figure 4b shows that the MPC controller, as a new optimal control method for variable-speed ACs, satisfactorily maintains the indoor air temperature between the upper and lower limits, although the temperature fluctuates due to the uncertain noise. The results also show that the ordinary MPC controller naturally implements the precooling strategy, which obviously improves thermal comfort at the beginning of occupancy. On Day 1, the prediction made by the MPC controller at 10:00 indicates that the indoor air temperature will exceed the upper boundary (i.e., 24 °C) at 12:00. Hence, the variable-speed AC begins to run at 10:00 to precool the room before it is occupied so that the indoor

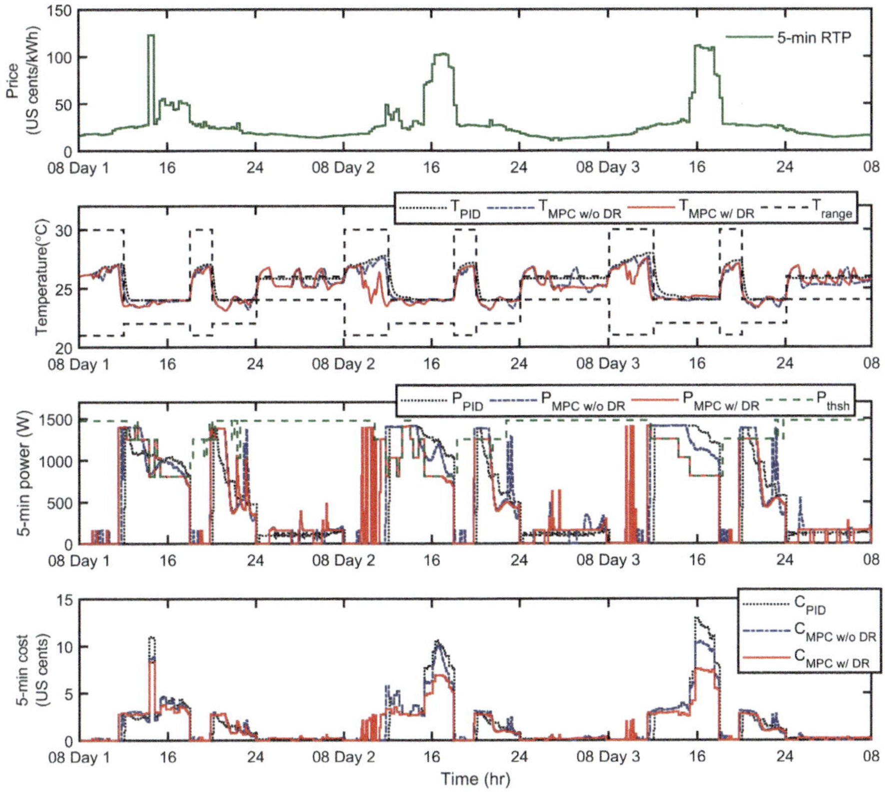

Fig. 4 **a** RTPs and system performances under PID control, ordinary MPC without DR function, and DR-enabled MPC, including **b** indoor air temperature, **c** power consumption, and **d** electricity costs

air temperature is at a comfortable level at the beginning of occupancy. Moreover, precooling is an optimal solution because the objective of the ordinary MPC is to minimize the energy consumption over the horizon. The ordinary MPC controller automatically implements precooling with different start-up times and durations on the three days because the weather conditions differ. In this way, the ordinary MPC controller solves the problem of determining the start-up time for precooling, which was challenging for most previous studies in which a precooling strategy was developed [14, 41]. Precooling may result in an increase in energy consumption, which can be observed in the test results shown in Fig. 5b. The AC under the ordinary MPC control consumes more energy than that under the PID control on these three days. However, the daily operating costs may be reduced, as shown in Fig. 5d, because the MPC controller allows a higher indoor air temperature than the PID controller.

Like the ordinary MPC controller without DR function, the DR-enabled MPC controller can implement automatic and optimal precooling. However, the DR-enabled MPC controller adopts longer and deeper precooling to limit power consumption

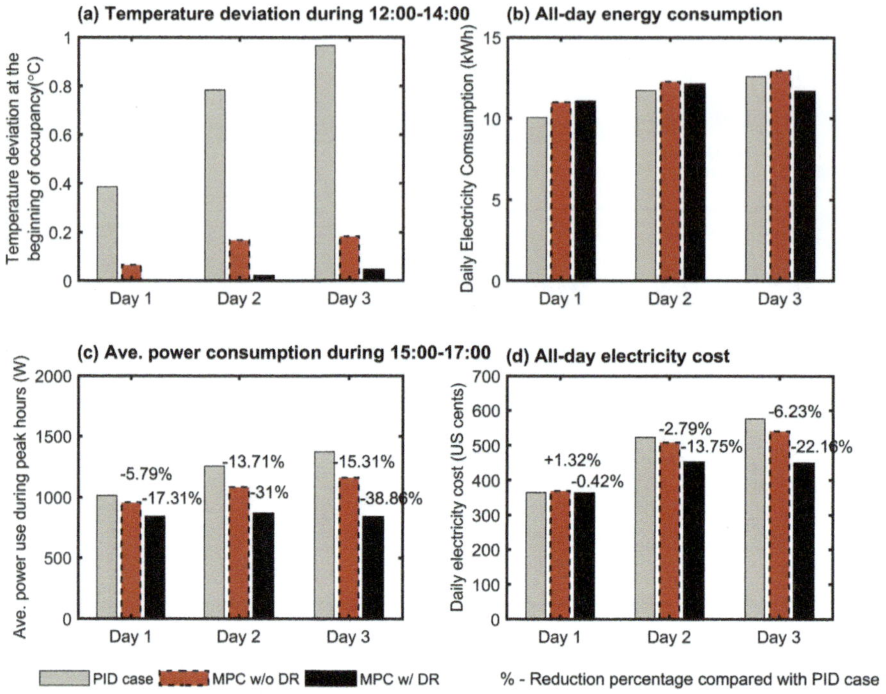

Fig. 5 System performances under PID control, ordinary MPC without DR function, and DR-enabled MPC

during peak demand periods. It starts the inverter AC to precool the room to a lower temperature so that the AC power consumption and operating frequency can be limited around 16:00, when the RTP is very high. Thermal comfort can be maintained due to the cooling energy stored in the thermal mass during the precooling period. As can be seen from Fig. 5c, compared with the PID case, the average power consumption from 15:00 to 17:00 is reduced by 17.31% on Day 1, 31.00% on Day 2, and 38.86% on Day 3, in the DR-enabled MPC control case. Because the DR-enabled MPC controller effectively shifts power consumption from high-RTP periods to low-RTP periods, the all-day electricity costs are reduced by 0.42% on Day 1, 13.75% on Day 2, and 22.16% on Day 3. The electricity cost saving is influenced by the weather conditions and RTP profiles. For example, the weather on Day 1 is cooler than that on Days 2 and 3, and the RTP curve is flatter, so the electricity cost saving is less. Meanwhile, as shown by Fig. 5, the daily energy consumption of the AC under the DR-enabled MPC may exceed that under the PID control. This is the price paid to achieve power reductions during the peak demand hours via precooling. From the perspective of power grids, it is more important to provide a stable power reduction as required.

2.1.2 Residential Buildings with Floor Heating Systems

Overview

Floor heating (FH) systems have attracted increasing interest and applications in recent years due to various advantages, including almost no noise, efficient use of space, uniform temperature distribution, energy saving potential, and use of environmentally friendly low-temperature energy sources [42, 43]. In this section, we aim to develop an MPC method for FH systems in response to dynamic electricity prices to shift peak power demands and provide energy flexibility. Like the MPC for variable-speed heat pumps in Sect. 2.1.1, the implementation structure of MPC for FH systems also includes two major parts: data-driven system model development and model predictive control. The system of interest in this study is an integrated building energy system, including a room with mechanical ventilation and an FH system.

Building thermal model and floor heating system

As shown in Fig. 6, the whole system model consists of an outdoor side, an indoor side, an underfloor side, and a water loop side (heating system). For the indoor side, it is divided into an exterior envelope ($T_{w,ext}$ and $T_{w,int}$), indoor air (T_{in}), and

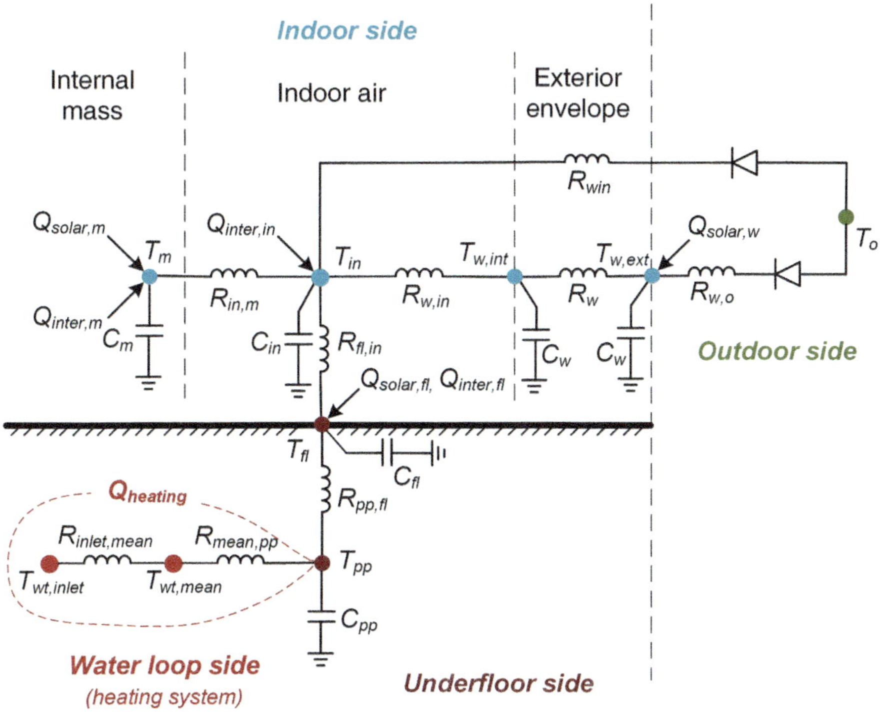

Fig. 6 Schematic of the gray-box RC model for FH systems

an internal thermal mass (T_m). The bulk of internal thermal mass Tm absorbs the heat from indoor heat sources, then gradually transfers the heat into the space by convection. The air in the interior zone is represented by a lumped node, which exchanges heat with the internal thermal mass, internal wall surface, outdoor air through windows, and the heated-up floor. Besides, the internal heat sources from occupants, lights, and equipment are assumed to be immediately absorbed by the indoor air.

The underfloor side provides the thermal interactions between the heating system (water loop side) and the indoor air. It consists of two critical nodes, i.e., the floor surface node, T_{fl}, and an assumed lumped node of the whole pipe plane, T_{pp}. The pipe plane is heated up by the hot water in the pipe and then gradually transfers the heat to the internal surface of the floor via construction elements. The heat transferred from the heating system to the pipe plane is denoted as $Q_{heating}$. On the indoor side and the underfloor side, the energy balance for the external and internal wall surfaces ($T_{w,ext}$ and $T_{w,int}$), the indoor air (T_{in}), the internal thermal mass (T_m), the floor surface (T_{fl}), and the core of the pipe plane (T_{pp}) are given by ordinary differential equations (ODEs). The ODEs can then be transformed into state-space models, which are commonly used for modeling multiple-input and multiple-output systems because of their advantage in explicitly expressing the relationships between the system outputs and inputs. The detailed process, including the development of ODEs, model transformation, discretization, and identification, can be found in our paper [16].

For the FH system, i.e., water loop side, it is critical to determine the amount of the transferred heat ($Q_{heating}$) and outlet water temperature ($T_{wt,outlet}$) given the mass flow rate, inlet hot water temperature ($T_{wt,inlet}$), and the core temperature of the pipe plane (T_{pp}). Detailed models for the FH system can also be found in [16].

Formulation and solution of optimization problems

The objective of the MPC controller for FH systems is to minimize the operating cost over the prediction horizon N while keeping the indoor air temperature in the pre-specified ranges. The optimization problem for MPC of FH systems is formulated as Eqs. (14)–(19).

$$\min_{u_1,u_2,\cdots,u_{N-1}} \sum_{k=1}^{N-1} (u_k/COP)\Delta t \cdot DAP_k + \rho_e e_k \tag{16}$$

$$\text{s.t.} \quad x_{k+1} = A_d x_k + B_d u_k + E_d d_k + w_k \tag{17}$$

$$y_k = C_d x_k \tag{18}$$

$$y_{lb,k} - e_k \leq y_k \leq y_{ub,k} + e_k \tag{19}$$

$$e_k \geq 0 \tag{20}$$

$$u_k = 0 \quad or \quad u_k = Q_{heating,k} \tag{21}$$

where N is the prediction horizon; Δt is the prediction interval; DAP_k is the day-ahead electricity prices (DAP) at the time step k; x_k, y_k, and d_k are the state vector, output vector, disturbance vector in the state-space system model, respectively; w_k

and v_k are the process and measurement noise, respectively, which are assumed to be independent, white and with normal distribution; A_d, B_d, C_d, E_d are the state-space matrices of the discrete-time state-space model; u_k is the input heating from inlet water to pipe plane at the time step k; COP is the coefficient of performance of the heat pump, which varies with the type, ambient medium temperature (air/ground) at the evaporate side, and the water temperature at the condenser temperature. Considering the ambient medium temperature and the water temperature at the condenser side almost remain constant under normal operating conditions, a typical constant value of COP [44], i.e., 3, is chosen for the case study in this paper. Considering the long thermal response time of the target room to the FH system, the prediction horizon N in this study is chosen as 12 h, which is sufficient for the FH to prepare for the upcoming disturbances. The prediction interval of the MPC controller in our study is set as 30 min, considering the thermal response time of the room.

Case study

The living room of a residential apartment in Nordhavn, Denmark, is chosen to carry out the simulation-based case study. The living room (L×W×H: 8.9m×6.22m×3m) has one south-facing exterior wall (8.9m×3m) and one east-facing exterior wall (6.22 m × 3 m). The window-wall ratios of the south-facing wall and the east-facing wall are 0.45 and 0.43, respectively. The residential building is well-insulated, and the overall heat transfer coefficients of the exterior wall and the window are 0.14 and 1.4 $W/(m^2 \cdot K)$, respectively. The apartment is heated by a water-based FH system. The outside diameter of the pipe and, spacing between pipes are 20 mm and 300 mm, respectively. The inlet water temperature to the FH system is regulated to be 35 °C by a manual thermostat, considering that higher temperatures could damage the wooden floor panels. The zone temperature is controlled using a thermostatic electro-valve. The driver actuator of the electro-valve adopts an on-off control signal. The set-point of the indoor air temperature is set as 22 °C when the target living room is occupied during 7:00–22:00.

For the MPC controller, the simulation time step in TRNSYS is set as 30 min, i.e., prediction interval, and at each time step, the MPC controller in MATLAB is called by TRNSYS and outputs the command signal to the FH system. Figure 7 shows the system performances including indoor air temperature, energy consumption, and electricity costs under the MPC control on seven winter days. For easy comparison, the performances under on-off control are also shown in the same figure.

As shown in Fig. 7b, the MPC, as a new control method for the FH system, can satisfactorily maintain the indoor air temperature between the upper and lower temperature bounds. Sometimes the temperature slightly exceeds the temperature bound because the prediction cannot be absolutely accurate, and some unknown random noises are added to the model of system dynamics, which is closer to the real case. Compared with the on-off control, the MPC is able to realize automatic and optimal preheating to improve the thermal comfort evaluated by the indoor air temperature at the beginning of occupancy. It makes a prediction at night time and realizes that the indoor air temperature at 7:00 will go below the lower bound, i.e., 22 °C. Hence, the FH system automatically preheats the room so that the indoor air temperature reaches

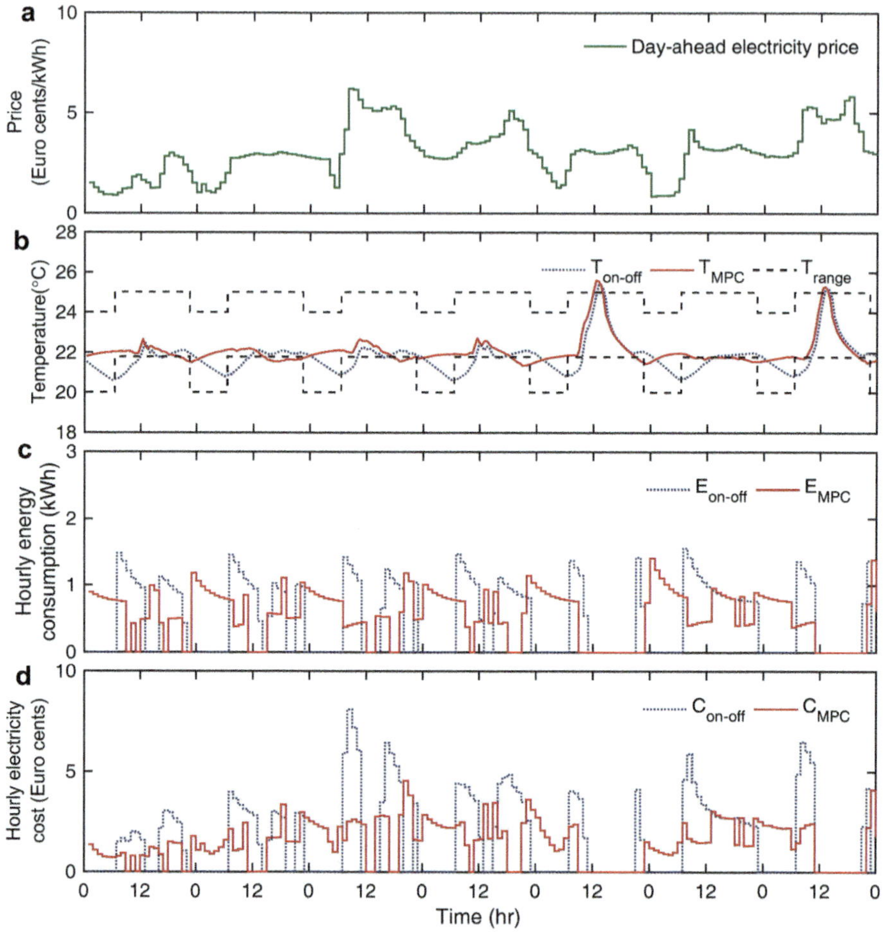

Fig. 7 **a** Day-ahead electricity prices and System performances under the on-off control and MPC on seven sequential days, including **b** indoor air temperature, **c** energy consumption, and **d** electricity costs

a comfortable level at the beginning of occupation. Moreover, the preheating action is optimal, which means the MPC controller can optimally determine the start-up time and duration of preheating based on the future weather conditions, price profile, and thermal characteristics of the room. Temperatures outside of the bounds are considered thermal comfort violations, which are evaluated by using $RMSE$ in the study. As shown in Fig. 8, the $RMSEs$ are reduced by 0.354 °C, 0.101 °C, 0.153 °C, 0.248 °C, 0.249 °C, 0.275 °C and 0.188 °C on Days 1–7, respectively.

As can be seen from Fig. 7, the MPC controller can also efficiently shift the energy demands from peak period (07:00–22:00) to off-peak period (23:00–6:00) and provide the energy flexibility for electric grids. The flexibility factor (FF), i.e.,

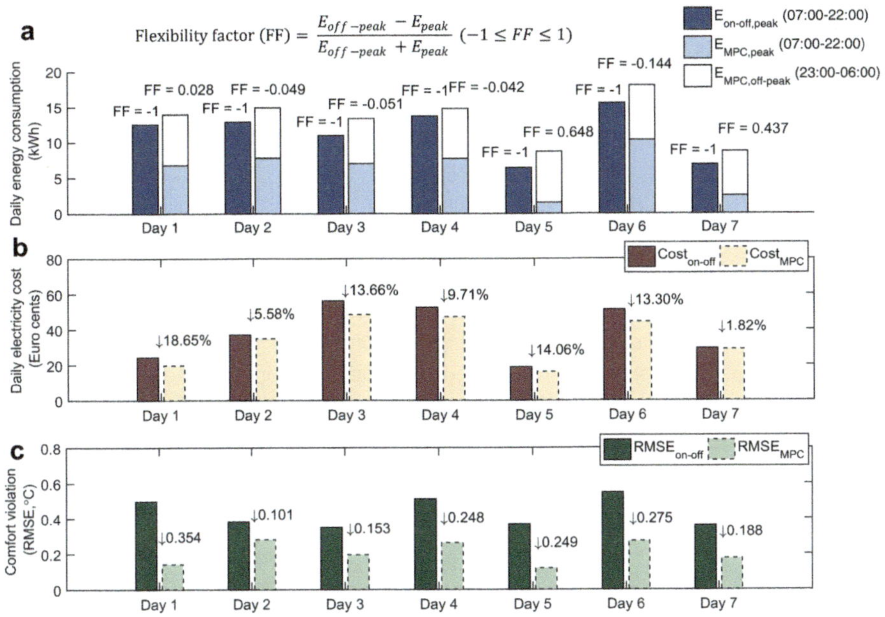

Fig. 8 Daily system performances under on-off control and MPC including **a** daily energy consumption, **b** daily electricity cost, and **c** thermal comfort violations

Eq. (22), is used as a performance indicator to assess the ability to shift the energy use from on-peak to off-peak periods.

$$Flexibility\ \ factor = \frac{E_{off-peak} - E_{peak}}{E_{off-peak} + E_{peak}} \quad (-1 \le FF \le 1) \qquad (22)$$

where $E_{off-peak}$ and E_{peak} refer to the total energy consumption during peak period (7:00–22:00) and off-peak period (23:00–6:00), respectively. The value of FF ranges between –1 and 1. The worst and optimal cases are when FF = –1 (i.e., energy is used only during peak period) and when $FF = 1$ (i.e., energy is used only during off-peak period), respectively. Figure 8a shows the energy flexibility is largely achieved by using the MPC, and FFs are increased from –1 to 0.028, –0.049, –0.051, –0.042, 0.648, –0.144 and 0.437 on Days 1–7, respectively, although the average daily energy consumption is increased by 15.81% in comparison to that under the on-off control. This is the price paid to achieve power reductions during the peak demand hours via preheating. From the perspective of power grids, providing a stable power reduction/shifting as required is more important than using less energy.

Due to the ability to shift energy use from high-price to low-price periods, the MPC helps to reduce the total daily electricity cost, which is influenced by the weather conditions and DAPs. As shown in Fig. 8b, the all-day electricity costs are reduced by 18.65%, 5.58%, 13.66%, 9.71%, 14.06%, 13.30% and 1.82% on Days 1–7, respectively.

2.2 *Individual Commercial Buildings*

In this section, a direct load control scheme is introduced in commercial buildings, which controls the building demand by limiting the supply side of thermal systems. Using a case study based in Hong Kong, the challenges are explained if without proper controls to deal with limited cooling/heating supplies, and a novel control strategy is developed accordingly to overcome the drawbacks of existing controls applied in buildings.

Current Controls and Challenges
In fact, adjusting the control states at the demand side of thermal systems is the most popular approach to realize the reshape of power demand in buildings by relying on the feedback mechanism of applied control strategies. However, these controls may experience inherent and inevitable delays in the control processes and therefore, cannot fulfill the needs of the grid's real-time operation without any pricing information in advance. Shutting down some of the operating chillers or limiting the cooling supply by chillers in buildings can achieve immediate demand reduction. Some utility companies have shown an interest in launching this direct load control program to automatically and remotely limit the cooling supply during peak demand periods [45].

Almost all the automatic control strategies commonly used today for air-conditioning systems in buildings are demand-based feedback control. The control mechanism of typical demand-based feedback control of an air-conditioning system in a building is shown in Fig. 9. The regulators, typically PID controllers, modulate the cooling intakes from their suppliers to maintain the states of the spaces served by the terminal units (or states of AHU outlets) at their set-points.

The control of cooling distribution in an air-conditioning system is based on the cooling demands of individual air-conditioned spaces, i.e., demand based. The cooling generated by chillers is delivered to the cooling demand side (i.e., indoor spaces) based on their cooling loads. Each AHU and its regulator form a local feedback control loop, which controls the cooling (i.e., chilled water) taken from its supply side (i.e., chillers) based on what is needed by the terminal units served by the AHU, typically controlling the supply air temperature at the expected value (set-point). Such control assumes that all AHUs can get what they need from chillers. Similarly, the cooling (i.e., air flow) taken by each terminal unit from its supply side (i.e., AHU) is also based on what is needed by the space it serves, typically controlling the space air temperature at its set-point. This control is based on the assumption that all

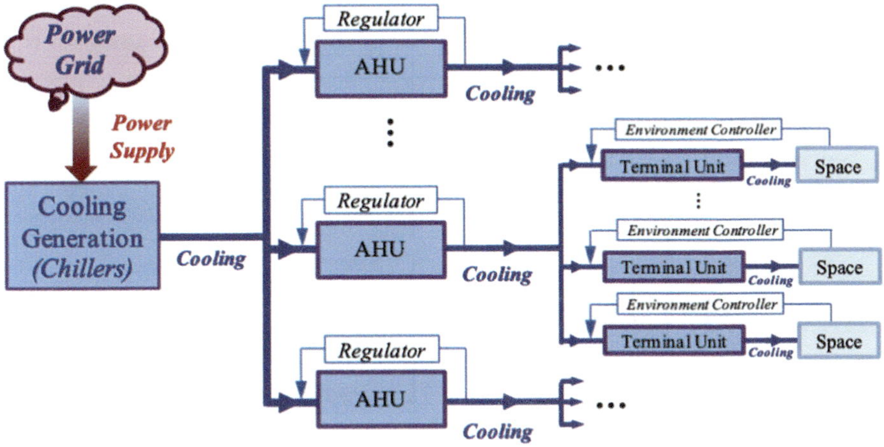

Fig. 9 Basic principle of demand-based feedback control

terminal units can get what they need from AHUs. The distribution of the cooling in a building is therefore managed based on those standalone demand-based feedback control loops.

The distribution of the cooling based on the demand-based feedback control loops can be managed properly in normal conditions. However, after limiting the cooling supply by chillers during DR events, the cooling supply is insufficient to meet the cooling demand of buildings. The problems with cooling distribution and system operation occur at the water and air sides of air-conditioning systems. With a limited cooling supply, all cooling devices at the demand side (e.g., AHUs and VAV boxes) will reach the maximum position to compete for the limited cooling supply. The modulating valves/dampers of most cooling devices at their demand side will be fully open quickly, as they cannot maintain the controlled variables at their set-points. The cooling distributions both among water-side users and air-side terminal units will not be evenly reduced due to hydraulic pressure loss across the pipelines/airducts. In addition, chilled water pumps and supply fans would operate at full speed because of the need to maintain the preset pressure set-points of pumps and supply fans in the condition of fully opened valves. The excessive speeding of water pumps and air fans would result in significantly increased power consumption and relieve the effects of DR control on power reduction.

Development of a Novel Control Method

A completely new control concept, supply-based feedback control strategy, is proposed to solve the inherent operation problems during the above DR events as shown in Fig. 10. At the chilled water side, the supply-based feedback control (i.e., global cooling distributor) manages the distribution of the total available chilled water (i.e., total cooling supply) among the individual zones based on the measured average space temperatures of all zones aiming at keeping space temperatures of all zones

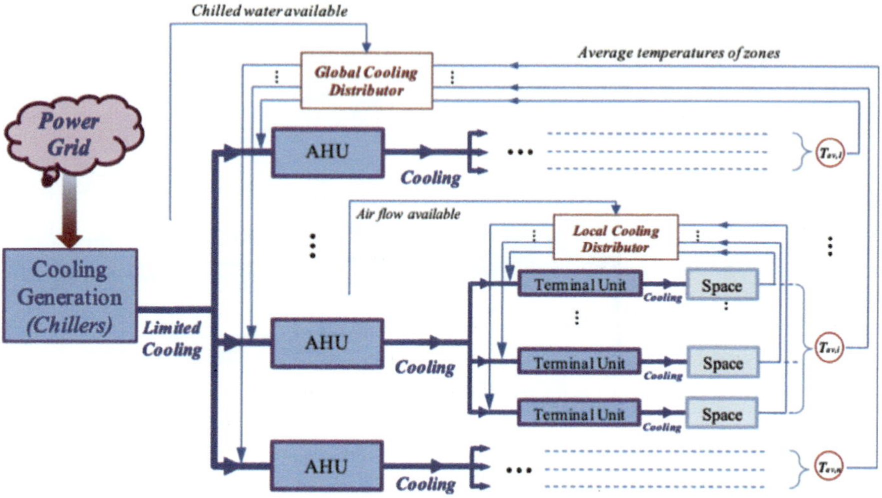

Fig. 10 Basic principle of supply-based feedback control

the same (rising gradually during DR events). Similarly, at the air side, the supply-based feedback control (i.e., local cooling distributor) manages the distribution of the total available cooling supply of an AHU among the individual spaces based on the measured space temperatures of all spaces in the zone, aiming at keeping temperatures of all spaces the same.

Same to the existing demand-based feedback control strategies widely used today, the supply-based control strategy is also based on the actual measurements as the feedback to ensure the accuracy, adaptiveness, and simplicity of cooling distribution as it is hard for a simple predictive control to achieve the same control performance due to the complexity and every change of conditions and disturbance factors. The major difference compared with the existing control strategies is that all the feedback is from the end-user side (i.e., space indoor temperature) rather than from their immediate outlets (e.g., outlet air temperatures of AHUs).

The analytical method of the cooling distribution optimizer is adopted from an economic concept, the utility function [46], to realize an expected cooling distribution under a limited cooling supply. The numerical procedure for cooling distribution at the water side is illustrated here as an example, and the procedure is the same for cooling distribution at the air side, except that the parameters and variables used at the water side are replaced by the corresponding ones.

Equation (23) shows the calculation of the utility value of a zone, where T_{band} is the deviation between T_i^k (measured indoor air temperature) and $T_{(set,i)}$ (original indoor set-point), which should be large enough to fully cover the whole possible indoor temperature range of zones in practical situations. Its value is set to be 10 °C.

U_i^k is the utility value, which is originally an economic concept adopted in this application.

$$U_i^k = 1 - \frac{|T_i^k - T_{set,i}|}{T_{band}} \quad U_i^k \in [0, 1] \tag{23}$$

The online calculation process for determining the chilled water flow set-points of individual zones is shown in Eqs. (24)–(26).

$$M_{U=1,i}^k = M_{w,i}^k - \sqrt{\frac{1 - U_i^k}{a_i}} \tag{24}$$

$$M_{U=1,i}^k = \lambda M_{U=1,i}^{k-1} + (1 - \lambda) M_{U=1,i}^k \tag{25}$$

$$M_{sp,i}^k = M_{U=1,i}^k + \sqrt{\frac{1 - \bar{U}_{sp}^k}{a_i}} \tag{26}$$

where $M_{U=1,i}^k$ is the required chilled water flow of ith zone to maintain the original utility value, that is, when the indoor air temperature is at its set-point of office hour. $M_{w,i}^k$ is the measured chilled water flow of ith zone at kth time step. λ is the forgetting factor selected to be 0.95 in this study. $\bar{U}_{sp}^k$ is the target utility value of all zones at kth time step, which is the expected utility value of all zones if the utility values of all zones are controlled to be the same. The value is determined based on the available cooling capacity (i.e., total chilled water flow in the primary loop). $M_{sp,i}^k$ is the chilled water flow set-point for ith zone at kth time step.

$$a_i = \frac{\Delta T_i}{T_{band} \times \Delta M_i^2} \tag{27}$$

$$\Delta T_i = (T_{out,i} - T_{set,i}) + 5 \tag{28}$$

Parameter a_i represents the thermodynamic characteristics of ith zone and the values are calculated by Eqs. (27) and (28) [47]. Where ΔT_i is the indoor air temperature rise (stabilized) of ith zone from its set-point, which can be obtained when the air-conditioning system is shut down at a selected working point in the morning. The value is assumed to be 5 °C higher than the design outdoor air temperature ($T_{out,i}$), as calculated by Eq. (28). ΔM_i is the chilled water flow rate of ith zone at the selected working point when the air-conditioning is shut down in the morning. This value can be obtained based on the chilled water flow of ith zone in the reference day, considering the current outdoor temperature, relative humidity, the average temperature of the building, and the type of day.

Case Study

A virtual simulation platform is constructed using dynamic models developed in TRNSYS to test and validate the effectiveness of the proposed control strategy. The central chiller plant simulated is a typical primary constant-secondary variable chilled water system, which is modified based on the air-conditioning system of a super-high-rise building in Hong Kong. It consists of six identical chillers, each with a rated capacity of 4080 kW. Each chiller is associated with a primary chilled water pump of constant speed (172.5 L/s). Two variable-speed chilled water pumps are employed in the secondary loop. The cooling source for the building comes from the chilled water circulating in the AHUs, which cools down the supply air temperature to a pre-defined set-point. After the heat exchanger in the AHU, the returned chilled water is distributed evenly to the operating chillers to be chilled again. For the air side, one zone served by one AHU is simulated, which contains a supply and a return fan with rated capacities of 34 kW and 32 kW, respectively. The zone consists of eight rooms with different cooling load profiles, and air duct resistances are considered. The design supply air static pressure is 650 Pa, and the fresh air flow set-point for a zone is set to be constant according to the ASHRAE Standard 62.1-2013. A typical summer day in Hong Kong is selected, and the office hours of the building are from 08:00 a.m. to 20:00 p.m. The indoor air temperature set-point during office hours is set as 24 °C.

The use of cooling distributors for the supply-based feedback control strategy not only effectively solves the unbalanced distribution problems that occur at the water side and air side of an air-conditioning system during DR events, but also achieves further power reduction by diminishing the operation problems of secondary water pumps and air delivery fans. In addition, the power rebound phenomenon right after DR events is also effectively relieved.

Figure 11 shows the comparison among the indoor air temperature profiles of the six zones using the conventional demand-based feedback control and the proposed supply-based feedback control strategies. In Fig. 11a, using the demand-based feedback control, the indoor air temperature profiles of the six zones are obviously different not only during the DR event, but also right after the DR event. This was mainly because right after the DR event, the cooling demand was very high, and individual zones competed for the chilled water again to push their comfort levels back to their original set-points. If the proposed control was released right after the DR event and the conventional control strategy was resumed accordingly, the imbalanced indoor air temperature rises among individual zones would happen again.

In Fig. 11b, it can be observed that the temperature profiles of the six zones are almost the same during the DR event using the proposed control strategy when the cooling supply from chillers is limited and have similar resume profiles after the DR event. The indoor air temperature at 16:00 pm had an obvious jump because there was a significant step change introduced in the solar radiation at 16:00 pm. This can demonstrate the robustness of the proposed control strategy based on an adaptive utility function when facing sudden significant changes in cooling loads. After the conventional feedback control was resumed when the indoor air temperature approached their original set-points, the profiles experienced little differences among individual zones as a result of the handover between the control strategies.

Figure 12 presents the actual chilled water flow rates distributed to six individual AHUs using the conventional demand-based feedback control and the supply-based feedback control strategies. During and right after the DR event, the proposed supply-based control strategy achieved stable and proper chilled water distribution among

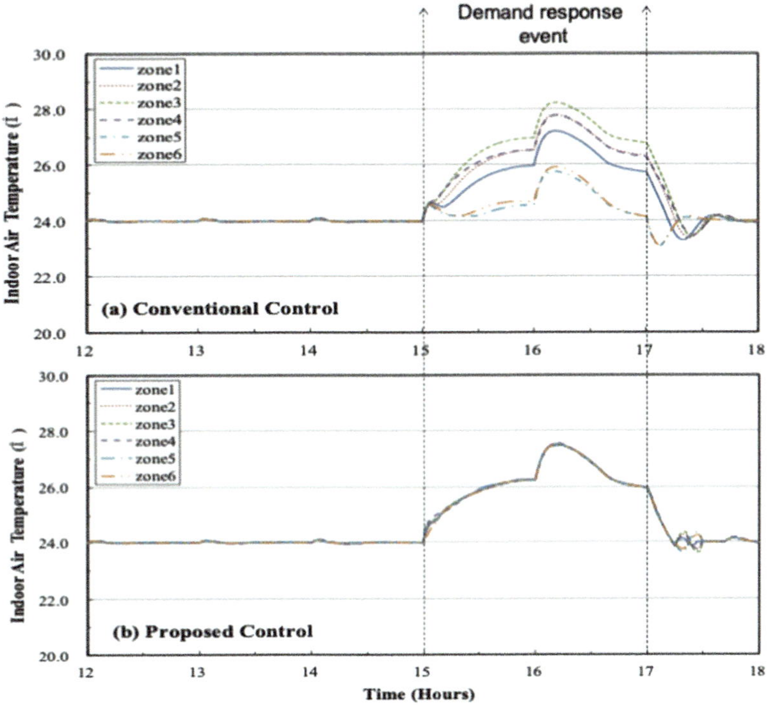

Fig. 11 Indoor air temperature profiles of zones in DR tests using conventional and proposed strategies

six zones, avoided the disordered water flow distribution effectively, and maintained the same indoor air temperature profiles among different zones, as shown in Fig. 11b. In Fig. 12a, using the conventional control strategy, the water flow rates were very high and out of control due to the competition among the zones. In Fig. 12b, the proposed control strategy could eliminate the deficit flow and keep the water flow rate in the by-pass line about zero during the DR event, while there was a serious deficit flow when using the conventional strategies.

The primary objective of the DR control is to achieve as much power reduction as possible. Shutting down some of the operating chillers directly might be a failure to realize a fast power reduction if the whole air-conditioning system is still controlled by the conventional demand-based feedback control strategies. Figure 13 compares the total power consumption using the two control strategies, including the chillers, the (primary and secondary) water pumps, and the (supply and return) fans. It can be observed that using the conventional demand-based feedback control, the extra power consumption caused by the full-speed operation of the secondary pumps and fans almost offsets the effect of shutting down chillers. During the DR event, the proposed control strategy could achieve a power reduction of about 1200 kW, accounting for 24% of the power consumption compared with that right before the start of

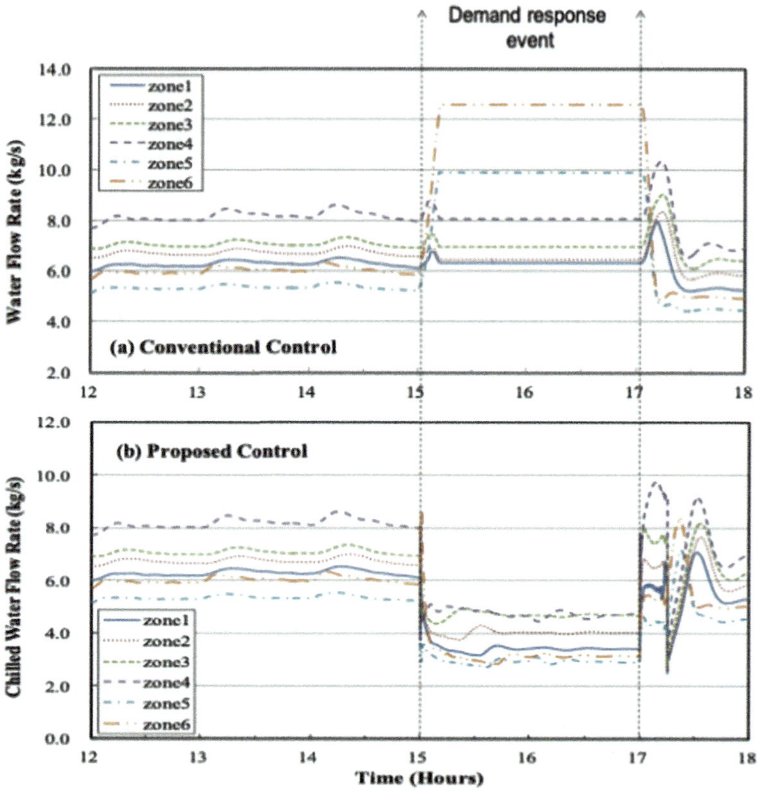

Fig. 12 Chilled water flow profiles of zones in DR tests using conventional and proposed strategies

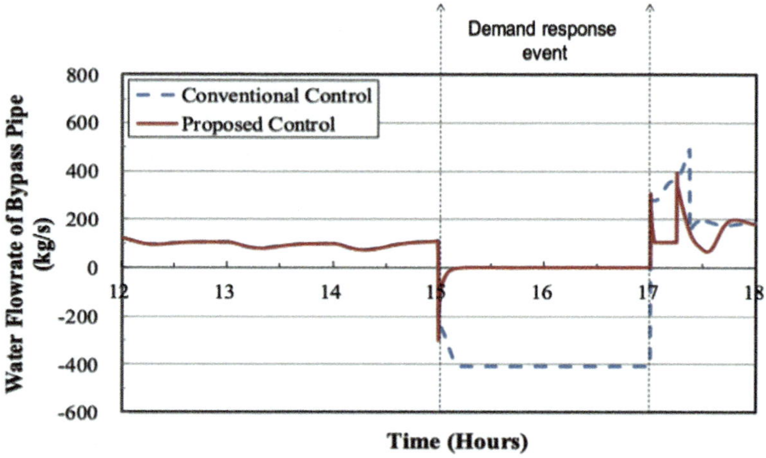

Fig. 13 Water flow rates in by-pass line in DR tests using conventional and proposed strategies

the DR event, while there was almost no power reduction (even increased) due to the compensation of increased power consumption of pumps and fans when using the conventional demand-based feedback control strategies. In addition, it is worth noticing that an obvious rebound phenomenon can be observed after the DR event when the conventional demand-based feedback control strategy is used. The proposed power-limiting control strategy was used during and after the DR event until the indoor air temperatures of all zones/spaces reached their original set-points.

3 Energy Flexibility in Building Clusters

3.1 Coordination and Negotiation in Building Clusters: A Review

All entities in residential microgrids, including in-home loads, energy storage systems, distributed generation, and utilities, are not independent but interconnected with each other. The conventional and separate demand-side management of multiple entities may result in detrimental effects on the grid reliability and the economic benefits for both utilities and consumers. For example, in a residential microgrid, random and unorganized responses of all demand response participants may result in peak "rebounds" at low-price periods, if they are given the same dynamic price profile and simultaneously schedule their loads during on-peak hours [48]. Harmony coordination among participating customers, therefore, is needed to solve the peak rebound issues and to improve the reliability of electrical grids.

Besides coordination in a cooperative sense, another social behavior, i.e., negotiation, is also desirable to solve the conflicts among multiple entities in a noncooperative environment [49, 50]. Effective negotiations can solve the conflicting goals among multiple entities, such as the conflicts between utility and end-use consumers (i.e., maximization of the payoffs of both sides) [51] and the competitions among multiple energy generators (i.e., maximization of their generations) [52, 53]. In summary, coordination and negotiation play a significant role in the management of energy flexibility in residential microgrids. The present study, therefore, focuses on investigating the existing technologies and techniques of coordination and negotiation in residential microgrids.

This section aims to review the state-of-the-art classification, architectures, and techniques of neighborhood-level coordination and negotiation in residential microgrids.

3.1.1 Classification of Coordination and Negotiation

Multiple coordination and negotiation behaviors can be found in a cyber-physical multi-entity residential microgrid, as illustrated in Fig. 14. Table 1 lists selected ref-

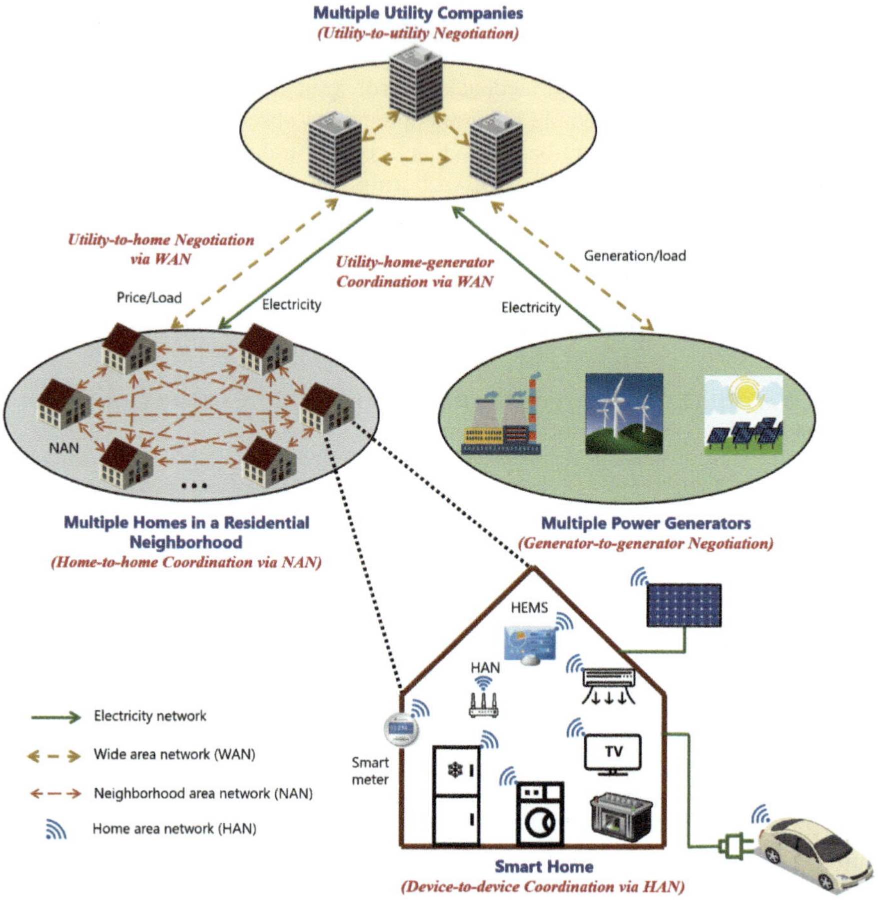

Fig. 14 Multiple types of coordination and negotiation in a cyber-physical multi-entity residential microgrid

Table 1 Selected references for coordination and negotiation behaviors in residential microgrids

Type of interaction		References
Coordination	Device-to-device	[54–57]
	Home-to-home	[58–62]
	Utility-home-generator	[63–66]
Negotiation	Utility-to-home	[51]
	Generator-to-generator	[52, 53]
	Utility-to-utility	[67]

erences for various coordination and negotiation behaviors in residential microgrids. Coordination can generally be classified into three types at different levels: (a) device-to-device coordination; (b) home-to-home coordination; and (c) Utility-home-generator coordination.

- **Device-to-device coordination**: Device-to-device or component-to-component coordination can be implemented in an individual home, mainly the optimal scheduling of the loads, distributed generations, and energy storage by using local coordinators in smart homes [54–57]. The device-to-device coordination is implemented via a home area network (HAN) in home energy management systems (HEMSs).
- **Home-to-home coordination**: Each home in a residential neighborhood is selfish in a sense and only interested in the minimization of its own electricity bill in nature. As observed in [58–60], if all homeowners are provided with the same dynamic price profile, the home energy management systems will simultaneously shift the loads to time periods with lower prices, which may result in some peak "rebounds" at low-price periods. In this regard, home-to-home coordination is an effective solution to solve the rebound issue by exchanging information and coordinating with neighboring homes [61, 62]. The home-to-home coordination is implemented via a neighborhood area network (NAN).
- Unlike device-to-device coordination and home-to-home coordination, **utility-home-generator coordination** is implemented at the microgrid level involving district power generations [63–66]. The utility-home-generator coordination is usually implemented via a wide area network (WAN).

Some interactions among the utility, home, and generator in the residential microgrids are in a non-cooperative and even conflicting environment. In this regard, negotiations are needed to effectively solve the conflicting goals among the involved entities. As shown in Fig. 14, there are three types of negotiation behaviors: (a) utility-to-home negotiation, (b) generator-to-generator negotiation, and (c) utility-to-utility negotiation.

- **Utility-to-home negotiation**: Utility-to-home negotiation is used to solve the conflicting goals between utility and homes, i.e., maximizing profits for both the utility side and consumer side [51].
- **Generator-to-generator negotiation**: Generator-to-generator negotiation is normally applied to deal with the conflict of maximizing generations for all generators [52, 53].
- **Utility-to-utility negotiation**: Utility-to-utility negotiation is used to solve the non-cooperative problems when there are multiple utility companies connected to a residential microgrid. These utility companies all aim to maximize their payoffs [67].

Overall, multiple types of coordination and negotiation behaviors are found in the multi-entity and complex residential microgrid systems. To deal with the complex interactions among multiple entities, various techniques have been proposed in the existing literature. In the following section, the bibliometric analysis is used to identify major technologies and techniques for coordination and negotiation problems in residential microgrids.

3.1.2 Architectures for Coordination and Negotiation Implementation

For effective coordination and negotiation in residential microgrids, management units are normally needed to determine the operating schedules of the demand-side flexible loads. The management units can be situated in utilities/aggregators (i.e., neighborhood energy management systems, NEMSs) or individual residential homes (home energy management systems, HEMSs). Various architectures were developed for implementing the coordination and negotiation of different parties in residential microgrids. In this study, the architectures are classified based on the following two criteria: (1) what entity determines the operating schedules of the demand-side flexible loads, and (2) whether and how neighboring homes share power consumption information, as illustrated in Fig. 15 and compared in Table 2.

If the decisions are made by utilities/aggregators via NEMSs, it is a centralized architecture; if the decisions are made by individual homes via HMESs, it is a decen-

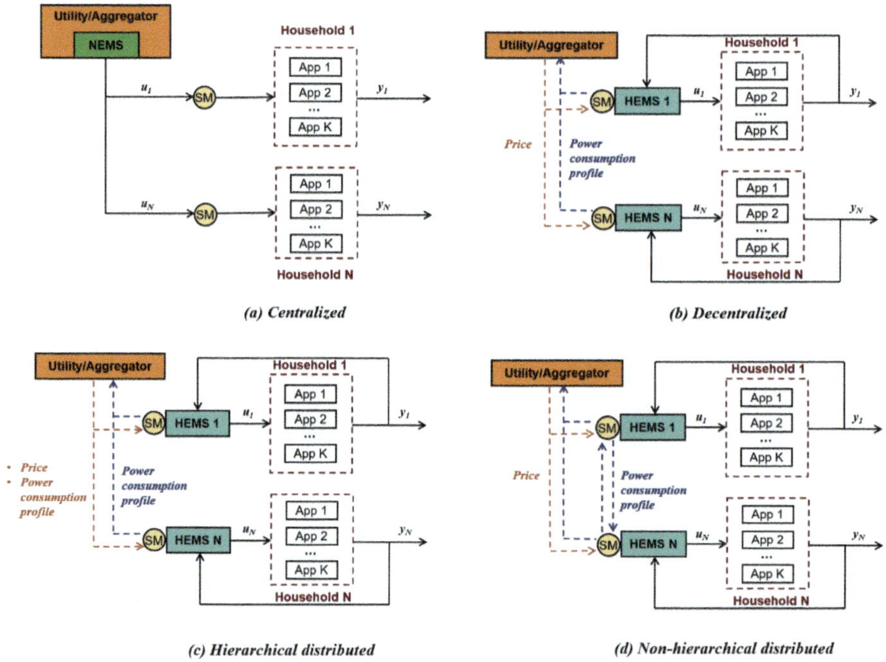

Fig. 15 Coordination architectures for demand-side flexibility in a residential neighborhood: **a** centralized, **b** decentralized, **c** hierarchical distributed, and **d** non-hierarchical distributed

Table 2 Features and selected existing studies for each coordination architecture

Architecture type	Decision-maker	Whether and how households share information?	Function	References
Centralized	NEMS at utility side	No	Coordination	[68–74]
Distributed (Hierarchical)	HEMS at home side	No	Merely cooperation	[75–77]
Distributed (Non-hierarchical)		Yes (via utility/aggregator at the higher layer)	Coordination	[58, 65, 66, 78, 79]
Distributed (Non-hierarchical)		Yes (via local HEMSs)	Coordination or negotiation	[60, 80–82]

tralized or distributed architecture. The power consumption information is shared in the distributed architecture, but not in the decentralized architecture. Based on how the power consumption information is shared, the distributed architecture can be further classified into hierarchical distributed type and non-hierarchical distributed type. If there is no direct communication of power consumption among neighboring homes and all the communication is done through unities/aggregators, it is a hierarchical distributed type; if neighboring homes can directly communicate with each other about their power consumption, it is a non-hierarchical distributed type.

3.1.3 Techniques of Coordination and Negotiation

This section focuses on the key coordination and negotiation techniques for managing the demand-side flexibility in a residential community. Two major techniques are discussed here: multi-agent systems and game theory. For the review of more techniques, such as optimization, please refer to [83].

Multi-agent System
Since the 1990s, agent-based modeling (ABM) and multi-agent system (MAS) based control/management have been widely applied in various fields such as economics, robotics, air traffic control, social science, and computer science due to their advantages in dynamic, multi-entity and complex environments [84, 85]. According to the definition in [84], an "agent" is an intelligent computational system, which has the ability to be autonomous, sociable, reactive, and pro-active.

Agent differs significantly from object/component. As highlighted in [86], agent-oriented approaches differ from object-oriented approaches over the following points: (a) objects are intrinsically passive; (b) objects don't encapsulate action choices; (c) object-oriented approach fails to mimic the behaviors in complex systems; and (d)

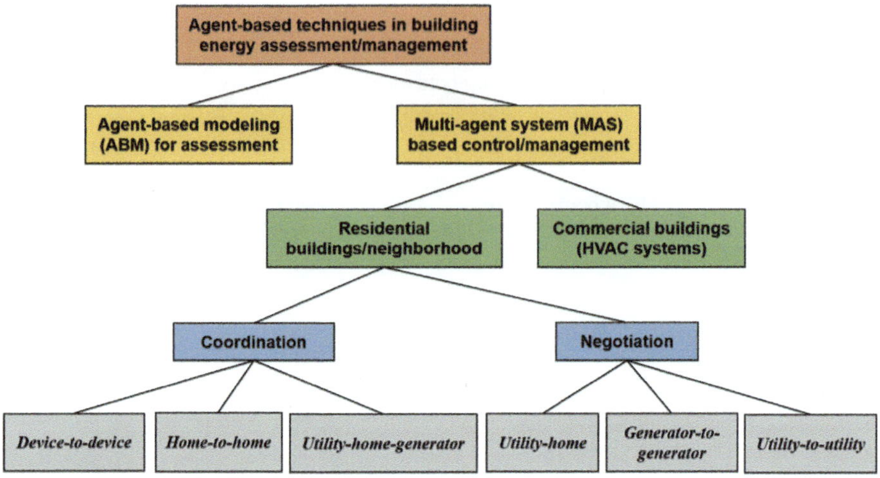

Fig. 16 Classifications of the applications of agent-based techniques to building energy assessment and management

object-oriented approach is not suited to characterize and manage organizational relationships. Due to those characteristics, ABM and MAS-based control techniques have been mainly used to solve problems in distributed, complex, and heterogeneous situations [87].

As shown in Fig. 16, over the last decade, agent-based modeling and MAS-based control techniques have been proven as effective solutions for energy assessment and management in both commercial buildings and residential buildings. Several studies [88–90], including a review work [91], focused on MAS-based energy management in commercial buildings. Our present work will focus on the use of MASs for energy flexibility management in residential neighborhoods integrated with multiple flexible resources. Various types of agents have been used for residential microgrids, including load agents for various household appliances, PHEV/EV agents, energy storage agents, distributed generation agents, HEMS agents, and central coordinator agents. By communicating and interacting with each other, these agents in residential microgrids majorly fulfill two types of social behaviors: (1) coordination behavior and (2) negotiation behavior, as shown in Table 3.

Game Theory

Game theory is a relatively new tool in the field of energy flexibility management. It is a study of analyzing strategic interactions among rational and self-interested players in given environments, i.e., games. In a game, multiple players interact with each other and choose the best strategy as an action in an attempt to achieve the best return. Game theory was originally developed by Von Neumann and Morgenstern in the 1940s for economics and social science [92]. In the last few decades, it has been increasingly used as a promising tool for analyzing negotiation and coordination problems in the framework of multi-agent systems [93]. MAS-based game theoretical

Table 3 Applications of MAS technique for various interactions in residential microgrids and selected references for each type of interaction

Type of interaction		References
MAS-based coordination	Device-to-device	[54–57]
	Home-to-home	[61, 62]
	Utility-home-generator	[63–66]
MAS-based negotiation	Utility-to-home	[51]
	Generator-to-generator	[52, 53]
	Utility-to-utility	[67]

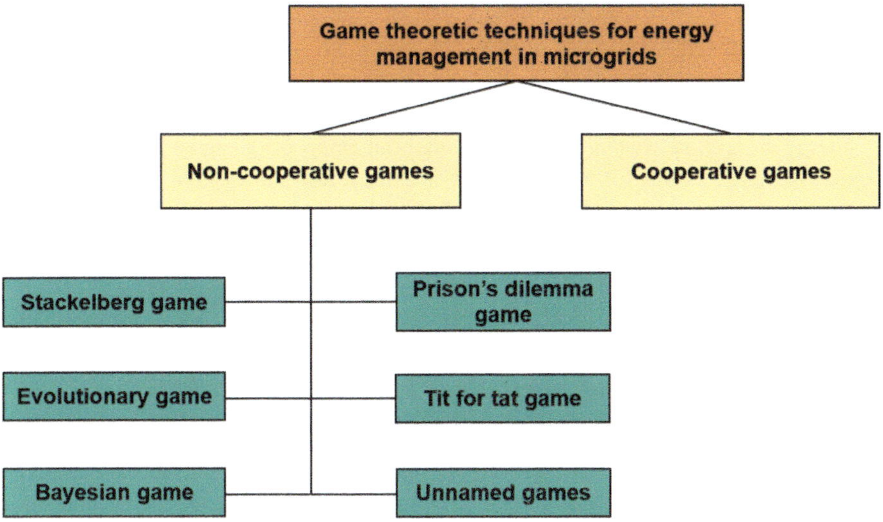

Fig. 17 Classifications of game-theoretic techniques for energy management in microgrids

techniques have been used in various research fields, such as traffic management, maintenance and repair chain, and electrical engineering. In the present study, the game theoretical techniques applied to energy management in residential microgrids are reviewed. As shown in Fig. 17, game-theoretic techniques for energy management in residential microgrids are categorized into two branches:

- **Cooperative game** is a game with cooperation between players, which is also called the team game. Cooperative game theory focuses on analyzing the joint actions and collective payoffs of all participating players.
- **Non-cooperative game** is a game where individual players interact and compete with each other. Non-cooperative game theory attempts to predict players' individual strategies and payoffs and to reach a desirable operating point, i.e., Nash equilibrium.

Table 4 Selected references for each type of game-theoretic technique for energy management in residential microgrids

Game type		References
Non-cooperative game	Stackelberg game	[67, 94–99]
	Evolutionary game	[100–102]
	Bayesian game	[103]
	Prison's dilemma game	[52]
	Tit for tat game	[104]
	Other unnamed games	[80–82, 105–108]
Cooperative game		[109, 110]

Table 4 lists some references for each type of game-theoretic technique for energy management in residential microgrids. It can be found that in the domain of energy flexibility management in residential microgrids, most studies employ non-cooperative games for negotiation problems; a few games focus on cooperative games for coalition and coordination problems. Moreover, among all non-cooperative games, 1-leader, N-follower Stackelberg game is the most used game type, where the utility and consumers are seen as the leader and followers, respectively [97, 111].

3.2 *Control Strategies for Energy Flexibility of Building Clusters: A Case Study*

Currently, most of the building demand management and demand response control strategies focus on individual buildings, i.e., individual level. Actually, the aggregated load of multiple buildings is always the main concern of the power grids. Gao and Sun [112] optimized the operation of central air-conditioning systems in cluster-level buildings from the viewpoint of their aggregated peak load. Shen et al. [113] investigated the electricity bill savings of cluster-level buildings when using different control strategies for their power demand management. They found that the demand management for cluster-level buildings conducted in a coordinated manner can achieve more savings on electricity costs than that conducted uncoordinatedly. Facing the cluster-level building demand management, the individual-level control strategies cannot deal with this coordination effectively, and hence, studies have started to pay attention to the cluster-level building demand management.

This section, therefore, aims to present the necessity of cluster-level building demand management based on a case study in a Hong Kong campus and introduce a decentralized method of optimizing the operation of buildings integrated with passive and active thermal storage.

3.2.1 Necessity of Cluster-Level Demand Management for Buildings

On-site data on building electricity collected from a Hong Kong campus are used to illustrate the problems when conducting demand management at the cluster level. The total electricity bill of the university is divided into four accounts, as charged by the power company. Each charging account is associated with several buildings regarded as a cluster to be counted together, due to historical and technical reasons. Hence, the electricity bill of an account will be influenced by the power usage of every building involved. In such a case, if the building demand management is conducted separately by individual buildings without considering the overall power use of the buildings at the cluster level, some problems and unexpected results will occur as described in the following.

The electricity tariff in Hong Kong considers energy consumption and peak demand simultaneously. The peak demand cost of a commercial building in Hong Kong contributes a great percentage to the electricity bill [114]. Therefore, an effective reduction of the peak demand in commercial buildings will bring significant economic benefits to building owners and also relieve the imbalance operation problem of power grids. However, when using individual-level methods, which consider the power demands of individual buildings separately, to address the peak demand issues of cluster-level buildings, the peak demand limiting control will have inferior effects and even fail. The occurrence times of individual building peak demands are always different from those of their aggregated peak demand associated with a charge account. As a result, using individual-level methods, the peak demand limited by some of the buildings would not contribute to reducing the peak demand of the building cluster and thus the electricity bill. Sometimes, it may even lead to a new peak in the power use of building clusters. In addition, such ineffective power demand controls would result in unnecessary indoor air temperature increases and thermal comfort sacrifices, as well as energy losses caused by the charging and discharging processes if thermal storages are deployed.

The on-site data of building electricity use in one charging account on the university campus were selected to illustrate the above problems. Four buildings with central air-conditioning systems were associated with this account. The data of a summer working day in 2017 collected from power meters installed in the buildings are presented in Fig. 18. The daily peak demands of buildings and the corresponding account (i.e., the aggregated demand of four buildings) were highlighted in the figure. The peak demand occurrence times of buildings and the account on the test day were inconsistent. If the peak demand limiting for the buildings was conducted in an uncoordinated manner, the effectiveness of limiting controls in Building 1 and Building 2 would not benefit the peak demand reduction of the account, and therefore, their corresponding efforts would be wasted. This was because the peak demands of building 1 and building 2 occurred at 16:00 pm and 9:00 am, respectively, but that of the (overall) account occurred at 15:00 pm. Therefore, the building demand management should be conducted from the viewpoint of the cluster level rather than the individual buildings to avoid the above problems.

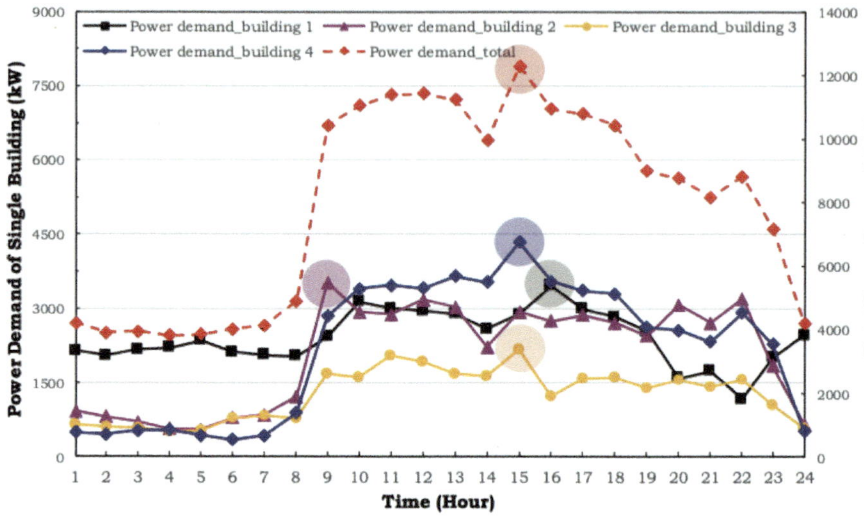

Fig. 18 Building power demands associated with one campus account on a test day

3.2.2 Needs of Cluster-Level Building Demand Management in a Decentralized Mechanism

In a commercial building, the operation of a central air-conditioning system is optimized and controlled by a building automation system. Following the same logic, a central control system collects all the required information from the concerned buildings in a cluster to conduct the peak demand optimization (called centralized method), as shown in Fig. 19. In principle, this central control system can effectively deal with the issue of cluster-level building demand management, and the best possible results can be obtained. However, in practical applications, such central systems often do not exist and are difficult to implement. As the systems scale up, the computational complexity is also concerning. Therefore, a decentralized control strategy is required to fully leverage the building automation systems equipped in individual buildings to enable the cluster-level management to achieve this in a distributed manner. Moreover, the decentralized mode is achieved by autonomous agents in individual buildings, and hence, the privacy of buildings can be protected if necessary, as little information needs to be shared. The distributed strategy allows the optimization of large systems or complex optimization problems to be divided into a few simple tasks, providing enhanced applicability and robustness in practical online applications.

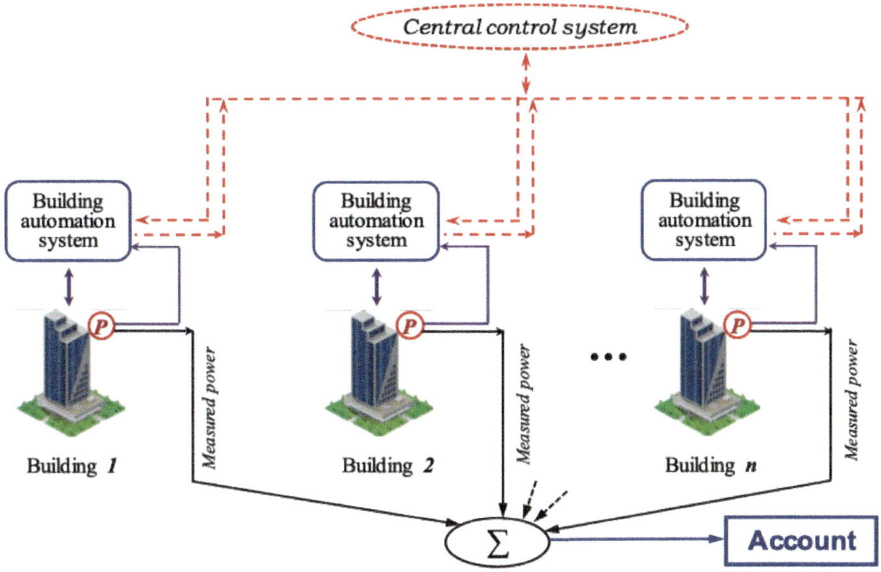

Fig. 19 Structure of control system for central air-conditioning systems in commercial buildings

3.2.3 Development of a Distributed Control Method

The schematic of the proposed decentralized optimal control strategy for cluster-level building demand management is presented in Fig. 20. The demand management of the building cluster is achieved by implementing and coordinating the supervisory control algorithms distributed in individual buildings. Compared with the aforementioned centralized optimal control strategies, the only required measurement to be shared among buildings associated with the account is the aggregated power demand of the building cluster. According to the aggregated power demand and required measurements of individual buildings themselves, the optimization is conducted individually in each building to achieve the global optimization of the building cluster. This decentralized approach is based on a game-theoretic principle of best response [115], that is, individual buildings simply follow what are the best decisions for them to determine their optimal controls.

Problem formulation

To illustrate the process of developing a distributed control for cluster-level buildings, a case study is conducted to minimize the total electricity bill of buildings, considering both energy charge and demand charge. The objective function is shown in Eqs. (29) and (30).

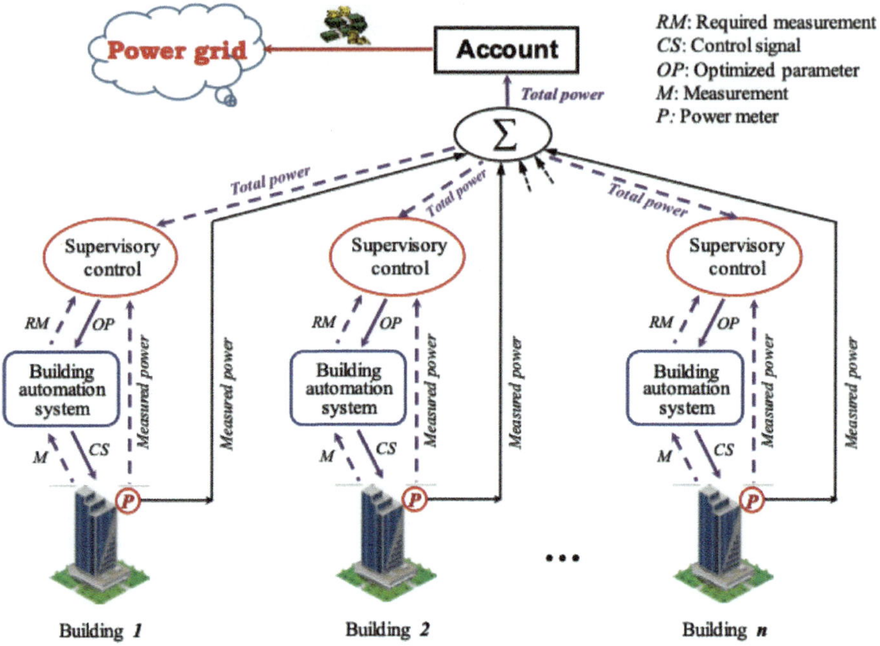

Fig. 20 Schematic of proposed decentralized optimal control strategy for cluster-level building demand management

$$\min C = \alpha \times PD_{tot} + \sum_{k=1}^{N}(\beta^k \times E_{tot}^k) \tag{29}$$

$$E_{tot}^k = \sum_{i \in \delta} P_i^k \times t \tag{30}$$

where C is the daily electricity bill of the account, in HKD. α and β are the unit prices for daily electrical peak demand and total energy consumption, respectively. PD_{tot} is the daily peak demand of the account, kW. E_{tot}^k is the total energy consumption (kWh) of the account at time slot k. N is the total number of equal-sized time slots in a day and is set to be 24 (i.e., each day is divided into 24 time slots). t is the duration in a time slot and is set as 1 h in this study. P_i^k is the power demand of building i at time slot k. The subscript i refers to the building i, and $i \in \delta$. δ refers to the building cluster.

Each building in the cluster is equipped with active cold storage in its central air-conditioning system in this study, as shown in Fig. 21. The indoor temperature can also be adjusted within an acceptable range, cooperating with thermal storage to provide the flexibility of power demand in buildings. The power demand in buildings under different settings of indoor temperature and thermal storage can be calculated by building dynamic thermal models, which can be referenced in [116].

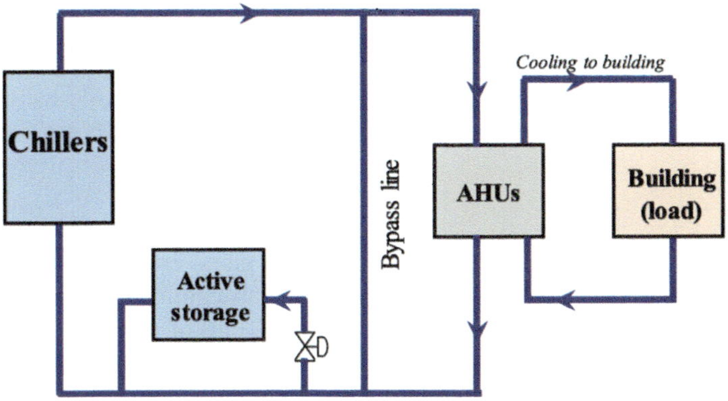

Fig. 21 Schematic of a central air-conditioning system integrated with an active cold storage

Game establishment and solution identification
In a game, each player may partially or totally conflict and cooperate with one another. They all want to maximize their own welfare by setting their strategies. Normally, a game $G = \sigma, S, U$ consists of three components, i.e., "player" set σ, "strategy" space S and "utility" set U. Each player i ($i \in \sigma$) selects its strategy s_i ($s_i \in S$) to maximize its utility/welfare u_i ($u_i \in U$) according to its payoff function. The solution of the game is the Nash equilibrium. The Nash equilibrium is a strategy profile S^* such that no player i can gain more profit by unilaterally deviating from their strategy s_i^*, assuming every other player follows the strategy s_{-i}^* [117]. As defined in Eqs. (31) and (32), a strategy vector S^* is a Nash equilibrium if and only if $\forall i \in \sigma \quad and \quad \forall s_i \in S_i$, where s_i is the strategy of player i and s_{-i} represents all of the "strategies" of other players.

$$U(S^*) \geq U(s_i, s_{-i}^*) \tag{31}$$

$$S^* = (s_i^*, s_{-i}^*) \tag{32}$$

In this case, the buildings associated with an account (a cluster) can be regarded as players, as shown in Eq. (33). The "strategy" set of each building in the game includes the indoor air temperature set-point ($T_{i,opt}$) and the charging/discharging control signals of the active storage (u_i). Where ϑ is the feasible set of the "strategy" that can satisfy the system constraints and indoor thermal comfort. The buildings associated with the concerned cluster share the same payoff function for calculating their utility, i.e., the total electricity bill, as shown in Eq. (37). When the Nash equilibrium of the established game is reached, all the buildings in the account will not change their decisions further (i.e., control signals on charging/discharging process of cold storage and indoor air temperature set-point are optimized) because their benefits have reached the maximum (i.e., the electricity bill of the account is minimum).

$$Player: \quad \sigma = \{i | i \in \delta\} \tag{33}$$

$$Strategy: \quad S = \{X | X \in \vartheta\} \tag{34}$$

$$X = \{x_i | i \in \delta\} \tag{35}$$

$$x_i = \{u_i^k, T_{i,opt}^k\} \quad \forall i \in \delta, k \in \gamma \tag{36}$$

$$Payoff: \quad U = -C(X) \tag{37}$$

To identify the Nash equilibrium of the above game established, the Nikaido-Isoda function is used in this study to transform an equilibrium problem into an optimization problem [118]. The optimal control signals of individual buildings are updated iteratively based on the relaxation algorithm till the Nikaido-Isoda function is nearly converged to zero. At the lth iteration, building i optimizes its power demand by adjusting the indoor temperature set-point and the charging/discharging rate of the active storage, based on the total power demand of the account, to maximize its own interest. Then the control decisions of building i are updated with new optimized values using the relaxation weighting factor. Such a process will be stopped until the Nikaido-Isoda function reaches its convergence within a pre-determined tolerance (ε).

3.2.4 Test and Validation

Figure 22 presents the optimized charging/discharging processes of active cold storage in four buildings associated with the account using different control strategies. In the baseline case, the active cold storage was not considered (i.e., the values were zero). Generally, the performance optimized by the centralized cluster-level control strategy would be the best among these four cases because all the required measurements of the four buildings were collected and used without restrictions. Compared with the case using the centralized strategy, a significantly different result was obtained if the building power demand was optimized individually in an uncoordinated manner. The aggregated demand of the account was almost the highest in the period of 15:00–16:00 pm on the test day, which was highlighted in Fig. 4. But during this period, the cold stored in the storage was not effectively discharged to reduce the high power demand, particularly at 15:00 pm when the peak demand of the account occurred using the individual-level control. At the time of the aggregated peak demand, almost no cooling was discharged by the storage in Building 1, Building 2, and Building 3. In contrast, using the decentralized game theory-based control, the cooling could be effectively discharged at the time of the aggregated peak demand, which was very close to the results using the centralized control strategy.

Figure 23 shows the optimized indoor air temperature set-points in four buildings associated with the account using different control strategies. In the baseline case, the indoor air temperatures of four buildings were maintained at 24 °C during office time. Using the individual-level control strategy, the optimized time schedule of indoor air temperature rise was significantly different compared with that using two cluster-

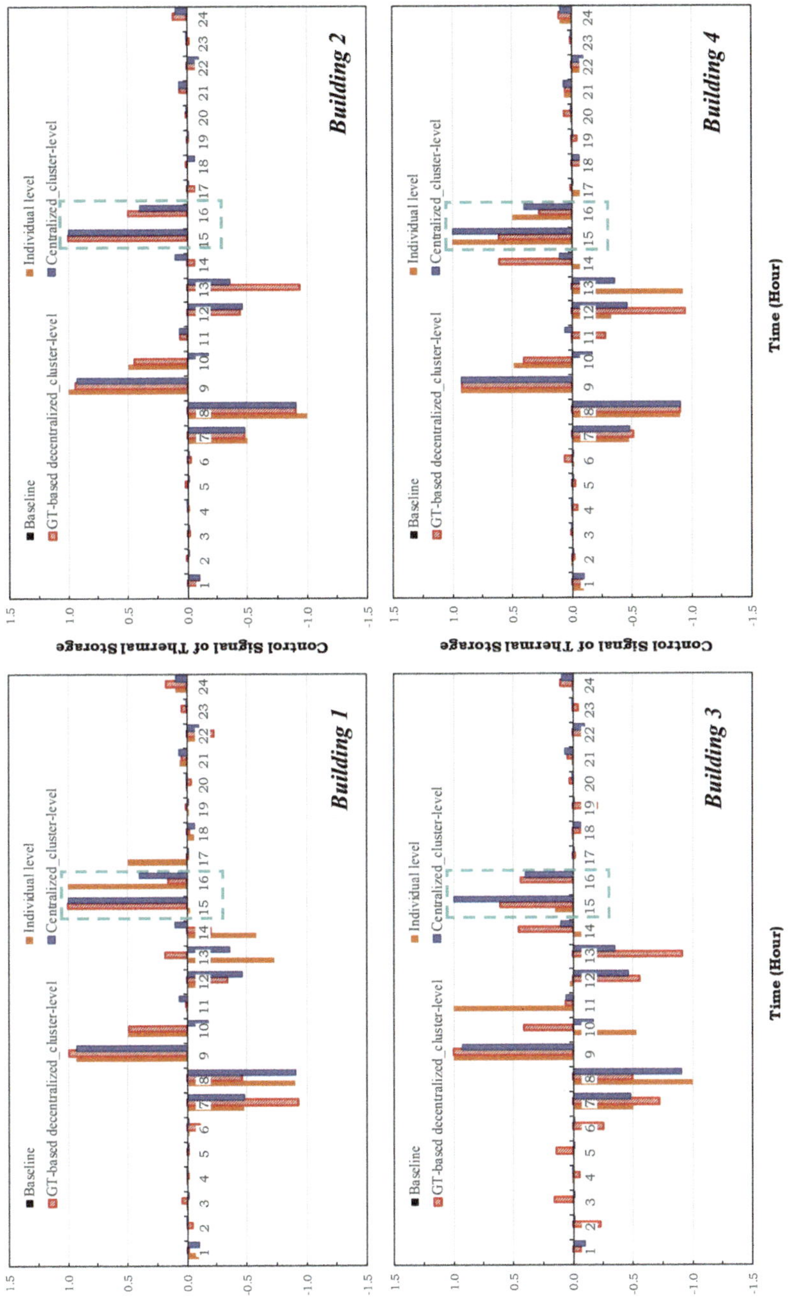

Fig. 22 Optimized charging/discharging of active cold storage for cluster-level building demand management using different control strategies

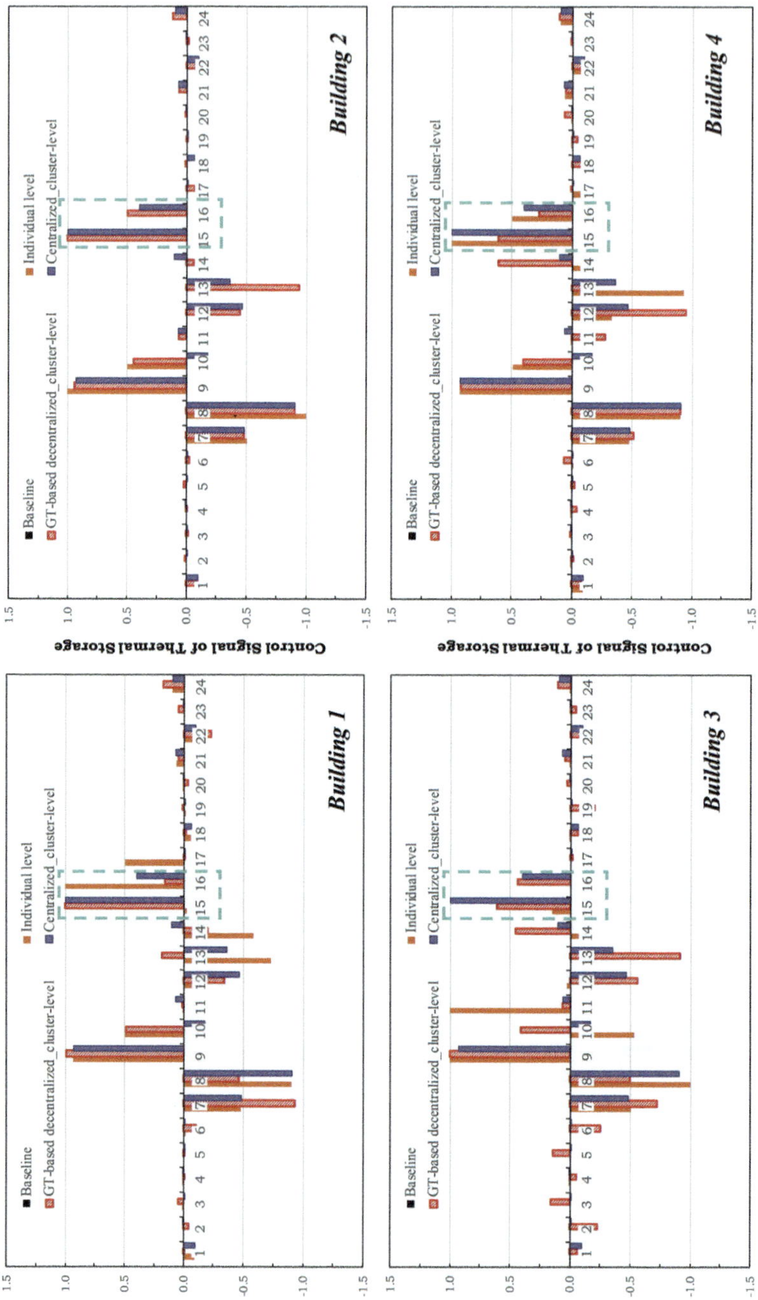

Fig. 23 Optimized indoor air temperature set-points for cluster-level building demand management using different control strategies

level control strategies (decentralized and centralized), particularly for Buildings 2, 3, and 4. This was mainly because the individual-level control strategy focuses on its own power demand only and hence cannot effectively deal with the aggregated power demand of the building cluster. Improper indoor temperature rise could waste the thermal comfort sacrifice using an individual-level strategy. Compared with the results using two cluster-level building demand management strategies, the indoor temperature rises occurred nearly at the same time, which demonstrated the effectiveness of the game theory-based control strategy.

Figures 24 and 25 show the optimized power demands of four individual buildings and the optimized aggregated power demand of the account using different control strategies. The peak demands of individual buildings were effectively reduced using the individual-level control strategy, and their values were even lower than the corresponding values optimized by two cluster-level control strategies, as shown in Fig. 24. However, as for the aggregated peak demand of the account, the peak demand reduction using individual-level strategy (i.e., 535 kW) was only half of that using two cluster-level control strategies (i.e., 1186 kW and 1254 kW), as shown in Fig. 25. As a result, some of the peak demand limiting controls conducted by the individual-level control strategy not only had almost no contribution to reducing the aggregated peak demand, but also resulted in extra charging/discharging of the storage and unnecessary sacrifice of indoor temperature rise. By contrast, the decentralized control strategy could achieve a similar result to that using the centralized control strategy.

4 Conclusion

This chapter investigates energy flexibility in both individual buildings and building clusters with the assistance of advanced control strategies. For individual residential buildings (Sect. 2.1), the model predictive control method is applied to two types of space heating and cooling systems, including variable-speed heat pumps in Hong Kong and floor heating systems in Denmark. The proposed model predictive control helps to improve thermal comfort at the beginning of occupancy, reduce power consumption during peak demand periods, and reduce daily electricity costs. The percentages of peak power reduction and cost savings are influenced by the room's thermal mass, the weather conditions, and the real-time electricity prices. For individual commercial buildings (Sect. 2.2), a cooling supply-based feedback control, employing global and local cooling distributors based on adaptive utility function, is proposed to properly distribute the chilled water/air flow among different zones/spaces to maintain uniform thermal comfort sacrifice during DR events, instead of the commonly used demand-based feedback control strategies. The proposed cooling supply-based feedback control strategy can facilitate the expected fast demand response and power-limiting control strategies to achieve fast power reduction when receiving demand response requests from smart grids.

At the building cluster level, we comprehensively review the state-of-the-art classification, architectures, and techniques for neighborhood-level coordination and

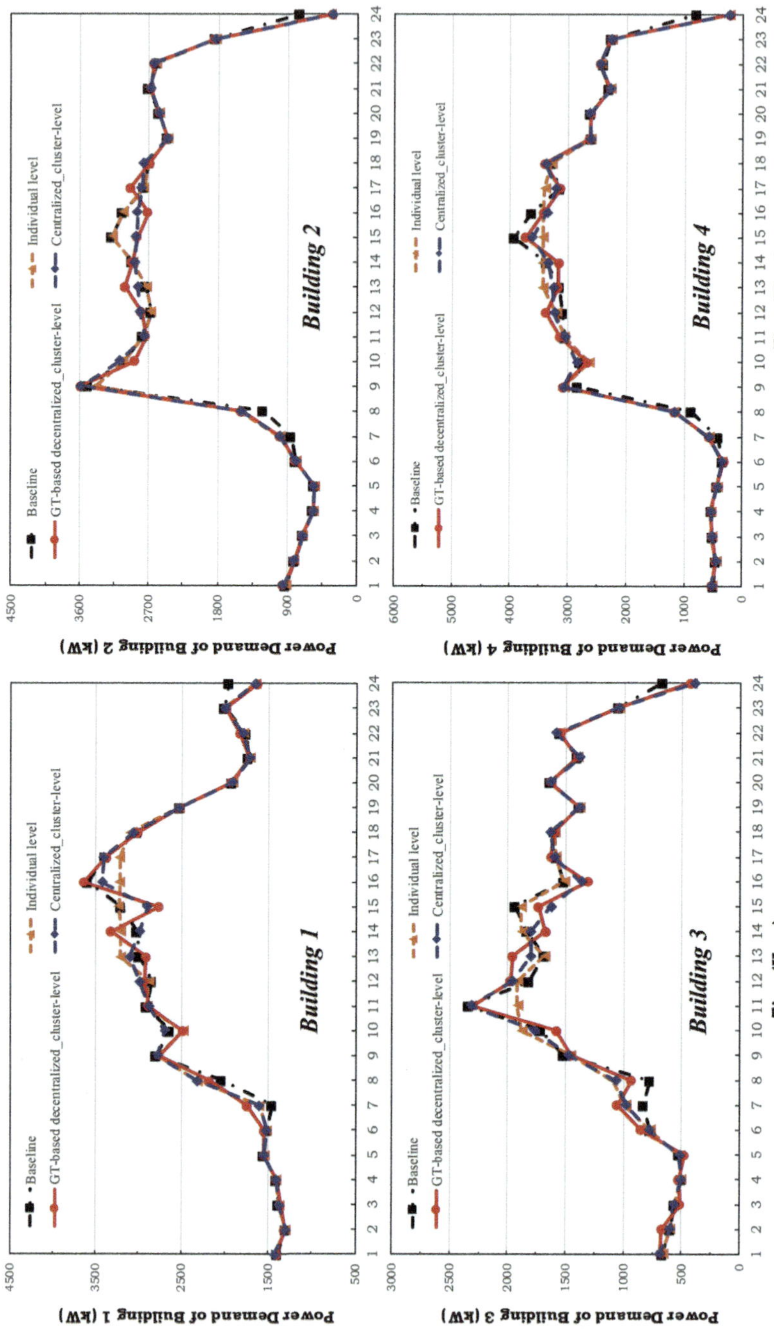

Fig. 24 Optimized power demands of four buildings associated with the campus account using different control strategies

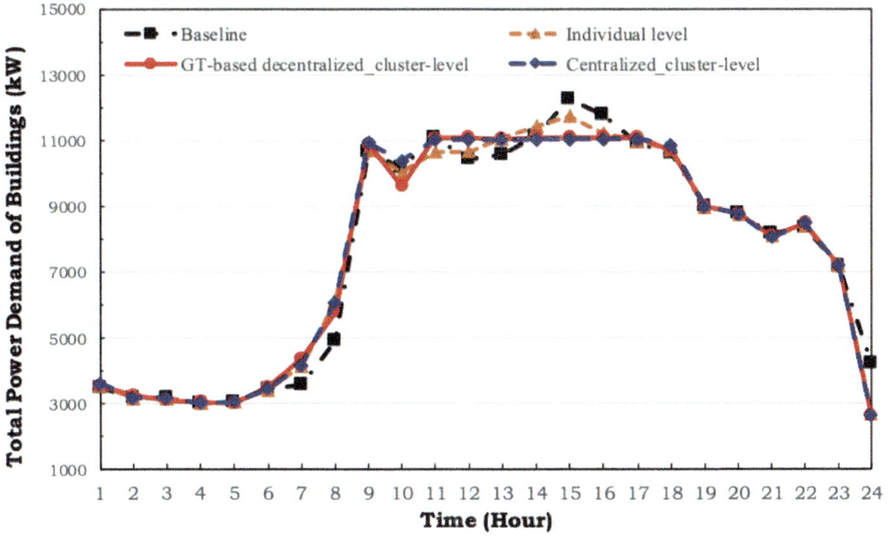

Fig. 25 Optimized aggregated power demand of the campus account using four different control strategies

negotiation in microgrids in Sect. 3.1. In Sect. 3.2, the necessity and importance of achieving cluster-level building demand management in a distributed way are explained. A game theory-based control is introduced as an effective solution to realize this distributed control in buildings. Test results show that the proposed game theory-based decentralized control strategy could effectively increase the aggregated (account) peak demand reduction and electricity cost saving with almost minimal sacrifice of indoor thermal comfort.

The development of energy flexibility in buildings is a complex socio-economic-technical challenge involving various stakeholders. This chapter mainly focuses on the technical solutions for the building end-users at the operational stage. Li et al. [27] pointed out that technical, social, commercial, and regulatory development was needed for the deployment of energy flexibility in buildings. Barriers to the development of aggregated building energy flexibility were also identified in [119] by development phase, including market and policy, early planning and design, and operation, and by different perspectives, including policy, economic, technical, professional, and social aspects.

References

1. Wang S, Xue X, Yan C (2014) Building power demand response methods toward smart grid. HVAC&R Res 20(6):665–687

2. Niknam T, Golestaneh F, Malekpour A (2012) Probabilistic energy and operation management of a microgrid containing wind/photovoltaic/fuel cell generation and energy storage devices based on point estimate method and self-adaptive gravitational search algorithm. Energy 43(1):427–437
3. Neves D, Brito MC, Silva CA (2016) Impact of solar and wind forecast uncertainties on demand response of isolated microgrids. Renew Energy 87:1003–1015
4. Chen Y et al (2018) Measures to improve energy demand flexibility in buildings for demand response (DR): a review. Energy Build 177:125–139
5. Global Alliance for Buildings and Construction (2019) International Energy Agency and the United Nations Environment Programme. In (2019) Global Status report for buildings and construction: towards a zero-emission, efficient and resilient buildings and construction sector. United Nations Environment Programme Nairobi, Kenya
6. US Department of Energy (2012) 2011 DOE building energy data book. (Book chapter)
7. Wang S (2016) Making buildings smarter, grid-friendly, and responsive to smart grids. Sci Technol Built Environ 22(6):629–632
8. Østergaard Jensen S et al (2017) IEA EBC annex 67 energy flexible buildings. Energy Build 155:25–34
9. Le Dréau J, Heiselberg P (2016) Energy flexibility of residential buildings using short term heat storage in the thermal mass. Energy 111:991–1002. (Sept 2016, Book chapter)
10. Shafie-khah M, Kheradmand M, Javadi S, Azenha M, de Aguiar JLB, Castro-Gomes J, Siano P, Catalão JPS (2016) Optimal behavior of responsive residential demand considering hybrid phase change materials. Appl Energy 163:81–92. (Feb 2016, Book chapter)
11. Finck C, Li R, Kramer R, Zeiler W (2018) Quantifying demand flexibility of power-to-heat and thermal energy storage in the control of building heating systems. Appl Energy 209:409–425. (Jan 2018, Book chapter)
12. Hu M, Xiao F, Wang L (2017) Investigation of demand response potentials of residential air conditioners in smart grids using grey-box room thermal model. Appl Energy 207:324–335
13. Yin R et al (2016) Quantifying flexibility of commercial and residential loads for demand response using setpoint changes. Appl Energy 177:149–164
14. Turner WJN, Walker IS, Roux J (2015) Peak load reductions: electric load shifting with mechanical pre-cooling of residential buildings with low thermal mass. Energy 82:1057–1067. (Mar 2015, Book chapter)
15. Hu M, Xiao F (2018) Price-responsive model-based optimal demand response control of inverter air conditioners using genetic algorithm. Appl Energy
16. Hu M, Xiao F, Jørgensen JB, Li R (2019) Price-responsive model predictive control of floor heating systems for demand response using building thermal mass. Appl Therm Eng. (Book chapter)
17. Grondzik WT, Kwok AG (2019) Mechanical and electrical equipment for buildings. John Wiley & Sons. (Oct 2019, Book chapter)
18. Braun J, Piette MA, Haves P, Xu P (2004) Peak demand reduction from pre-cooling with zone temperature reset in an office building. Citeseer. (Book chapter)
19. Xue X, Wang S, Sun Y, Xiao F (2014) An interactive building power demand management strategy for facilitating smart grid optimization. Appl Energy 116:297–310. (Mar 2014, Book chapter)
20. Fukai J, Hamada Y, Morozumi Y, Miyatake O (2003) Improvement of thermal characteristics of latent heat thermal energy storage units using carbon-fiber brushes: experiments and modeling. Int J Heat Mass Transf 46(23):4513–4525. (Nov 2003, Book chapter)
21. Sharma A, Tyagi VV, Chen CR, Buddhi D (2009) Review on thermal energy storage with phase change materials and applications. Renew Sustain Energy Rev 13(2):318–345. (Feb 2009, Book chapter)
22. Henze GP, Felsmann C, Knabe G (2004) Evaluation of optimal control for active and passive building thermal storage. Int J Therm Sci 43(2):173–183. (Feb 2004, Book chapter)
23. Hajiah A, Krarti M (2012) Optimal control of building storage systems using both ice storage and thermal mass–Part I: simulation environment. Energy Convers Manag 64:499–508. (Dec 2012, Book chapter)

24. Hajiah A, Krarti M (2012) Optimal controls of building storage systems using both ice storage and thermal mass–Part II: parametric analysis. Energy Convers Manag 64:509–515. (Dec 2012, Book chapter)
25. de Gracia A, Navarro L, Castell A, Ruiz-Pardo Á, Álvarez S, Cabeza LF (2013) Thermal analysis of a ventilated facade with PCM for cooling applications. Energy Build 65:508–515. (Oct 2013, Book chapter)
26. Tyagi VV, Kaushik SC, Tyagi SK, Akiyama T (2011) Development of phase change materials based microencapsulated technology for buildings: a review. Renew Sustain Energy Rev 15(2):1373–1391. (Feb 2011, Book chapter)
27. Li R et al (2022) Ten questions concerning energy flexibility in buildings. Build Environ 223:109461
28. Hu M, Xiao F (2020) Quantifying uncertainty in the aggregate energy flexibility of high-rise residential building clusters considering stochastic occupancy and occupant behavior. Energy 194:116838. (Mar 2020, Book chapter)
29. Wang A, Li R, You S (2018) Development of a data driven approach to explore the energy flexibility potential of building clusters. Appl Energy 232:89–100. (Dec 2018, Book chapter)
30. Vigna I, Pernetti R, Pasut W, Lollini R (2018) New domain for promoting energy efficiency: energy flexible building cluster. Sustain Cities Soc 38:526–533. (Apr 2018, Book chapter)
31. Taniguchi A, Inoue T, Otsuki M, Yamaguchi Y, Shimoda Y, Takami A, Hanaoka K (2016) Estimation of the contribution of the residential sector to summer peak demand reduction in Japan using an energy end-use simulation model. Energy Build 112:80–92. (Jan 2016, Book chapter)
32. Perfumo C, Kofman E, Braslavsky JH, Ward JK (2012) Load management: model-based control of aggregate power for populations of thermostatically controlled loads. Energy Convers Manag 55:36–48. (Mar 2012, Book chapter)
33. US Energy Information Administration (2018) What's new in how we use energy at home: results from EIA's 2015 residential energy consumption survey. RECS, 2018
34. Administration Energy Information U. S. (2016) Electric power annual 2015. Department of Energy: Washington DC, USA, p 2016
35. Federal Energy Regulatory Commission (2009) A national assessment of demand response potential. Federal Energy Regulatory Commission, 2009
36. Insight A (2021) Residential DER set for long-term growth, Guidehouse insights. (Book chapter)
37. Oldewurtel F, Parisio A, Jones CN, Gyalistras D, Gwerder M, Stauch V, Lehmann B, Morari M (2012) Use of model predictive control and weather forecasts for energy efficient building climate control. Energy Build 45:15–27. (Feb 2012, Book chapter)
38. Hu M, Xiao F, Jørgensen JB, Wang S (2019) Frequency control of air conditioners in response to real-time dynamic electricity prices in smart grids. Appl Energy 242:92–106 (May 2019, Book chapter)
39. Khoshnevisan L, Momeni HR, Ashraf-Modarres A (2012) Introduction of a robust fault detection filter model based on continuous Kalman filter for a real drum boiler system. Adv Model Optim. (Book chapter)
40. Simon D (2006) Optimal state estimation: Kalman, H infinity, and nonlinear approaches. John Wiley & Sons. (Jun 2006, Book chapter)
41. Sun Y, Wang S, Xiao F, Huang G (2012) A study of pre-cooling impacts on peak demand limiting in commercial buildings. HVAC&R Res 18(6):1098–1111. (Dec 2012, Book chapter)
42. Xu X, Wang S, Wang J, Xiao F (2010) Active pipe-embedded structures in buildings for utilizing low-grade energy sources: a review. Energy Build 42(10):1567–1581. (Oct 2010, Book chapter)
43. Olesen BW, Ashrae F (2022) Radiant floor heating in theory and practice. legalett.ca. (Book chapter)
44. Halvgaard R, Poulsen NK, Madsen H, et al (2012) Economic model predictive control for building climate control in a smart grid. In: 2012 IEEE PES. (Book chapter)

45. Shan K, Wang S, Yan C, Xiao F (2016) Building demand response and control methods for smart grids: a review. Sci Technol Built Environ 22(6):692–704. (Aug 2016, Book chapter)
46. Varian HR (1992) Microeconomic analysis. secure.nodebox.net. (Book chapter)
47. Wang S, Tang R (2017) Supply-based feedback control strategy of air-conditioning systems for direct load control of buildings responding to urgent requests of smart grids. Appl. Energy 201:419–432. (Sept 2017, Book chapter)
48. Cui B, Gao D-C, Wang S, Xue X (2015) Effectiveness and life-cycle cost-benefit analysis of active cold storages for building demand management for smart grid applications. Appl Energy 147:523–535. (Jun 2015, Book chapter)
49. Kraus S (1997) Negotiation and cooperation in multi-agent environments. Artif Intell 94(1):79–97. (Jul 1997, Book chapter)
50. Xiao Y, Plasters C, Seagull FJ, Mackenzie C, Kobayashi M, Fussell S, Kiesler S (2005) Negotiation and coordination: a preliminary field study of conflict management in large scale collaboration. In: Proceedings of the human factors and ergonomics society annual meeting, vol 49, no. 3, pp 539–543. (Sept 2005, Book chapter)
51. Wang Z, Wang L (2013) Adaptive negotiation agent for facilitating bi-directional energy trading between smart building and utility grid. IEEE Trans Smart Grid 4(2):702–710. (Jun 2013, Book chapter)
52. Khan B, Reyasudin M, Jidin R, Pasupuleti J (2016) Multi-agent based distributed control architecture for microgrid energy management and optimization. Energy Convers Manag 112:288–307. (Mar 2016, Book chapter)
53. Karavas C-S, Arvanitis K, Papadakis G (2017) A game theory approach to multi-agent decentralized energy management of autonomous polygeneration microgrids. Energies 10(11):1756. (Nov 2017, Book chapter)
54. Rajeev T, Ashok S (2015) Dynamic load-shifting program based on a cloud computing framework to support the integration of renewable energy sources. Appl Energy 146:141–149. (May 2015, Book chapter)
55. Wang Z, Wang L, Dounis AI, Yang R (2012) Multi-agent control system with information fusion based comfort model for smart buildings. Appl Energy 99:247–254. (Nov 2012, Book chapter)
56. Fernandes F, Morais H, Vale Z, Ramos C (2014) Dynamic load management in a smart home to participate in demand response events. Energy Build 82:592–606. (Oct 2014, Book chapter)
57. Bünning F, Sangi R, Müller D (2017) A Modelica library for the agent-based control of building energy systems. Appl Energy 193:52–59. (May 2017, Book chapter)
58. Safdarian A, Fotuhi-Firuzabad M, Lehtonen M (2016) Optimal residential load management in smart grids: a decentralized framework. IEEE Trans Smart Grid 7(4):1836–1845. (Jul 2016, Book chapter)
59. Li Y, Ng BL, Trayer M, Liu L (2012) Automated residential demand response: algorithmic implications of pricing models. IEEE Trans Smart Grid 3(4):1712–1721. (Dec 2012, Book chapter)
60. Chang T-H, Alizadeh M, Scaglione A (2013) Real-time power balancing via decentralized coordinated home energy scheduling," *IEEE Trans. Smart Grid*, vol. 4, no. 3, pp. 1490–1504, Sep 2013, *Book chapter*
61. Rahman MS, Oo AMT (2017) Distributed multi-agent based coordinated power management and control strategy for microgrids with distributed energy resources. Energy Convers Manage 139:20–32. (May 2017, Book chapter)
62. Rahman MS, Mahmud MA, Oo AMT, Pota HR, Hossain MJ (2016) Agent-based reactive power management of power distribution networks with distributed energy generation. Energy Convers Manage 120:120–134. (Jul 2016, Book chapter)
63. Anvari-Moghaddam A, Rahimi-Kian A, Mirian MS, Guerrero JM (2017) A multi-agent based energy management solution for integrated buildings and microgrid system. Appl Energy 203:41–56. (Oct 2017, Book chapter)
64. Ju L, Zhang Q, Tan Z, Wang W, Xin H, Zhang Z (2018) Multi-agent-system-based coupling control optimization model for micro-grid group intelligent scheduling considering autonomy-cooperative operation strategy. Energy 157:1035–1052. (Aug 2018, Book chapter)

65. Ramchurn S, Vytelingum P, Rogers A, Jennings N (2011) Agent-based control for decentralised demand side management in the smart grid, pp 5–12. (Book chapter)
66. Roche R, Siddharth S, Hansen TM, Kiliccote S, Miraoui A (2015) A multi-agent model and strategy for residential demand response coordination. In: 2015 IEEE Eindhoven PowerTech, 2015. (Book chapter)
67. Maharjan S, Zhu Q, Zhang Y, Gjessing S, Basar T (2013) Dependable demand response management in the smart grid: a stackelberg game approach. IEEE Trans Smart Grid 4(1):120–132. (Mar 2013, Book chapter)
68. Paterakis NG, Erdinc O, Pappi IN, Bakirtzis AG, Catalao JPS (2016) Coordinated operation of a neighborhood of smart households comprising electric vehicles, energy storage and distributed generation. IEEE Trans Smart Grid 7(6):2736–2747. (Nov 2016, Book chapter)
69. Nguyen DT, Le LB (2014) Joint optimization of electric vehicle and home energy scheduling considering user comfort preference. IEEE Trans Smart Grid 5(1):188–199. (Jan 2014, Book chapter)
70. Logenthiran T, Srinivasan D, Shun TZ (2012) Demand side management in smart grid using heuristic optimization. IEEE Trans Smart Grid 3(3):1244–1252. (Sept 2012, Book chapter)
71. Niro G, Salles D, Alcântara MVP, da Silva LCP (2013) Large-scale control of domestic refrigerators for demand peak reduction in distribution systems. Electric Power Syst Res 100:34–42. (Jul 2013, Book chapter)
72. Tushar MHK, Assi C, Maier M, Uddin MF (2014) Smart microgrids: Optimal joint scheduling for electric vehicles and home appliances. IEEE Trans Smart Grid 5(1):239–250. (Jan 2014, Book chapter)
73. Hansen TM, Roche R, Suryanarayanan S, Maciejewski AA, Siegel HJ (2015) Heuristic optimization for an aggregator-based resource allocation in the smart grid. IEEE Trans Smart Grid 6(4):1785–1794. (Jul 2015, Book chapter)
74. Ouammi A (2016) Optimal power scheduling for a cooperative network of smart residential buildings. IEEE Trans Sustain Energy 7(3):1317–1326. (Jul 2016, Book chapter)
75. Molitor C, Marin M, Hernandez L, Monti A (2013) Decentralized coordination of the operation of residential heating units. IEEE PES ISGT Europe 2013. (Book chapter)
76. Cole WJ, Rhodes JD, Gorman W, Perez KX, Webber ME, Edgar TF (2014) Community-scale residential air conditioning control for effective grid management. Appl Energy 130:428–436. (Oct 2014, Book chapter)
77. Sarker MR, Ortega-Vazquez MA, Kirschen DS (2015) Optimal coordination and scheduling of demand response via monetary incentives. IEEE Trans Smart Grid 6(3):1341–1352. (May 2015, Book chapter)
78. Guo Y, Pan M, Fang Y, Khargonekar PP (2013) Decentralized coordination of energy utilization for residential households in the smart grid. IEEE Trans Smart Grid 4(3):1341–1350. (Book chapter)
79. Chavali P, Yang P, Nehorai A (2014) A distributed algorithm of appliance scheduling for home energy management system. IEEE Trans Smart Grid 5(1):282–290. (Jan 2014, Book chapter)
80. Mohsenian-Rad A-H, Wong VWS, Jatskevich J, Schober R, Leon-Garcia A (2010) Autonomous demand-side management based on game-theoretic energy consumption scheduling for the future smart grid. IEEE Trans Smart Grid 1(3):320–331. (2010, Book chapter)
81. Deng R, Yang Z, Chen J, Asr NR, Chow M-Y (2014) Residential energy consumption scheduling: a coupled-constraint game approach. IEEE Trans Smart Grid 5(3):1340–1350. (May 2014, Book chapter)
82. Chen H, Li Y, Louie RHY, Vucetic B (2014) Autonomous demand side management based on energy consumption scheduling and instantaneous load billing: an aggregative game approach. IEEE Trans Smart Grid 5(4):1744–1754. (Jul 2014, Book chapter)
83. Hu M, Xiao F, Wang S (2021) Neighborhood-level coordination and negotiation techniques for managing demand-side flexibility in residential microgrids. Renewable Sustain Energy Rev 135:110248

84. Wooldridge M, Jennings NR (1995) Intelligent agents: theory and practice. Knowl Eng Rev 10(2):115–152. (Jun 1995, Book chapter)
85. Lewis FL, Zhang H, Hengster-Movric K, Das A (2013) Cooperative control of multi-agent systems: optimal and adaptive design approaches. Springer Science & Business Media. (Dec 2013, Book chapter)
86. Jennings NR, Bussmann S (2003) Agent-based control systems. IEEE Control Syst. 23(3):61–74
87. Barbati M, Bruno G, Genovese A (2012) Applications of agent-based models for optimization problems: A literature review. Expert Syst. Appl. 39(5):6020–6028
88. Cai J, Kim D, Jaramillo R, Braun JE, Hu J (2016) A general multi-agent control approach for building energy system optimization. Energy Build. 127:337–351
89. Constantin A, Löwen A, Ponci F, Müller D, Monti A (2017) Design, implementation and demonstration of embedded agents for energy management in non-residential buildings. Energies 10(8):1106
90. Jiang L, Yao R, Liu K, McCrindle R (2017) An Epistemic-Deontic-Axiologic (EDA) agent-based energy management system in office buildings. Appl Energy 205:440–452
91. Windham A, Treado S (2016) A review of multi-agent systems concepts and research related to building HVAC control. Sci Technol Built Environ 22(1):50–66
92. Von Neumann J, Morgenstern O (1944) Theory of games and economic behavior. Princeton University Press
93. Parsons S, Wooldridge M (2002) Game theory and decision theory in multi-agent systems. Auton Agent Multi Agent Syst 5(3):243–254
94. Meng F-L, Zeng X-J (2013) A Stackelberg game-theoretic approach to optimal real-time pricing for the smart grid. Soft Comput 17(12):2365–2380
95. Soliman HM, Leon-Garcia A (2014) Game-theoretic demand-side management with storage devices for the future smart grid. IEEE Trans Smart Grid 5(3):1475–1485
96. Yoon S-G, Choi Y-J, Park J-K, Bahk S (2016) Stackelberg-game-based demand response for at-home electric vehicle charging. IEEE Trans Veh Technol 65(6):4172–4184
97. Yu M, Hong SH (2016) Supply-demand balancing for power management in smart grid: a stackelberg game approach. Appl Energy 164:702–710
98. Salyani P, Abapour M (2019) Stackelberg based optimal planning of DGs and electric vehicle parking lot by implementing demand response program. Sustain Cities Soc 51:101743
99. Gazijahani FS, Salehi J (2019) Game theory based profit maximization model for microgrid aggregators with presence of EDRP using information gap decision theory. IEEE Syst J 13(2):1767–1775
100. Chai B, Chen J, Yang Z, Zhang Y (2014) Demand response management with multiple utility companies: a two-level game approach. IEEE Trans Smart Grid 5(2):722–731
101. Gong C, Tang K, Zhu K, Hailu A (2016) An optimal time-of-use pricing for urban gas: a study with a multi-agent evolutionary game-theoretic perspective. Appl Energy 163:283–294
102. Cheng X, Long R, Chen C, Yang J (2019) Does social interaction have an impact on residents' sustainable lifestyle decisions? A multi-agent stimulation based on regret and game theory. Appl Energy 251:113366
103. Xu Q, Zhang N, Kang C, Xia Q, He D, Liu C, Huang Y, Cheng L, Bai J (2016) A game theoretical pricing mechanism for multi-area spinning reserve trading considering wind power uncertainty. IEEE Trans Power Syst. 31(2):1084–1095
104. Sofana Reka S, Ramesh V (2016) A demand response modeling for residential consumers in smart grid environment using game theory based energy scheduling algorithm. Ain Shams Eng J 7(2):835–845
105. Atzeni I, Ordóñez LG, Scutari G, Palomar DP, Fonollosa JR (2013) Noncooperative and cooperative optimization of distributed energy generation and storage in the demand-side of the smart grid. IEEE Trans Signal Process 61(10):2454–2472
106. Yang P, Tang G, Nehorai A (2013) A game-theoretic approach for optimal time-of-use electricity pricing. IEEE Trans Power Syst 28(2):884–892

107. Nwulu NI, Xia X (2015) Implementing a model predictive control strategy on the dynamic economic emission dispatch problem with game theory based demand response programs. Energy 91:404–419
108. Qi Y, Liu Y, Wu Q (2017) Non-cooperative regulation coordination based on game theory for wind farm clusters during ramping events. Energy 132:136–146
109. Han X, Yu X, Liang Y, Li J, Zhao Z (2018) A game theory-based coordination and optimization control methodology for a wind power-generation hybrid energy storage system. Asian J Control 20(1):86–103
110. Zhang N, Yan Y, Su W (2015) A game-theoretic economic operation of residential distribution system with high participation of distributed electricity prosumers. Appl Energy 154:471–479
111. Tang R, Wang S, Li H (2019) Game theory based interactive demand side management responding to dynamic pricing in price-based demand response of smart grids. Appl Energy 250:118–130
112. Gao D-C, Sun Y (2016) A GA-based coordinated demand response control for building group level peak demand limiting with benefits to grid power balance. Energy Build 110:31–40
113. Shen L, Li Z, Sun Y (2016) Performance evaluation of conventional demand response at building-group-level under different electricity pricings. Energy Build 128:143–154
114. Tang R, Wang S, Shan K, Cheung H (2018) Optimal control strategy of central air-conditioning systems of buildings at morning start period for enhanced energy efficiency and peak demand limiting. Energy 151:771–781
115. Nisan N, Roughgarden T, Tardos E, Vazirani VV (2007) Algorithmic game theory. Cambridge University Press, Cambridge, UK
116. Tang R, Li H, Wang S (2019) A game theory-based decentralized control strategy for power demand management of building cluster using thermal mass and energy storage. Appl Energy 242:809–820
117. Basar T, Olsder GJ (1999) Dynamic noncooperative game theory. Academic Press, SIAM, New York
118. Nikaidô H, Isoda K (1955) Note on non-cooperative convex games
119. Le Dréau J et al (2023) Developing energy flexibility in clusters of buildings: a critical analysis of barriers from planning to operation. Energy Build 300:113608

Building the Resilience of Local Power Systems Against Climate Change

Weijia Yang, Pu Yang, and Stephanie Hirmer

Abstract As extreme weather events become more frequent and severe, energy infrastructure resilience to these disruptions is crucial to ensure stability and continuity of energy service delivery. This chapter examines the critical issue of enhancing the resilience of local power systems in response to growing challenges, specifically climate change. The chapter begins by outlining the dual impacts of climate change: the long-term shifts in energy production, distribution, and consumption, as well as the acute challenges posed by extreme weather events like hurricanes, extreme heat, and floods. It explores various resilience strategies to strengthen energy systems against these threats. Hard resilience strategies discussed include infrastructure enhancements such as fortifying power lines, while soft resilience strategies cover technological and operational innovations like smart grid technologies and advanced monitoring systems. Finally, this chapter discusses the importance of robust frameworks that facilitate the implementation of resilience measures. It emphasises the role of advanced digital tools in monitoring and forecasting to prevent, manage, and mitigate potential disruptions. Through a blend of theory, case studies, and real-world applications, the chapter illustrates the practical applications and benefits of integrating climate resilience into power systems planning and operations.

1 Introduction

In the face of increasingly extreme weather events, this chapter aims to provide a comprehensive understanding of the importance of building resilience in local energy systems, with a focus on power networks. In 2015, the Paris Agreement was adopted

W. Yang · P. Yang (✉) · S. Hirmer
Department of Engineering Science, University of Oxford, Oxford, UK
e-mail: pu.yang@eng.ox.ac.uk

W. Yang
e-mail: weijia.yang@eng.ox.ac.uk

S. Hirmer
e-mail: stephanie.hirmer@eng.ox.ac.uk

W. Hua et al. (eds.), *Digitalisation of Local Energy Systems*, Green Energy and Technology, https://doi.org/10.1007/978-3-031-77833-9_7

by 196 Parties at the UN Climate Change Conference (COP21), setting the goal to "strengthen the global response to the threat of climate change by keeping a global temperature rise this century well below 2 degrees Celsius above pre-industrial levels and to pursue efforts to limit the temperature increase even further to 1.5 degrees Celsius". In order to achieve the long-term goals of the Paris Agreement, nationally determined contributions (NDCs) are critical to the agreement. NDCs provide specific climate commitments that each country sets based on its own national circumstances, priorities, and capabilities, allowing for tailored approaches for reducing greenhouse gas emissions and adapting to climate change. However, current NDC efforts are not able to maintain the 2 °C targets, let alone the 1.5 °C [1, 2]. As a result of human-induced climate change, extreme weather events will become more frequent and intense as global temperatures rise above pre-industrial levels. Therefore, understanding the complexities of these events and their impact on our energy systems is essential.

This chapter begins by introducing the long-term and short-term impacts of climate change on local power systems. Despite the fact that such events have always been a part of Earth's climatic history, human-induced climate change has dramatically increased their frequency and intensity. Literature has increasingly focused on the long-term effects of climate change and the challenges it poses to the power system. A few examples of these changes include changes in energy demand patterns, impacts on infrastructure over time, and an increase in uncertainty about renewable energy potentials. Furthermore, we examined the short-term impacts of climate change on the local power system. These impacts are mostly induced by various extreme weather events, such as hurricanes, typhoons, and heat waves.

With an understanding of the impact of extreme weather events, the next section focuses on the various definitions of resilience, explaining how it differs from reliability and can be measured from various aspects. Resilience is a multidimensional concept that has gained increasing importance in the field of systems engineering, particularly in the context of critical infrastructure systems and their ability to withstand and recover from disruptive events. In contrast, reliability is a narrower and well-defined concept that refers to a system's performance during normal operation.

The final section of the chapter discusses three main categories of methods to enhance energy system resilience, including hard, soft, and digital strategies. Hard strategies involve upgrading electrical equipment, while soft strategies are conventional system-wide methods to protect systems from extreme events, such as microgrids and dynamic prices. Digital tool-based strategies take advantage of emerging Artificial Intelligence methods such as Neural Networks to improve the accuracy of weather and energy forecasts, aiming to assist grid operators in making proper decisions during periods of extreme stress.

2 Climate Change and Its Impact on Local Power Systems

Climate change influences local power systems through both long-term transformations and short-term disruptions. Over time, climate change gradually modifies electricity production and consumption, leading to hydroelectric power generation deficiency due to changing precipitation patterns [3] and shifts in heating and cooling demands [4]. Additionally, extreme weather events' increased frequency and severity pose immediate challenges, requiring swift recovery and adaptation of energy systems [5]. This chapter is structured to first address the long-term effects of climate change on local energy infrastructure in Sect. 2.1, then to discuss the short-term impacts resulting from extreme weather events in Sect. 2.2. Understanding these impacts is essential for developing effective strategies for resilient local power systems, ensuring they can withstand and adapt to the evolving climate landscape [6].

2.1 Long Term Impact of Climate Change

Local power systems are profoundly affected by climate change, which manifests primarily through three interrelated aspects: changes to renewable energy potentials, ill-equipped infrastructure, and changing energy demand patterns. These factors affect the operational efficiency and resilience of local power systems, demanding that current energy strategies be re-evaluated to mitigate the adverse effects of climate change.

2.1.1 Changes in Renewable Energy Supply

Climate change influences the potential and efficiency of renewable energy sources. The generation capacities of hydroelectric, solar, and wind power facilities will be significantly affected by changes in precipitation patterns, water levels, solar irradiance, and wind patterns because of climate change. It has been shown that decreased water availability during prolonged droughts may adversely affect hydroelectric power generation, while increased cloud cover and atmospheric dust may reduce solar panel efficiency. For example, Zambia relies heavily on hydropower for electricity generation, with the Kariba dam generating more than 10,000 GWh (57%) of Zambia's electricity generation each year. However, prolonged droughts pose a significant threat to the country's energy supply. The recent drought that began in January 2024 has resulted in 8–10 h of power outages a day [7]. Additionally, shifting wind patterns may decrease the efficiency of existing wind farms. To address these challenges, adaptive strategies and innovations (e.g. AI-based renewable generation forecast tools) are necessary to ensure the resilience and sustainability of renewable energy sources in the face of climate change.

2.1.2 Equipped Infrastructure

Many local power systems rely on ill-equipped infrastructure, including power plants, transmission lines, and distribution networks. These systems were designed for historical temperature and weather patterns instead of the increasingly extreme conditions we witness today. Rising temperatures can reduce the efficiency of thermal power plants, as they require cooler water for optimal operation. Additionally, extreme weather events like storms and floods can damage critical infrastructure, leading to outages and disruptions.

In addition, high temperatures also pose safety concerns. Transformers and power lines are engineered to operate within specific temperature ranges to ensure safe operation and reliability. However, exceeding these limits can compromise their functionality. An overheated transformer, for example, can malfunction, significantly increasing the risk of a fire and power outage. Similarly, excessive heat can cause power lines to sag, leading to short circuits and equipment failure (real-life examples of such equipment failures are detailed in the Heatwave section of Sect. 2.2).

2.1.3 Energy Demand Change

Increased global temperatures have directly impacted energy consumption patterns, notably increasing the need for cooling [4]. According to the International Energy Agency (IEA), the global demand for space cooling may triple by 2050, posing significant challenges for local power grids [8]. This shift not only stresses existing energy supply systems but also exacerbates energy inequality and increases greenhouse gas emissions from fossil fuel-based power generation.

2.2 How Extreme Weather Events Affect Local Power Systems

This section reviews how extreme weather affects power systems from the perspectives of generation, transmission, distribution, and demand behaviour. In addition, while extreme weather events curtail grid performance, some electrical faults can also initiate extreme weather, e.g. wildfires. The bidirectional impact between power systems and extreme weather is discussed in this section.

Different types of extreme weather events may damage various parts of an electricity grid. Apart from physical damage, generation characteristics, and demand behaviours can also be indirectly affected, increasing the power system dispatch complexity. According to a U.S.-based study [9], 62.1% of 8+ hour outages co-occur with extreme weather events, particularly heavy precipitation, anomalous heat, and tropical cyclones. In turn, some electrically caused faults initiate extreme weather events, resulting in significant economic and social costs.

Typical extreme weather events that weaken the safety of power grid operations include floods, ice storms, tropical cyclones (wind), heatwaves, droughts, and

Table 1 Extreme weather effects on power grids–major concerns and examples

Weather type	Major concerns	Example
Flood	Inundated substation	A four-day power cut in Lancaster, UK, due to an inundated substation in 2016 [7]
Extratropical cyclone (wind)	High-voltage/extra high-voltage transmission network (e.g. lines contact failure & pole collapse)	Storm Arwen & Eunice hit the UK in 2021 & 2022 causing power outages that disrupted over one million customers' power connections [11]
Ice storm	Frozen generators; Surging demand	Texas's Power Failure [12]
Heatwave	Reduced line capacity; surging demand	The worst heat wave clipped 80% of power generation capacity in Sichuan Province, [13]
Drought	Hydro power shortage, additional transboundary fossil fuel generation needed	Drought-induced power shortage led to 65% more fossil fuel-based electricity consumed in the Western U.S., which was imported from other regions [14]
Wildfire	Direct damage on grid equipment & indirect damage: heat, smoke, and particle matter related faults	2019-2020 devastating wildfire seasons, resulting in over 20,000 household disconnections in Victoria, [15]

wildfires [10]. Each weather type affects various aspects of the network. The major concern for each is summarised in Table 1. Explanations for how each extreme weather type affects power grid operation are discussed in detail after Table 1.

Major concerns about power faults associated with various extreme weather types are discussed in the paragraphs below. Specific examples are provided for each weather type following the order listed in Table 1.

2.2.1 Flooding

Extreme flood events caused by climate change can affect different aspects of the electricity network. Amongst all impacts, the submergence of power substations are the most common and significant events during floods. Power substations act as the community-level power hub, stepping down electricity from the high-voltage transmission network to the distribution and residential use levels. When the drainage system reaches its capacity, the substation is submerged, leading to power interruption and widespread blackouts. In addition, a large amount of water, mud, and rust left trapped can make the repair work lengthy, complex, and costly [10]. As mentioned in Table 1, a four-day power cut happened in Lancaster, UK, due to an inundated substation in 2016, affecting more than 100,000 people's energy use [7].

2.2.2 Tropical and Extratropical Cyclone

To effectively manage the dual threat of high winds and flooding from tropical cyclones, adaptation strategies should integrate complementary approaches for both flood and wind hazards. However, upgraded cyclones, storms, and hurricanes (or typhoons) may lead to more severe damage to electricity networks. When the maximum sustained wind speed reaches 39 miles per hour, the tropical cyclones become tropical storms. The tropical storm will upgrade as a hurricane or typhoon when the maximum sustained wind speed is equal to or greater than 74 miles per hour [16].

The major concerns of wind-related power faults are line and pole failures due to fallen trees or contact-caused short circuits. It is noticeable that there is an evident correlation between the proximity of trees to distribution network lines and the vulnerability of power lines to high winds. Moreover, when the wind speed exceeds a certain threshold (60 miles per hour), the risk of tree-related faults will increase exponentially [10]. In [17], a Texas case study of power utility poles' resilience was conducted, demonstrating that the non-resilient poles doubled once the wind condition increased to 160 miles per hour from normal conditions. Short circuit faults due to line contacts are also a great concern under high wind conditions. For example, different phases of power lines may swing together, resulting in power outages.

Storm Arwen and Storm Eunice are the two most severe and damaging storms affecting England and Wales since 2014 [7, 11]. Over a million homes were left without power during Storm Eunice as strong winds brought down trees, with ongoing power cuts lasting several days in February 2022. It is reported that both Arwen and Eunice caused substantial damage to distribution networks. However, the possibility of customers impacted by a fault generally scales as the voltage level increases, with High Voltage (HV) and Extra High Voltage (EHV) affecting most customers. The distribution of the storm-affected power faults and the customers impacted by each fault type are displayed in Fig. 1. While most faults happen in distribution networks, only 2–3% of the customers lost power due to Low Voltage (LV) faults. As the pre-action cost (resilient strategies) of transmission networks is usually lower than the maintenance cost, reinforcement methods to build robust networks will be reviewed in Sect. 4 to enhance network resilience in a more financially viable way.

2.2.3 Ice Storm

Ice or snow accumulation on power lines can cause power outages with issues similar to those caused by high winds. The impact of winter storms on both the generation and demand sides is increasingly severe and cannot be ignored. Millions of Texans

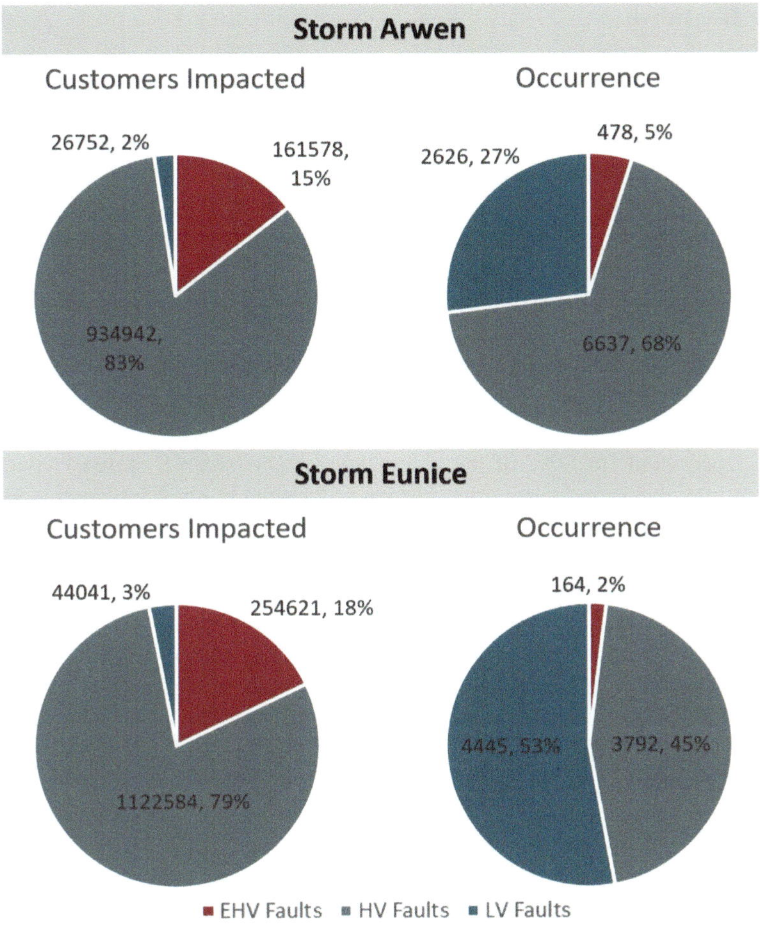

Fig. 1 Power fault classifications, occurrences, and customers affected by fault types for Storm Arwen and Storm Eunice [18]

experienced power outages at a time when electricity was most needed due to the unprecedented and devastating winter storm in 2021. More than 30 GW of generation was curtailed, mainly due to offline frozen wind turbines. Meanwhile, the natural gas available for electricity generation was largely diverted to heat homes in frigid conditions. On the end-user side, peak demand hit a record of more than 69 GW due to the extreme freezing conditions [12].

2.2.4 Heatwave

Heatwaves are defined as periods of at least five consecutive days with a maximum temperature of at least 5°C higher than the climatological norm for the same calendar day over the period 1961–1990, according to the World Meteorological Organisation [10]. The scorching temperature can reduce the transmission system efficiency. A study projected that by the mid-century (2040–2060), the average summertime transmission capacity in the U.S. will be reduced by 1.9–5.8% relative to the 1990-2010 reference period, due to the rising ambient temperature [19]. Moreover, the impact of heatwaves on generation and demand is becoming overwhelming. For instance, the worst heatwave in 60 years struck Sichuan Province in China from July to August in 2022. The National Meteorological Centre of China raised a red alert, and the power system was on the brink of collapse. On the generation side, electricity produced from hydropower was most affected by the heatwave. While hydropower usually accounted for 80% of the electricity generation in Sichuan Province, the hydro source, the Yangtze River, was severely affected by the heatwaves [13]. Apart from hydropower, the remaining dispatchable thermal power plant generation was also impacted by the heatwave due to insufficient water for the cooling process [20]. On the demand side, Sichuan's residential demand almost doubled from July 2021 to July 2022. Further, in August, the excessive heat continued to raise the demand, with a 78.5% increase compared to 2021. The Yangtze River's level change and the Sichuan Province residential demand trend are shown in Fig. 2.

2.2.5 Drought

Drought also affects grid performance in many countries. For example, the western U.S. has experienced severe droughts in recent decades. A study indicates that extreme drought can increase fossil fuel generation by up to 65% to compensate for the hydropower shortage, based on the 2001-2021 operational data in the Western U.S. Over 54% of this drought-induced generation is transboundary, leading to the fossil fuel-induced polluted emission upwards in other regions that provide energy export [14].

2.2.6 Wildfire

Most electrical faults related to wildfire disasters are triggered in transmission and distribution networks. Large fires can lead to physical damage to grid-supporting infrastructures (i.e. poles and towers), resulting in the collapse of power lines. However, the heat, smoke, and particle matter in a wildfire can threaten normal grid operation, even without direct physical damage to the structure [21]. For instance, soot accumulation on the insulation can create a conducting path for leakage currents, leading to line shutdowns. The smoke may change the dielectric property of the surrounding air, causing line-to-line or line-to-ground arcing [22]. The line capacity

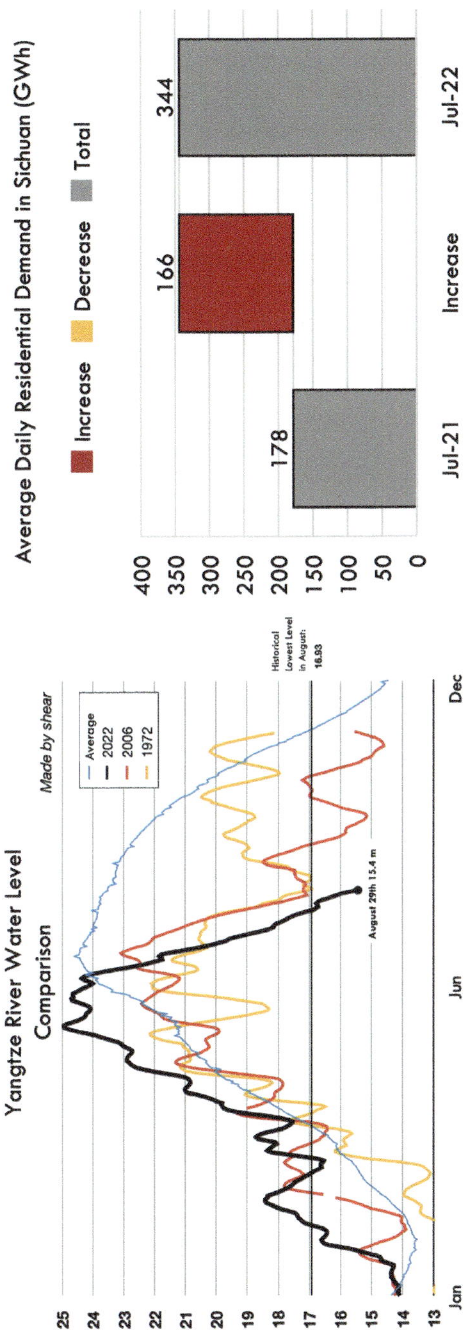

Fig. 2 Yangtze River water level changes and average daily residential demand in Sichuan, July 2021 & July 2022 [13]

can be affected by the fire retardant dumped from firefighting aircrafts. Even without overload, the line sagging caused by high fire temperatures may violate safety clearance and decrease the safe loading capacity of lines [23]. In [24], ten indices from three aspects of meteorological factors, environmental factors, and line factors were selected to build a comprehensive evaluation model of grid wildfire disaster using the BP neural network, demonstrating potential methods to quantify fire risk impact on grid operation.

2.2.7 Electrically Induced Extreme Weather Events (Wildfire)

While high wildfire risk conditions may lead to grid outages and losses, electrical faults sometimes can, in turn, cause wildfire ignitions or increase the probability of wildfire initiations. For instance, in summer 2020, a transformer explosion in Utah Valley (US) ignited a wildfire, leading to utility property damage and mass evacuation of people [25]. In Muhs et al. [26], the probability of fire igniting in a power distribution network was characterised using a sensitivity analysis, which included the impact of wind speed, line congestion, and grid protection. They found that mid-range wind speeds, line overloading, and long arc duration due to a lack of protection could increase the probability of line-induced wildfires.

3 Definition and Measurement of Power System Resilience

3.1 Definition of Power System Resilience

In Bie et al. [27], power system resilience is defined as the capacity to anticipate, withstand, adapt to, and promptly recover from grid disruptions. Reliability constitutes a conventional notion in power grids, assessing the quality and quantity of electricity delivered to satisfy the demand [28]. These two principles are pivotal for evaluating power grid efficacy, each delineating a distinct aspect. While grid resilience underscores the capability to withstand and recuperate from infrequent and severe incidents, grid reliability denotes the constancy of routine system functioning.

Figure 3 illustrates a conceptual resilience curve depicting the systemic states throughout an extreme event. The resilience levels over time concerning the disturbance event are divided into various zones [29]. Before the event occurs at t_e, the system needs to be robustly prepared to withstand the initial attack. As the event progresses from t_e to t_{pe}, the system enters the degraded state after the system resilience level decreases from R_o to R_{pe}, where R represents a metric to measure the resilience (load shedding rate, number of customers affected, etc.). The start of the system restoration is marked with t_r. With multiple recuperation methods, the system recovers to the post-restoration state R_{pr}, where R_{pr} can be equal to or less than R_o, depending on the effectiveness of the actions. Furthermore, it may take longer for

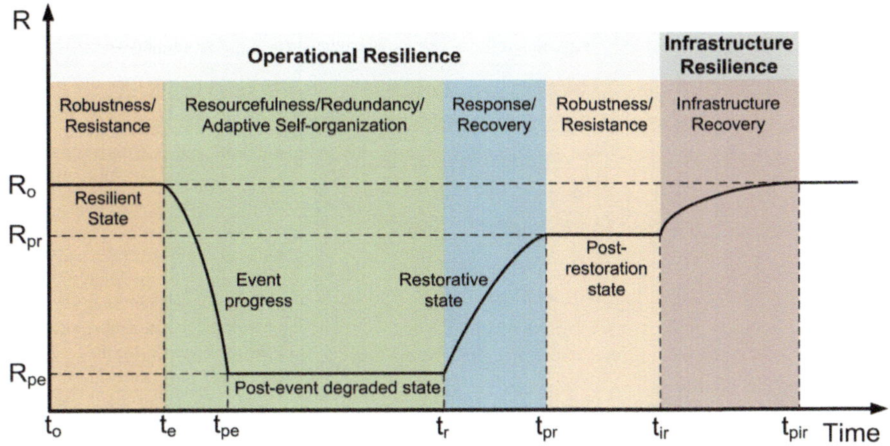

Fig. 3 A proposed resilience trapezoid used for assessing the resilience of a power system [29]

the system to fully recover to the pre-event infrastructure resilience level, from t_{ir} to t_{pir}. The whole process can be summarised as five main states: prevention, degradation, degraded, restoration, and adaptation. Infrastructure resistance and operational recoverability can be measured before and after the system's minimal function operational period to quantify the system's resilience.

A highly resilient system is not necessarily reliable, and vice versa. For instance, rolling brownouts may be acceptable in a resilient power system as long as the power supply can be quickly restored [28]. As extreme weather is a critical condition in this chapter, grid resilience will be the primary standard to evaluate grid performance. Reliability is still an essential parameter as it reflects the consistency of safe grid operation.

3.2 Measurement of Power System Resilience

Both qualitative and quantitative metrics to evaluate power grid reliability and resilience have been discussed in many existing studies. In this section, common metrics for assessing grid reliability and resilience are reviewed in Tables 2 and 3 [27, 30]. The metrics provide a standardised way to measure grid performance improvement after taking certain actions and compare grid resilience levels amongst various power networks.

Table 2 records common metrics to evaluate grid reliability during normal operating conditions (excluding extreme events). In Table 3, not only traditional ways of measuring resilience are covered, i.e. from the technical, economic, and social perspectives, but also topological resilience is introduced to calculate grid resilience

Table 2 Reliability indices–definitions and equations

Name	Equation	Definition and comment
System Average Interruption Duration Index (SAIDI) [31]	$SAIDI = \frac{\sum U_i N_i}{N_T}$ (Hour)	Average annual total duration of outages per customer. i: the location, U_i: the annual outage time, N_i: the customer numbers at the ith location, N_T: the total number of the customers served
Loss Of Load Expectation (LOLE) [32]	$\sum_{i=1}^{N} P[C_i < L_i]$ (Hour)	Expected number of days/hours in a pre-defined period that demand is not met by supply. N: the number of days/hours, P: the probability function, C_i: the available capacity on the ith day/hour, L_i: the daily/hourly peak demand [33]
System Average Interruption Frequency Index (SAIFI) [34]	$SAIFI = \frac{\sum N_{ci}}{\sum N}$	Average annual total number of outages per customer. N_{ci}: the total number of customer interruptions, N: the total number of served customers
Customer Average Interruption Duration Index (CAIDI) [34]	$CAIDI = \frac{\sum U_i N_i}{l_i N_i}$ (Hour)	Average time to recover once an outage occurs for a specific customer. U_i: yearly outage time, N_i: the number of customers, l_i: failure rate, i: the location
Customer Average Interruption Frequency Index (CAIFI) [30]	$CAIFI = \frac{\sum N_{ci}}{\sum N_{dc}}$	Frequency of power supply interruption for customers who have at least experienced one power outage. N_{ci}: the total number of customer interruptions, N_{dc}: the number of distinct customers interrupted (customers who experienced at least one power outage)
Customer Total Average Interruption Duration Index (CTAIDI) [30]	$CTAIDI = \frac{\sum U_i N_i}{N_{io}}$ (Hour)	Cumulative average interruption duration for customers who have at least experienced one power outage. i: the location, U_i: yearly outage time, N_i: the number of customers, N_{io}: the customer number that were interrupted
Average Service Availability Index (ASAI) [35]	$ASAI = \frac{\sum N_i \times 8760 - \sum U_i N_i}{\sum N_i \times 8760}$	Ratio between total available customer hours and demanded hours. i: the location, U_i: yearly outage (hours), N_i: the number of customers
Average Service Unavailability Index (ASUI) [30]	$ASUI = \frac{\sum U_i N_i}{\sum N_i \times 8760} = 1 - ASAI$	Ratio between total unavailable customer hours and demanded hours. i: the location, U_i: yearly outage (hours), N_i: the number of customers
Momentary Average Interruption Frequency Index (MAIFI) [30]	$MAIFI = \frac{\sum N_{ci}}{\sum N_c}$	Cumulative interruption amount experienced by a customer in a defined period. N_{ci}: the total number of customer interruptions (momentary–less than a specific time), N_c: the total number of customers served

Table 3 Resilience indices–definitions and equations

Name	Equation	Definition and comment
Resilience triangle (resilience impact) [37]	$I = \int_{t_E}^{t_R} [Q_N - Q(t)]dt$	I is the quantified resilience impact. t_R: time of restoration completion, t_E: starting time of the event, Q_N: grid normal functionality, $Q(t)$: grid functionality at time t
Availability-based Resilience Index [38]	$\rho = \frac{A_1}{n ln(t1)} \sum_{i=1}^{n} \frac{A_2^i A_3^i}{ln(t_3^i - t_2^i)}$	ρ is the ultimate availability-based resilience index. A_1, A_2, A_3: the availability at the steady-state, the after-shock transient state, and the new balance state. t_1: steady-state time. $t_3 - t_2$: measured static post shock time. $i \in [1, n]$, n is the number of shocks
Resilience trapezoid [39]	$FLEP(\Phi \Lambda E \Pi)$ $\Phi = \frac{R_{pdo} - R_{0_o}}{t_{ee} - t_{oe}}$ (MW/hour) $\Lambda = R_{0_o} - R_{pdo}$ (MW) $E = t_{or} - t_{ee}$ (hour) $\Pi = \frac{R_{0_o} - R_{pdo}}{T_{or} - t_{or}}$ (MW/hour)	A group of resilience metrics to measure the system performance during various phases of the trapezoid. $R_{0_o} - R_{pdo}$: the degree of resilience deterioration. $t_{ee} - t_{oe}$: period of disaster. $t_{or} - t_{ee}$: period of the post-threat degraded state. $T_{or} - t_{or}$: period of the restoration state
Resilience–economic [40]	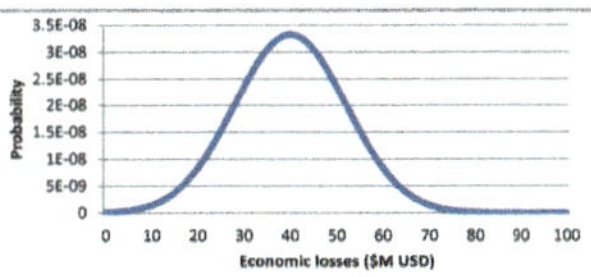 (M USD)	Probability Density Function (PDF) of the economic impact attributed to weather-related risk
Resilience–social impact (life at risk) [40]	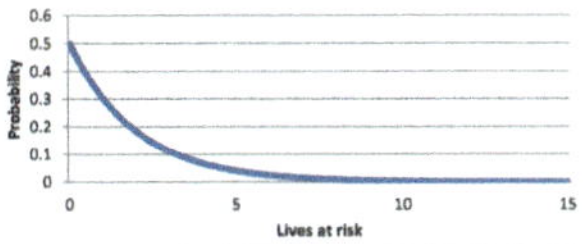	PDF describing life at risk attributed to weather-related risk
The power ratio[41]	$P_\omega = \frac{\sum_{m=1}^{M} \frac{1}{N} \sum_{n=1}^{N} P_d(m,n)}{\sum_{m=1}^{M} P_{(m)}} \times 100\%$ (%)	P_ω: percentage of power supply. $m \in M$: the area of services. $n \in N$: simulation numbers. $P_d(m, n)$: power output of the area m under the nth simulation. $P_{(m)}$: power output of the area m under normal condition
Topological resiliency [42]	$G = (N, E, W)$ $D = \frac{2\|E\|}{\|N\|(\|N-1\|)}$ $l_G = \frac{\sum_{i \neq j} d(n_i, n_j)}{N(N-1)}$ $C_n = \frac{2e_n}{k_n(k_n-1)}$	Graph G comprises of a set of nodes N and edges E. W: weights (power flow of each section). D: graph diameter. l_G: average path length of the graph G. $d(n_i, n_j)$: distance between the node i and j. C_n: clustering coefficient. e_n: number of connected pairs between all neighbours of a node n. k_n: number of neighbours of node n

(continued)

Table 3 (continued)

Name	Equation	Definition and comment
Network [43]	$R_\alpha = \sum_\beta P(\phi(x) \geq d\|\alpha, \beta)P_\beta$	R_α: expected resiliency in the presence of α failure. β: unique non-repeating combination. d: network performance requirement. $\phi(x)$: network capacity under system state vector x
Time-to-restoration [44]	N/A	Based on historical outage and recovery records

based on the connection structure. The definitions, equations, and variables of system reliability and resilience are detailed in Tables 2 and 3.

4 Strategies for Enhancing Resilience

In the study by Bie et al. [27], a comprehensive review was conducted on methods to enhance grid resilience, predominantly categorised as system hardening and operational resilience strategies, commonly called hard and soft resilience. With the development of Artificial Intelligence (AI), digital tools based on data-driven methods, including monitoring, forecasting, and pattern analysis, have emerged. Section 4 will discuss techniques for bolstering power grid resilience across three domains: hard resilience strategies, soft resilience strategies, and AI-driven digital tools.

4.1 Hard Resilience Strategies

Hard resilience strategies, which refer to physical alterations to infrastructure, include undergrounding lines, upgrading poles with stronger materials, adding grid redundancy, tree trimming, and elevating substations. Hard resilience strategies can improve the durability of transmission and distribution networks. For instance, a snowstorm in Southern China in 2008 resulted in 129 power line faults and power loss for 14.66 million households [36]. Since then, higher ice-resistant transmission lines, towers, and ice-melting equipment have been widely adopted to improve grid resilience in snowstorm-prone areas [27].

Amongst all hard resilience solutions, undergrounding is one of the most effective methods to avoid physical damage to lines. Therefore, the pros and cons of undergrounding are discussed as the focus here. While undergrounding high-voltage electricity lines can protect transmission networks from extreme weather events and mitigate the probability of electrical faults, there are some issues related to this approach [45].

- *Technical Issues*: Undergrounding requires larger electrical cables than the commonly used overhead lines (OHL) due to different materials. To match the thermal performance of a 400kV double circuit OHL, as many as 12 separate cables in four separate trenches are needed. This arrangement allows for heat dissipation, provides physical insulation for the cable, and ensures the protection of surrounding people while in service. In addition, while undergrounding is one of the most effective ways to reduce the frequency of electrical outages, maintenance time increases due to the difficulty of detecting underground lines (especially during a flood) [46].
- *Financial Issues*: The cost of underground cables can be up to 10 times higher than that of OHL, and this increased cost will ultimately fall on the bill payer. According to Pacific Gas and Electric (PG&E), the costs for undergrounding existing overhead distribution infrastructure can range anywhere between US$1.85 million to US$6.07 million per mile, accounting for trenching, substructures, cables & connections, conduits, metre panel modifications, cutover work, and the final removal from poles & wires services [47].
- *Environmental and Social Issues*: The installation of underground cables can lead to significant environmental and social issues. Potential costs related to threats to sensitive habitats and damage to archaeological heritage and biodiversity may arise [48].

4.2 Soft Resilience Strategies

In the previous section, issues related to hard resilience strategies were outlined, including technical restrictions, financial costs, and environmental impacts. Since hard resilience strategies may only be suitable for a single type of extreme event (e.g. extreme heat), soft strategies such as emerging smart grid technologies should be incorporated alongside hard measures. Examples of these soft strategies include microgrid islanding operations and dynamic transmission line control strategies. This section reviews various soft resilience strategies designed for electricity networks.

The Distribution System Resilient Load Restoration Framework was proposed by Bie et al. [27] as a soft grid resilience strategy using reconfiguration and Distributed Generation (DG) islanding. Islanding in microgrid operation refers to a condition where a portion of the electric grid, typically a microgrid, continues to operate and generate power independently even when it is disconnected from the main grid. DG technologies include multiple renewable generation and storage techniques. Solar Photovoltaic (PV) systems, wind turbines, small-scale hydroelectric generators, and biogas digesters are commonly considered DG technologies that assist microgrid islanding by taking advantage of clean and local resources. Attributed to the inherent intermittency of renewable generation, proper smart storage control techniques are selected to cooperate with the operation of DGs to guarantee a continuous and sufficient supply. There are a variety of storage technologies emerging in recent years, including Battery Energy Storage Systems (BESS), Pumped Hydro Storage (PHS),

Compressed Air Energy Storage (CAES), Flywheel Energy Storage (FES), Thermal Storage, and Hydrogen Storage. With the combined optimised control of DGs and storage, microgrids can achieve greater energy independence, resilience, and sustainability, especially during islanding events or when connected to unreliable main grids. More information about optimising distribution system performances by DG and load reconfiguration can be found in [49, 50].

Soft resilience strategies have also been applied to transmission networks. Specifically, geographical mapping methods have been used to develop dynamic grid control methods considering weather risk. Yang et al. [15] proposed a novel method to manage wildfire risks in transmission networks without resorting to blackouts. Victoria, Australia, was selected as the case study region to study how grid resilience can be enhanced during extreme wildfire events. The Fire Weather Index (FWI) is a crucial indicator of global fire risk both spatially and temporally, as proved by robust analyses in many previous studies. FWI is applied to a Wildfire-Energy Nexus System for the first time in [15]. The wildfire risk index and the power network model were considered simultaneously based on the geographical information of the transmission lines. In Figs. 4 and 5, the transmission line map of Victoria State and the corresponding FWI risk are demonstrated, which provide guidance on system dispatch in the Nexus model. It has been proven that 68% of system costs can be recuperated by implementing networked microgrid solutions in the worst-case scenarios.

In addition to the approach described earlier, dynamic pricing can also help decrease peak load by adjusting electricity rates based on demand. Several dynamic pricing methods have been suggested, with three primary strategies being widely debated: Time-of-Use, Real-Time Pricing, and Critical Peak Pricing. These measures aim to encourage consumers to shift their energy consumption to off-peak hours, thus lowering peak demand and preventing extensive investments in additional capacity.

Time-of-Use pricing schemes are designed to reflect the cost of electricity at different times of the day [51]. They charge more during high-demand peak hours and less during off-peak hours, encouraging users to shift their usage to times when electricity is cheaper and the grid is less strained.

Real-Time Pricing goes beyond dynamic pricing by providing prices that fluctuate hourly according to current market conditions [52]. This approach gives consumers precise signals, aligning with the actual production and delivery costs in real time. Real-Time Pricing can effectively encourage consumers to adjust their energy usage in response to price fluctuations, potentially resulting in substantial decreases in peak demand.

Critical Peak Pricing is a form of pricing where electricity rates are significantly higher during critical peak periods than normal rates [53]. These periods are often when the system is at its maximum capacity or during a grid emergency. Critical Peak Pricing is designed to curb energy use during these vital times, ensuring system reliability and preventing blackouts. Based on studies conducted in North America, Critical Peak Pricing has proven to be the most effective strategy to curtail loads during event periods automatically [54].

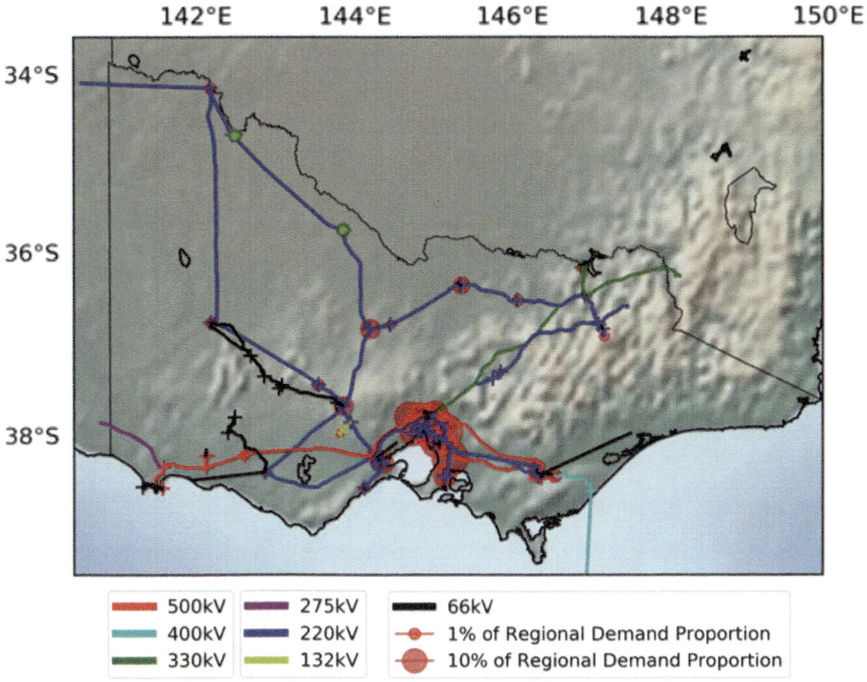

Fig. 4 High-voltage transmission network in Victoria, Australia, where '+' represents substations, the size of red circles denotes energy demand proportion at each node, and the two green spots indicate an example 220kV transmission line from Red Cliffs to Kerang

4.3 Digital Tools for Monitoring and Forecasting

In addition to traditional hard and soft resilience strategies, advancements in digital technologies are emerging and becoming promising solutions to enhance grid resilience. Real-time monitoring of weather conditions and grid performance plays a critical role in this approach. In addition, accurate energy forecasting tools considering climate risks are essential to help grid operators dispatch electricity in advance to get prepared for extreme events.

4.3.1 Real-Time Monitoring of Weather Conditions

In response to extreme climate events, the ability to capture local weather data in real time has become crucial to enhancing the resilience of local energy systems. By providing a constantly updated picture of atmospheric conditions, real-time weather monitoring empowers local energy providers to make informed decisions that optimise energy production, consumption, and overall grid stability. This section delves into the numerous benefits real-time weather monitoring offers, explores the tech-

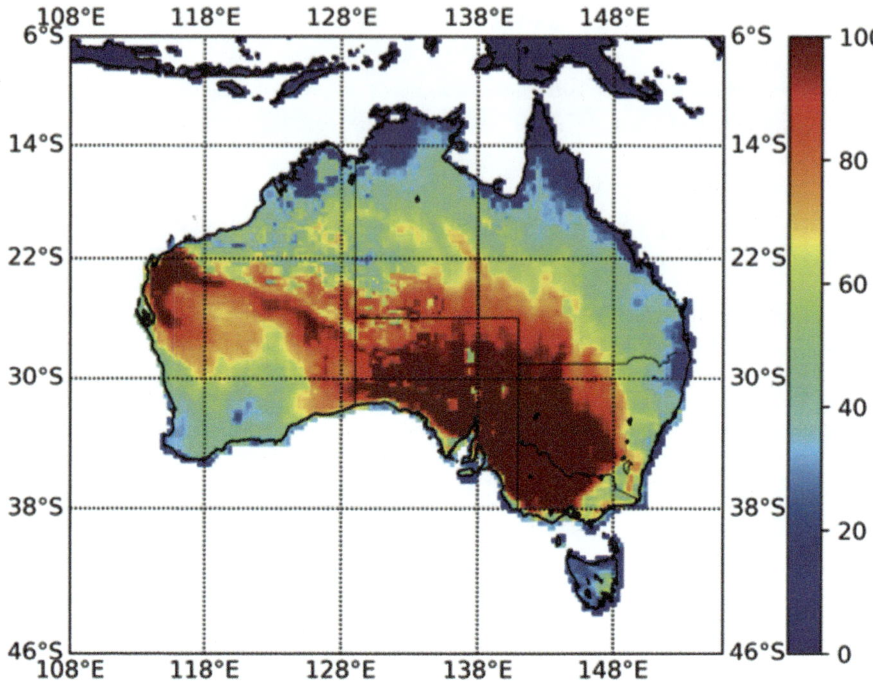

Fig. 5 Australian Fire Weather Index (FWI) colourmap on 30th Dec 2019

nologies employed for data collection, and examines how this data is utilised to strengthen local energy systems.

4.3.2 Enhanced Forecasting Resolution and Accuracy

Traditional weather forecasting models rely on data collected from a network of widely spaced meteorological stations. While these models provide valuable insights into large-scale weather patterns, they often lack the granularity necessary for accurate short-term forecasts at the local level, which is particularly important for assessing local climate impacts and identifying appropriate resilience strategies. Real-time weather monitoring bridges this gap by capturing hyper-local weather data. Ground-based weather stations, strategically placed within the local energy system's footprint, provide crucial information on temperature fluctuations, wind speed and direction, precipitation levels, and humidity. This detailed data can be fed into advanced weather prediction models that account for local microclimates and topographic variations. Previous studies showcase the effectiveness of integrating ground-based stations with real-time data transmission, enabling the generation of highly accurate short-term forecasts specifically tailored to the local energy system's needs [55]. With a clearer picture of impending weather changes, local energy providers can proactively adjust their strategies. For instance, anticipating a sudden drop in solar irradiance due to cloud cover allows for the activation of backup power generation or the implementation of demand response programmes to minimise strain on the grid.

4.3.3 Optimised Dispatch Strategies

Local energy systems often comprise a diverse mix of renewable and non-renewable energy sources. Real-time weather data empowers operators to optimise the dispatch of these resources based on anticipated weather patterns. For example, knowing that high wind speeds are expected within the next few hours allows for maximising wind turbine output. Conversely, a forecast of reduced solar irradiance due to heavy cloud cover necessitates adjustments in energy production. Devaraju [56] designed a wireless portable microcontroller-based weather monitoring station for local energy applications. This system enables real-time monitoring of wind speed and direction, enabling local energy providers (distribution level) to optimise the utilisation of wind turbines and potentially reduce reliance on fossil fuel backups during periods of low solar power generation. Similarly, Murthy [57] proposed a real-time weather monitoring system that integrates data on solar irradiance and temperature. This data can be used to predict solar power generation potential, guiding the activation of supplementary power sources or the implementation of demand-side management programmes to ensure grid stability during periods of reduced solar output.

4.3.4 Early Warning Systems for Extreme Events

Extreme weather events such as hurricanes, thunderstorms, and heat waves pose a significant threat to the stability of local energy systems. Real-time weather data plays a crucial role in triggering early warning systems for these events. By analysing data on wind speed, precipitation patterns, and atmospheric pressure changes, local energy providers can anticipate the potential impact of extreme weather events. A range of studies have explored the development and effectiveness of Early Warning Systems (EWS) for extreme events. Alfieri et al. [58] and Neußner et al. [59] both emphasised the importance of reliable and consistent EWS, with the former proposing a flood early warning system based on ensemble numerical predictions and the latter advocating for standardised warning schemes. Lenton et al. [60] highlighted the potential of EWS for reducing the impact of earthquakes and environmental shocks, while Mayhorn [61] provided a global perspective on risk communication in the context of EWS. These studies collectively underscore the need for robust, standardised, and multi-hazard EWS that can effectively communicate risk and minimise the impact of extreme events.

4.3.5 Importance of Incorporating Climate Resilience into Energy Forecasting Tools

After going through the digital applications in weather monitoring and forecasting, this subsection discusses how Artificial Intelligence tools are utilised in the field of electricity monitoring and forecasting. With the increasing occurrence of power outages attributed to climate change, there is a growing necessity for climate-resilient energy forecasting to facilitate prompt network reconfiguration during extreme weather events. Climate-resilient energy forecasting encompasses the prediction of power generation [62], electricity load [63, 64], and the system's net demand error [64]. In recent years, pertinent forecasting models have predominantly advanced along two dimensions to improve forecast accuracy: the complexity of the model structure and the diversity of factors involved.

As we delve into the realm of research aimed at enhancing the structure of forecasting models, we encounter two primary method types: traditional statistical methods grounded in time-series analysis and Artificial Intelligence (AI) based approaches [65, 66]. Hong and Fan [67] provide a comprehensive overview of several widely adopted statistical forecasting techniques, encompassing Multiple Linear Regression (MLR), Semi-parametric Additive Models (SAM), Autoregressive Integrated Moving Average (ARIMA), and Exponential Smoothing Models (ESM). Presently, statistical forecasting methods represent the most utilised tools. However, these methods, despite their widespread use, may encounter difficulties in capturing demand fluctuations induced by extreme weather events owing to their implicit non-linear characteristics. This challenge has led to the emergence of data-driven approaches, such as Neural Networks (NN), as promising solutions [68, 69]. Amongst NN-based models, Recurrent Neural Networks (RNNs) have proven particularly adept at handling time-series datasets [69], incorporating architectures like Long Short-Term Memory (LSTM) and Gated Recurrent Unit (GRU). Convolutional Neural Network (CNN) is another branch of NN methods, specialising in handling image processing issues and capturing spatial patterns [70–72]. While CNN is not commonly used for time-series load forecasting methods, some modified CNN methods were developed to solve sequential tasks [73, 74]. Figure 6 illustrates typical structures for load forecasting.

In addition to the comparison and adaptation of various forecast models, the incorporation of multiple factors is crucial for enhancing forecast accuracy. Few studies have delved into the internal relationship between various factors and load behaviours before integrating them as model inputs. For instance, previous studies [71, 75] typically utilised 24-hour ahead load data as the default input owing to its daily periodicity, yet had not explored the optimised design of input structures. Calendar effects have been applied to load forecasting studies through various classification methods, such as distinguishing working and non-working day characteristics [76]. Whilst most studies adhere to a specific method for incorporating calendar effects, there remains scope to explore and compare different approaches.

Moreover, various climate-related factors, including temperature, humidity, wind speed, and precipitation, have been previously considered. Daily mean or instantaneous temperature are those most commonly employed [77, 78]. However, the development of an energy forecasting model that encompasses a broader spectrum of climate factors is still in its nascent stage, diverging from traditional climate variables. In [63, 78], a load forecasting model was designed to exhibit robustness and resilience to a specific extreme weather event—wildfires—by integrating fire-related factors. The model outlined in [79] attained an average forecast accuracy of 97.05%, ensuring that reconfiguration measures mitigate rather than exacerbate high load conditions during periods of extreme stress. Future energy forecasting models could incorporate additional weather-related risk indices and long-term climate conditions (such as El Niño or La Niña patterns) to enhance the resilience of energy systems to climate variations.

5 Conclusion

As climate change threats intensify, local energy systems worldwide face unprecedented challenges. Extreme weather events, such as hurricanes, floods, and heatwaves, are becoming more frequent and severe, posing significant risks to the stability and functionality of power systems. This chapter offered a comprehensive analysis of how these challenges can be navigated through strategic resilience-enhancing measures in power networks.

The discussion delineates the dual impact of climate change on energy systems with practical examples, encompassing both long-term transformations and short-term disruptions. Long-term effects are considered alterations in electricity production, distribution, and consumption patterns. Short-term disruptions, typically triggered by extreme weather events, necessitate rapid and effective adaptation and

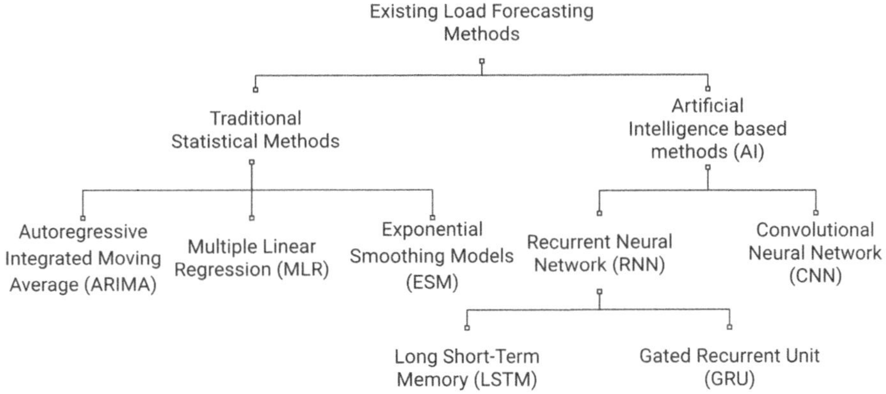

Fig. 6 Common electricity load forecasting methods

recovery strategies not only to ensure the stability and functionality of the energy system but also to increase its capacity to withstand future shocks.

After reviewing how various extreme weather events damage different aspects of the power grid, the two main evaluation systems, reliability and resilience, are defined and compared. Common metrics to assess power grid reliability and resilience have been studied to quantify grid performance during extreme stress periods.

To address these challenges, the chapter examines both hard and soft resilience strategies. Hard strategies encompass physical enhancements to infrastructure, such as fortifying power lines and stations, upgrading materials, and implementing redundant systems to maintain stability during extreme events. On the other hand, soft resilience strategies concentrate on operational and technological innovations. These include advanced grid technologies such as demand response programmes, improved weather forecasting, real-time monitoring systems, and integrating renewable energy sources with smart grid technologies.

Furthermore, advancements in digital technologies play a pivotal role in enhancing the resilience of energy systems. Real-time monitoring and forecasting tools facilitate the proactive management of energy systems in response to changing weather conditions. These tools enable energy providers to adjust operations preemptively, optimising the balance between supply and demand and mitigating potential disruptions caused by extreme weather.

However, with the increasing frequency of extreme weather events, it has been found that current power systems need to be regularly updated to maintain their resilience. This includes not only upgrading physical infrastructure to withstand more severe conditions but also integrating advanced technologies and adaptive management practices, e.g. grid-weather integrated models and AI-based renewable generation forecasting tools. Regular assessments and proactive enhancements are essential to ensure that power systems remain robust, flexible, and capable of handling both current and future challenges posed by a changing climate.

Throughout the chapter, case studies are used to illustrate the practical applications of resilience strategies. For example, the integration of microgrids in wildfire-prone areas, the use of predictive analytics for flood risk management, and the reinforcement of lines against ice storms showcase how specific tools and strategies can be effectively employed to mitigate the impacts of extreme weather events on local energy systems.

While the chapter outlines numerous strategies to enhance resilience, it also acknowledges the challenges involved. These include the high costs associated with upgrading infrastructure, the technical difficulties of integrating new technologies, and the complexities of coordinating policies and actions across different governmental and regulatory frameworks. However, these challenges also present opportunities for innovation, research, and collaboration amongst stakeholders to develop cost-effective and sustainable solutions.

Looking ahead, the chapter suggests several areas for further research and development. These include the exploration of new materials and technologies that can withstand extreme weather conditions, the development of more accurate and localised prediction models for weather and electricity, the design of a more comprehensive evaluation system for grid resilience under extreme weather risks, and the investigation of socio-economic factors that influence the resilience of energy systems. Furthermore, fostering a deeper understanding of the interdependencies within energy systems and the broader socio-economic landscape can facilitate the roll-out of resilience strategies.

References

1. Yang P, Mi Z, Yao YF, Cao YF, Coffman DM, Liu LC (2021) Solely economic mitigation strategy suggests upward revision of nationally determined contributions. One Earth 4:1150–1162. https://doi.org/10.1016/j.oneear.2021.07.005
2. Yang P, Mi Z, Wei YM, Hanssen SV, Liu LC, Coffman D et al (2023) The global mismatch between equitable carbon dioxide removal liability and capacity. Natl Sci Rev 10. https://doi.org/10.1093/NSR/NWAD254
3. Dallison RJH, Patil SD (2023) Impact of climate change on hydropower potential in the UK and Ireland. Renew Energy 207:611–628. https://doi.org/10.1016/J.RENENE.2023.03.021
4. Miranda ND, Lizana J, Sparrow SN, Zachau-Walker M, Watson PAG, Wallom DCH et al (2023) Change in cooling degree days with global mean temperature rise increasing from 1.5°C to 2.0°C. Nat Sustain 6:1326–1330. https://doi.org/10.1038/s41893-023-01155-z
5. Schaeffer R, Szklo AS, Pereira de Lucena AF, Moreira Cesar Borba BS, Pupo Nogueira LP, Fleming FP et al (2012) Energy sector vulnerability to climate change: a review. Energy 38:1–12 (2012). https://doi.org/10.1016/J.ENERGY.2011.11.056
6. IPCC (2014) Key economic sectors and services. In: Climate change 2014–impacts, adaptation and vulnerability: part a: global and sectoral aspects: working group II contribution to the IPCC fifth assessment report: volume 1: global and sectoral aspects. Cambridge University Press, Cambridge, pp 659–708. https://doi.org/10.1017/CBO9781107415379.015
7. Harvey F (2016) Power cuts caused by flooding "major danger" to British lives, engineers warn. Guardian
8. IEA (2018) The future of cooling
9. Vivian D, Heather M, Nina MF, Alexander JN, Jeffrey S, Mathew VK et al (2023) Spatiotemporal distribution of power outages with climate events and social vulnerability in the USA. Nat Commun
10. Abi-Samra NC, Malcolm WP (2011) Extreme weather effects on power systems. In: 2011 IEEE power and energy society general meeting, 2011
11. Met Office (2022) Storms Dudley, Eunice and Franklin, Feb 2022
12. Xie L, Barteau M, Singh C, Pistikopoulos E (2021) What went wrong with Texas' power failure and how to fix it. Texas A&M
13. Chen Z, Gao S (2022) From Sichuan to California, extreme heat is pushing the grid to its limit. Rocky Mountain Institute
14. Minghao Q, Nathan R, Inés MLA, Noah SD, Marshall B (2023) Drought impacts on the electricity system, emissions, and air quality in the western United States. In: PNAS. https://doi.org/10.1073/pnas.2300395120
15. Yang W, Sparrow SN, Ashtine M, Wallom DCH, Morstyn T (2022) Resilient by design: preventing wildfires and blackouts with microgrids. Appl Energy 313:118793. https://doi.org/10.1016/J.APENERGY.2022.118793
16. National Oceanic and Atmospheric Administration (2023) What is the difference between a hurricane and a typhoon?
17. Alam MM, Tokgoz BE, Hwang S (2019) Framework for measuring the resilience of utility poles of an electric power distribution network. Int J Disaster Risk Sci

18. Department for Business, Energy & Industrial Strategy (2022) Energy emergencies executive committee storm Arwen review
19. Bartos M, Chester M, Johnson N, Gorman B, Eisenberg DA, Linkov I et al (2013) Impacts of rising air temperatures on electric transmission ampacity and peak electricity load in the United States. Environ Res Lett
20. American Nuclear Society (2021) Report: extreme weather is affecting nuclear power's reliability
21. Allen-Dumas BKC, Colin I, Cunliff MR (2019) Extreme weather and climate vulnerabilities of the electric grid: a summary of environmental sensitivity quantification methods
22. Sathaye J, Dale L, Larsen P, Fitts G, Koy K, Lewis S et al (2011) Estimating risk to California energy infrastructure from projected climate change
23. POWERLINK Queensland (2015) Fire and high voltage transmission line safety-information sheet
24. Wu C, Zhong Y, Liu H, Wang Y, Niu D, Gao F (2016) A novel application of BP neural networks to evaluate the safety of power grid in wildfire disasters. Int J Simul Syst Sci Technol. https://doi.org/10.5013/IJSSST.a.17.27.32
25. ExpertsNRG (2021) Transformer explosion causes a forest fire. https://expertsnrg.com/transformer-explosion-causes-a-forest-fire/. Accessed 26 July 2021
26. Muhs J, Parvania M, Nguyen HT, Palmer JA (2020) Characterizing probability of wildfire ignition caused by power distribution lines. IEEE Trans. Power Deliv. https://doi.org/10.1109/TPWRD.2020.3047101
27. Bie Z, Lin Y, Li G, Li F (2017) Battling the extreme: a study on the power system resilience. Proc IEEE
28. Clark-Ginsberg A (2021) What's the difference between reliability and resilience? ResearchGate. https://www.researchgate.net/publication/320456274_What's_the_Difference_between_Reliability_and_Resilience. Accessed 22 July 2021
29. Panteli M, Trakas DN, Mancarella P, Hatziargyriou ND (2016) Boosting the power grid resilience to extreme weather events using defensive islanding. IEEE Trans Smart Grid (2016)
30. Hossain E, Roy S, Mohammad N, Nafiu N, Dipta DR (2021) Metrics and enhancement strategies for grid resilience and reliability during natural disasters. Appl Energy
31. The World Bank (2021) Getting electricity: system average interruption duration index (SAIDI) (DB16-20 methodology). https://datacatalog.worldbank.org/getting-electricity-system-average-interruption-duration-index-saidi-db16-20-methodology. Accessed 22 July 2021
32. Panteli M, Mancarella P (2017) Modeling and evaluating the resilience of critical electrical power infrastructure to extreme weather events. IEEE Syst J 11:1733–1742. https://doi.org/10.1109/JSYST.2015.2389272
33. Ibanez E, Milligan M (2014) Comparing resource adequacy metrics
34. Bourezg A, Meglouli H (2014) Reliability assessment of power distribution systems using disjoint path-set algorithm. J Ind Eng Int 11:45–57. https://doi.org/10.1007/S40092-014-0083-5
35. National Association of Regulatory Utility Commissioners (2021) Reliability. NARUC. https://www.naruc.org/servingthepublicinterest/about/reliability/. Accessed 23 July 2021
36. Lin Y, Bie Z, Qiu A (2018) A review of key strategies in realizing power system resilience. Global Energy Interconnect 1:70–78. https://doi.org/10.14171/J.2096-5117.GEI.2018.01.009
37. Bruneau M, Chang SE, Eguchi RT, Lee GC, O'Rourke TD, Reinhorn AM et al (2003) A framework to quantitatively assess and enhance the seismic resilience of communities. Earthq Spectra 19:733–752. https://doi.org/10.1193/1.1623497
38. Cai B, Xie M, Liu Y, Liu Y, Feng Q (2018) Availability-based engineering resilience metric and its corresponding evaluation methodology. Reliab Eng Syst Saf 172:216–224. https://doi.org/10.1016/j.ress.2017.12.021
39. Panteli M, Mancarella P, Trakas DN, Kyriakides E, Hatziargyriou ND (2017) Metrics and quantification of operational and infrastructure resilience in power systems. IEEE Trans Power Syst 32:4732–4742. https://doi.org/10.1109/TPWRS.2017.2664141

40. Watson J-P, Guttromson R, Silva-Monroy C, Jeffers R, Jones K, Ellison J et al (2015) Conceptual framework for developing resilience metrics for the electricity, oil, and gas sectors in the united states
41. Shinozuka M, Chang SE, Cheng T-C, Feng M, O'rourke TD, Saadeghvaziri MA et al (2003) Resilience of integrated power and water systems
42. Chanda S, Srivastava AK (2016) Defining and enabling resiliency of electric distribution systems with multiple microgrids. IEEE Trans Smart Grid 7:2859–2868. https://doi.org/10.1109/TSG.2016.2561303
43. Whitson JC, Ramirez-Marquez JE (2006) Resiliency as a component importance measure in network reliability. Reliab Eng Syst Saf 94:1685–1693. https://doi.org/10.1016/J.RESS.2009.05.001
44. Maliszewski PJ, Perrings C (2012) Factors in the resilience of electrical power distribution infrastructures. Appl Geogr 32:668–679. https://doi.org/10.1016/J.APGEOG.2011.08.001
45. Evennett H (2024) 'Undergrounding' electrical transmission cables
46. Davis G, Snyder AF, Mader J (2014) The future of distribution system resiliency. In: 2014 Clemson university power systems conference. IEEE, pp 1–8. https://doi.org/10.1109/PSC.2014.6808134
47. California Public Utilities Commission (2024) CPUC undergrounding programs description
48. Richardson ML, Wilson BA, Aiuto DA, Crosby JE, Alonso A, Dallmeier F, Golinski GK (2017) A review of the impact of pipelines and power lines on biodiversity and strategies for mitigation. Biodivers Conserv 26:1801–1815
49. Zhang M, Chen J (2015) Islanding and scheduling of power distribution systems with distributed generation. IEEE Trans Power Syst 30:3120–3129. https://doi.org/10.1109/TPWRS.2014.2382564
50. Dong X, Yang W, Wu L, Tian X, Xie X, Chu W et al (2018) Subsynchronous oscillation characteristic study of wind-thermal power bundling and EHV AC-DC hybrid transmission system. 2018 International conference on power system technology (POWERCON). IEEE, pp 1995–2000. https://doi.org/10.1109/POWERCON.2018.8602352
51. Çelebi E, Fuller JD (2012) Time-of-use pricing in electricity markets under different market structures. IEEE Trans Power Syst 27:1170–1181. https://doi.org/10.1109/TPWRS.2011.2180935
52. Barbose G, Goldman C, Neenan B (2004) A survey of utility experience with real time pricing. https://doi.org/10.2172/836966
53. Herter K (2007) Residential implementation of critical-peak pricing of electricity. Energy Policy 35:2121–2130. https://doi.org/10.1016/J.ENPOL.2006.06.019
54. Newsham GR, Bowker BG (2010) The effect of utility time-varying pricing and load control strategies on residential summer peak electricity use: a review. Energy Policy 38:3289–3296 (2010). https://doi.org/10.1016/j.enpol.2010.01.027
55. Calderon-Cordova C, Jaramillo A, Tinoco C, Quinones M (2016) Design and implementation of an architecture and methodology applied to remote monitoring of weather variables. In: Iberian conference on information systems and technologies (CISTI), vol 2016-July. IEEE Computer Society. https://doi.org/10.1109/CISTI.2016.7521465
56. Devaraju JT, Suhas KR, Mohana HK, Patil VA (2015) ireless portable microcontroller based weather monitoring station. Measurement 76:189–200. https://doi.org/10.1016/J.MEASUREMENT.2015.08.027
57. Sreerama Murthy M, Ram Kumar RP, Saikiran B, Nagaraj I, Annavarapu T (2023) Real time weather monitoring system using IoT. In: E3S web of conferences, vol 391, EDP sciences. https://doi.org/10.1051/E3SCONF/202339101142
58. Alfieri L, Salamon P, Pappenberger F, Wetterhall F, Thielen J (2012) Operational early warning systems for water-related hazards in Europe. Environ Sci Policy 21:35–49. https://doi.org/10.1016/J.ENVSCI.2012.01.008
59. Neußner O (2021) Early warning alerts for extreme natural hazard events: a review of worldwide practices. Int J Disaster Risk Reduct 60. https://doi.org/10.1016/J.IJDRR.2021.102295. (Elsevier Ltd)

60. Lenton TM (2013) What early warning systems are there for environmental shocks? Environ Sci Policy 27:S60–S75. https://doi.org/10.1016/J.ENVSCI.2012.06.011
61. Mayhorn CB, McLaughlin AC (2014) Warning the world of extreme events: a global perspective on risk communication for natural and technological disaster. Saf Sci 61:43–50. https://doi.org/10.1016/J.SSCI.2012.04.014
62. Mesa-Jiménez JJ, Tzianoumis AL, Stokes L, Yang Q, Livina VN (2023) Long-term wind and solar energy generation forecasts, and optimisation of power purchase agreements. Energy Rep 9:292–302. https://doi.org/10.1016/j.egyr.2022.11.175
63. Yang W, Sparrow SN, Wallom DCH (2023) Optimising multi-factor assistance in a deep learning-based electricity forecasting model with climate resilience: an Australian case study. In: 2023 IEEE PES innovative smart grid technologies Europe (ISGT EUROPE). IEEE, pp 1–5. https://doi.org/10.1109/ISGTEUROPE56780.2023.10407524
64. Yan X, Abbes D, Francois B (2017) Uncertainty analysis for day ahead power reserve quantification in an urban microgrid including PV generators. Renew Energy 106:288–297. https://doi.org/10.1016/j.renene.2017.01.022
65. Eskandari H, Imani M, Moghaddam MP (2021) Convolutional and recurrent neural network based model for short-term load forecasting. Electric Power Syst Res 195:107173. https://doi.org/10.1016/J.EPSR.2021.107173
66. Wang O, Zhou S, Li GY (2024) ADM: batch ADMM for deep learning. https://doi.org/10.48550/arXiv.2407.01640
67. Hong T, Fan S (2016) Probabilistic electric load forecasting: a tutorial review. Int J Forecast 32:914–938. https://doi.org/10.1016/j.ijforecast.2015.11.011
68. Wang O, Zhou S, Li GY (2023) Few-shot learning for new environment adaptation. In: 2023 IEEE global communications conference (GLOBECOM). IEEE, pp 351–356. https://doi.org/10.1109/GLOBECOM54140.2023.10437273
69. Xiao H, Pei W, Wu L, Ma L, Ma T, Hua W (2023) A novel deep learning based probabilistic power flow method for Multi-Microgrids distribution system with incomplete network information. Appl Energy 335:120716. https://doi.org/10.1016/j.apenergy.2023.120716
70. Wang O, Zhou S, Li GY (2023) New environment adaptation with few shots for OFDM receiver and mmWave beamforming. https://doi.org/10.48550/arXiv.2310.12343
71. Kumar S, Hussain L, Banarjee S, Reza M (2018) Energy load forecasting using deep learning approach-LSTM and GRU in spark cluster. In: 2018 Fifth international conference on emerging applications of information technology (EAIT). IEEE, pp 1–4. https://doi.org/10.1109/EAIT.2018.8470406
72. Wang O, Gao J, Li GY (2023) Learn to adapt to new environments from past experience and few pilot blocks. IEEE Trans Cogn Commun Netw 373–385. https://doi.org/10.1109/TCCN.2022.3228536. (IEEE)
73. Wei X, Juang B-H, Wang O, Zhou S, Li GY (2024) Accretionary learning with deep neural networks with applications. IEEE Trans Cogn Commun Netw 10:660–673. https://doi.org/10.1109/TCCN.2023.3342454
74. Wang O, Zhou S, Li GY (2023) Effective adaptation into new environment with few shots: applications to OFDM receiver design. 2023 IEEE 33rd international workshop on machine learning for signal processing (MLSP). IEEE, pp 1–6. https://doi.org/10.1109/MLSP55844.2023.10285904
75. Yu B, Li J, Liu C, Sun B (2022) A novel short-term electrical load forecasting framework with intelligent feature engineering. Appl Energy 327:120089. https://doi.org/10.1016/j.apenergy.2022.120089
76. Hu X, Li B, Shi J, Li H, Liu G (2021) A novel forecasting method for short-term load based on TCN-GRU model. In: 2021 IEEE international conference on energy internet (ICEI). IEEE, pp 79–83. https://doi.org/10.1109/ICEI52466.2021.00020
77. Wang Y, Chen J, Chen X, Zeng X, Kong Y, Sun S et al (2021) Short-term load forecasting for industrial customers based on TCN-LightGBM. IEEE Trans Power Syst 36:1984–1997. https://doi.org/10.1109/TPWRS.2020.3028133

78. Yang W, Sparrow SN, Wallom DCH (2023) A generalised multi-factor deep learning electricity load forecasting model for wildfire-prone areas. https://doi.org/10.48550/arXiv.2304.10686
79. Yang W, Sparrow SN, Wallom DCH (2024) A comparative climate-resilient energy design: wildfire resilient load forecasting model using multi-factor deep learning methods. Appl Energy 368:123365. https://doi.org/10.1016/j.apenergy.2024.123365

Robust Multi-energy System Planning for Zero Energy Buildings

Zhi Wu and Qirun Sun

Abstract Zero Energy Buildings (ZEBs) will encounter timescale uncertainties during their life cycle. This chapter investigates a robust planning method for designing a multi-energy system (MES) in ZEBs. In order to ensure the efficient multi-energy supplies and achievement of ZEBs' yearly zero energy objective, uncertainties in both the short-term and long-term timescales are considered and addressed in a tri-level planning structure. The primary objective at the upper level is to determine an optimal device sizing strategy within a MES. In the middle level, a set of representative scenarios is generated from future temperature forecasts, which aims to tackle the long-term temperature changes. In the lower level, a hybrid stochastic and robust optimization (HSRO) method is applied to deal with source-demand uncertainties in a short-term timescale. A reformulation method is introduced to convert the tri-level model into a tractable form. Simulation results illustrate the effectiveness of the proposed method in optimizing the deployment of energy equipment in ZEBs and mitigating the impact of uncertainties across different timescales.

Nomenclature

ω	Index for weather condition scenarios, from 1 to N_ω
s	Index for seasonal typical day scenarios, from 1 to N_s
t	Index for intraday dispatch periods, from 1 to N_t
c	Index for candidate equipment installation capacities
i	Index for rooms in the ZEBs
$(\hat{\cdot})/(\tilde{\cdot})$	Index for short-term forecast/uncertain variables
$(\bar{\cdot})/(\underline{\cdot})$	Index for upper/lower bounds of uncertainty

Z. Wu (✉) · Q. Sun
School of Electrical Engineering, Southeast University, Nanjing, Jiangsu, China
e-mail: zwu@seu.edu.cn; 101012079@seu.edu.cn

Q. Sun
e-mail: 230218326@seu.edu.cn

W. Hua et al. (eds.), *Digitalisation of Local Energy Systems*, Green Energy and Technology, https://doi.org/10.1007/978-3-031-77833-9_8

ψ	Index for all the energy equipment
ψ_1	Index for energy storage equipment, including BS/HS/SHS/TS
c_{ψ}^{inv}	Unit investment cost of energy equipment
c_{ψ}^{om}	Unit operation and maintenance cost of energy equipment
$c_{st}^{\mathrm{grid+}}/c^{\mathrm{grid-}}$	Seasonal retail price/fixed feed-in tariff of electricity
$c_{\mathrm{on/off}}^{\mathrm{chp/ed}}$	Startup/shutdown cost of CHP/ED
$c_{\mathrm{bs/chp/ed}}^{\mathrm{deg}}$	Degradation cost of BS/CHP/ED
$A_{i,j}^{\mathrm{wall}}/A_{i,j}^{\mathrm{win}}$	Area of the wall/window
$C_i^{\mathrm{room}}/C_{i,j}^{\mathrm{wall}}$	Thermal capacity of the room/wall
Cap_c^{ψ}	Candidate installed capacity of energy equipment
$Cap_{\max}^{\psi}$	Maximum installed capacity of energy equipment
$d_{\langle w,w'\rangle}$	Euclidean distance between weather condition scenarios
D_s	Number of days in scenario s in one year
M	A relatively large positive number
N_y	Number of forecast data
l_S	Coefficient related to N_y
$N_{\min}^{\mathrm{chp/ed,on}}$	Minimum on-state time periods of CHP/ED
$N_{\min}^{\mathrm{chp/ed,off}}$	Minimum off-state time periods of CHP/ED
$N_{\max}^{\mathrm{chp/ed,on}}$	Maximum on-state time periods of CHP/ED
$P_{\max}^{\mathrm{grid}}$	Maximum exchange active power with power grid
$Q_{\max}^{\mathrm{grid}}$	Maximum exchange reactive power with power grid
$R_{i,j}^{\mathrm{wall}}/R^{\mathrm{win}}$	Thermal resistance of wall/window
$Ra_{\max}^{\mathrm{chp/ed}}$	Maximum ramping rate of CHP/ED
$T_{\mathrm{min/max}}^{\mathrm{room}}$	Minimum/maximum indoor temperature
ϕ	Present worth factor
σ/Y_{N}	Interest rate/length of system life cycle
Δt	Duration of intraday dispatch period
$\hat{\pi}_w$	Reference probability of future weather condition scenario w
$\theta_{i,j}^{\mathrm{wall}}/\theta_i^{\mathrm{win}}$	Heat absorption coefficient of wall/window
$\upsilon_{i,j}^{\mathrm{rad}}$	Identifier of sunlit wall
κ^{chp}	Heat recovery ratio of CHP
κ^{hp}	Heating power allocation ratio of HP
$\delta^{\mathrm{chp/ed}}$	Minimum operating ratio of CHP/ED
$\delta_{\mathrm{min/max}}^{\psi_1}$	Minimum/maximum operating ratio of energy storage equipment
$\varphi_{\mathrm{el/pv/wt}}$	Power factor of electric demand/PV/WT
ξ_{st}^{wt}	Forecast output factor of WT
$\eta^{\mathrm{ach/chp/ed/pv/st}}$	Efficiency of ACH/CHP/ED/PV/ST
$\eta_{\mathrm{h/c}}^{\mathrm{hp}}$	Heating/cooling efficiency of HP
$\eta^{\psi_1,+/-}$	Charge/discharge efficiency of energy storage equipment
η^{ψ_1}	Self-discharge rate of energy storage equipment
μ^{ψ_1}	Power-capacity ratio of energy storage equipment

Γ_s^{el}	Deviation budget of electric demand under scenario s
$x_c^{\mathrm{ach/chp/ed/hp/wt}}$	Binary, 1 if candidate cth ACH/CHP/ED/HP/WT is installed (kW), 0 otherwise
$x_c^{\mathrm{bs/hs/shs/ts}}$	Binary, 1 if candidate cth BS/HS/SHS/TS is installed (kWh), 0 otherwise
$x_c^{\mathrm{pv/st}}$	Binary, 1 if candidate cth PV/ST is installed (m^2), 0 otherwise
C^{inv}	Annual investment cost in ETH-MES
$C^{\mathrm{om/grid/deg}}$	Annual operation and maintenance/electricity trading/ degradation cost
$E_{wst}^{\psi_1}$	Energy stored in energy storage equipment
$I_{\mathrm{on/off}}^{\mathrm{chp}}$	Binary, 1 for CHP startup/shutdown, 0 otherwise
$I_{\mathrm{on/off}}^{\mathrm{ed,wst}}$	Binary, 1 for ED startup/shutdown, 0 otherwise
$P_{w,s,t}^{\mathrm{bs/hs/shs,+}}$	Charge power of BS/HS/SHS
$P_{w,s,t}^{\mathrm{bs/hs/shs,-}}$	Discharge power of BS/HS/SHS
$P_{w,s,t}^{\mathrm{chp,in/out}}$	Input hydrogen power/output electric power of CHP
$P_{w,s,t}^{\mathrm{ed,in/out}}$	Input electric power/output hydrogen power of ED
$P_{w,s,t}^{\mathrm{el}}$	Active power demand in ZEBs
$P_{w,s,t}^{\mathrm{grid+/-}}$	Active power buy from/sold to the power grid
$P_{w,s,t}^{\mathrm{hp/pv/wt}}$	Active power of HP/PV/WT
$Q_{w,s,t}^{\mathrm{ach/in/out}}$	Input heating/output cooling power of ACH
$Q_{w,s,t}^{\mathrm{chp/out}}$	Output heating power of CHP
$Q_{w,s,t}^{\mathrm{grid}}$	Exchanged reactive power with the power grid
$Q_{w,s,t}^{\mathrm{hl/cl}}$	Space heating/cooling demand
$Q_{w,s,t}^{\mathrm{hp,h/c}}$	Output heating/cooling power of HP
$Q_{w,s,t}^{\mathrm{hw}}$	Hot water demand
$Q_{w,s,t}^{\mathrm{pv/wt}}$	Reactive power of PV/WT
$Q_{w,s,t}^{\mathrm{rad},i,j}$	Radiative heating power on the wall
$Q_{w,s,t}^{\mathrm{st}}$	Heating power of ST collector
$Q_{w,s,t}^{\mathrm{ts+/-}}$	Charge/discharge heating power of TS
$Ra_{w,s,t}^{\mathrm{chp/ed}}$	Ramping ratio of CHP/ED
$S_{w,s,t}^{\mathrm{rad}}$	Solar radiation
$T_{w,s,t}^{\mathrm{room},i}$	Indoor temperature in room i
$T_{w,s,t}^{\mathrm{wall},i,j}$	Temperature of the wall between node i and j
$T_{w,s,t}^{\mathrm{out}}$	Outdoor temperature
$\epsilon_{\mathrm{wst}}^{\mathrm{chp/ed}}$	Binary, 1 for CHP/ED on-state, 0 otherwise
$\epsilon_{w,s,t}^{\mathrm{grid,+/-}}$	Binary, 1 for buying/selling power from/to power grid, 0 otherwise
$\epsilon_{w,s,t}^{\mathrm{shs,+/-}}$	Binary, 1 for SHS charge/discharge in one day, 0 otherwise
π_w	Actual probability of future weather condition w
$r_w^{+/-}$	Auxiliary variables for measuring the positive/negative deviation

$v_{ww'}^{+/-}$	Auxiliary variables for measuring the positive/negative transfer quantity from w to w'
$\zeta_{\mathrm{wst}}^{el+/-}$	Intraday uncertainty deviation sign for electric demand

1 Introduction

To mitigate energy consumption and reduce carbon emissions in the building sector, Zero Energy Buildings (ZEBs) have gained prominence globally [1]. ZEBs are green buildings that utilize renewable energy sources (RES) and trade zero net-metering electricity with the grid over a year [2]. Enhancing energy efficiency and optimizing the utilization of RES are crucial for successful ZEB design. Specifically, the integration of various energy equipment plays a pivotal role in enhancing the operation efficiency, making multi-energy systems (MES) a popular choice for addressing the energy challenges in ZEBs [3]. Notably, the optimal sizing of multiple energy equipment is vital to ensure cost-effective energy supplies for ZEBs.

During the life cycle, ZEBs encounter uncertainties across various timescales, including temperature fluctuations in the long-term timescale as well as RES and multi-energy demand variations in the short-term timescale [4, 5]. These above uncertainties profoundly impact ZEBs' energy performance and may hinder compliance with the yearly zero energy target. In daily operation, ZEBs contend with uncertainties and fluctuations such as electric demand, RES output, and ambient temperature [6]. Over the long term, the heating and cooling demand of climate-sensitive ZEBs are expected to be significantly influenced under the background of global warming [7].

Failure to comprehensively account for multi-timescale uncertainties can lead to significant disparities between the designed and actual performance of energy equipment [8]. Consequently, ZEBs might not achieve the yearly zero energy objective during their life cycle. This mismatch can cast doubts on the credibility of the ZEB concept and have far-reaching implications, undermining overall efforts in this direction [9]. Therefore, a comprehensive approach that addresses both short-term and long-term uncertainties is imperative to ensure the sustained success of ZEB initiatives [10].

This chapter presents a tri-level hybrid robust approach for planning MES in ZEBs. In particular, both long-term uncertainty and short-term uncertainty are handled in this model to ensure a reliable and robust sizing scheme. This scheme is crucial for providing a reliable energy supply and meeting the yearly zero energy objective across the buildings' life cycle. At the long-term level, a distributionally robust optimization approach is employed. This approach utilizes forecast datasets to model uncertainties related to long-term temperature changes. Concurrently, the short-term level addresses uncertainties from RES and multi-energy demand under various seasonal and daily patterns using a hybrid stochastic and robust optimization (HSRO) approach. To tackle the computational complexity, the proposed tri-level model is then reconstructed to a tractable and identical form and solved by an enhanced C&CG-AOP algorithm.

2 Modeling of ZEBs and Tri-Level Planning Framework

2.1 ZEB Structure with ETH-MES

As is shown in Fig. 1, the electric, thermal, and hydrogen integrated MES (ETH-MES) functions as a localized energy infrastructure to a ZEB community [11]. Its core components comprise diverse energy generation, conversion, and storage equipment. Electrically, the system incorporates photovoltaic (PV) panels, wind turbines (WT), and fuel cell-based μCHP enhance the availability of electrical power. In terms of thermal flow, the system integrates a heat pump (HP), heating recovery (HR) equipment, and a solar thermal (ST) panel to meet the thermal demand. Absorption chiller (ACH) and HP can convert the electrical and thermal energy to meet the cooling demands. Additionally, hydrogen can power the μCHP system. The electrolyzer device (ED) facilitates hydrogen production using surplus renewable energy sources (RES). The produced hydrogen is then stored in hydrogen storage equipment. In the ETH-MES, diverse energy storage equipment including battery storage (BS), thermal storage (TS), hydrogen storage (HS), and seasonal hydrogen storage (SHS) is strategically equipped to improve the reliability and flexibility of the energy system. Hydrogen serves as the medium of RES; it can mitigate the RES fluctuations in daily operations [12].

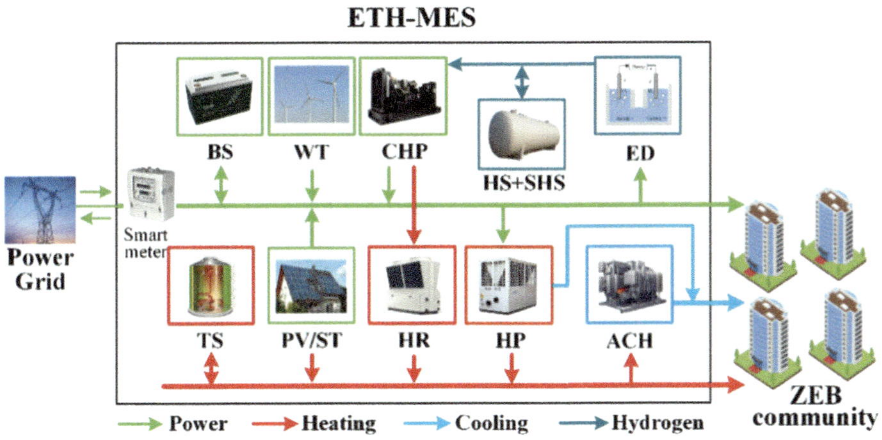

Fig. 1 Structure of the ETH-MES for a ZEB community

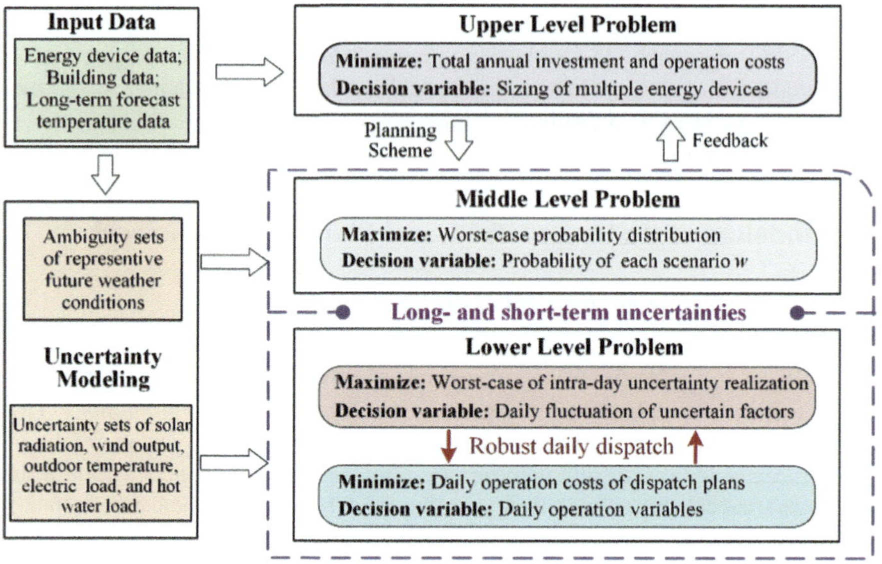

Fig. 2 Tri-level hybrid robust planning framework

The presented planning framework in Fig. 2 adopts a tri-level hybrid robust structure. The objective function of the planning model is to minimize the present value of the total planning costs, including investment and operation costs. Specifically, at the upper level, the model makes strategic decisions related to the allocation of capacities for each energy equipment. It aims to immunize the system against the impact of both long-term uncertainty and short-term uncertainty. In the middle level, the long-term forecast dataset related to yearly ambient temperature is utilized to address the uncertain future weather conditions. In the lower level, it deals with the operational challenges of ZEBs under short-term operation uncertainties.

$$\min_{x} C^{\text{inv}} + \sup_{\pi \in \mathcal{P}} \mathbb{E}_{\pi_w} \left[\max_{\boldsymbol{u}_w \in \mathcal{U}_w} \min_{\boldsymbol{y}_w, \boldsymbol{z}_w \in \Omega(x, \boldsymbol{u}_w)} \left(C_w^{\text{om}} + C_w^{\text{grid}} + C_w^{\text{deg}} \right) \right] \tag{1}$$

Upper Level: In Eq. (2), $\boldsymbol{x}$ denotes the investment decisions for energy equipment.

$$\boldsymbol{x} = \left(\boldsymbol{x}^{\text{ach}}, \boldsymbol{x}^{\text{bs}}, \boldsymbol{x}^{\text{chp}}, \boldsymbol{x}^{\text{ed}}, \boldsymbol{x}^{\text{hp}}, \boldsymbol{x}^{\text{hs}}, \boldsymbol{x}^{\text{pv}}, \boldsymbol{x}^{\text{wt}}, \boldsymbol{x}^{\text{shs}}, \boldsymbol{x}^{\text{st}}, \boldsymbol{x}^{\text{ts}} \right) \tag{2}$$

Middle Level: The planning model utilizes the forecast dataset to create a typical weather condition scenario set, which can capture both typical and extreme future weather conditions. In the middle level, the objective is to minimize the operation cost under uncertain weather conditions in the system life cycle. The term "sup" is used to resist the worst situation of the probability distribution π of the scenario

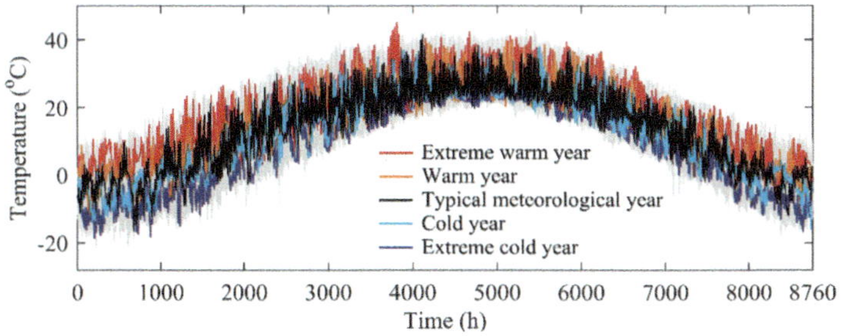

Fig. 3 Hourly temperature profiles of typical scenarios.

set. Specifically, hourly temperature during the system life cycle under four representative evolution pathways is adopted as the reference dataset. From this, five typical scenarios ($N_w = 5$) are derived, i.e., extreme warm year, warm year, typical meteorological year, cold year, and extreme cold year. Figure 3 illustrates the hourly temperature profiles sourced from the dataset, with the light-gray lines representing the original dataset, and the highlighted lines indicating the generated representative scenarios.

A data-driven DRO method [13] is employed to model long-term uncertainty of temperature change. This approach enables the fluctuations of the uncertain probability distribution within a confidence set.

$$P = \left\{ \pi \in [0,1]^{N_w} \;\middle|\; \begin{array}{c} \sum_w \pi_w = 1, \\ L_1 = \sum_w (r_w^+ + r_w^-) \le \alpha_1,\; L_\infty = \max_w (r_w^+ + r_w^-) \le \alpha_\infty, \\ \hat{\pi}_w + r_w^+ - r_w^- = \pi_w,\; \hat{\pi}_w + \sum_{w'} (v_{ww'}^+ - v_{ww'}^-) = \pi_w, \forall w, \\ L_W = \sum_{(w,w')} d_{<w,w'>} \cdot (v_{ww'}^+ + v_{ww'}^-) \le \alpha_W, \\ r_w^+, r_w^-, v_{ww'}^+, v_{ww'}^- \ge 0, \forall w, w' \end{array} \right\} \tag{3}$$

In Eq. (3), the first constraint ensures the sum of probabilities is equal to 1. The second to fourth constraints employ non-negative variables to quantify the 1-norm and ∞-norm distance. Additionally, in the fifth and sixth equations, the Wasserstein distance is restricted. These constraints impose a stricter limit on π, effectively avoiding the overestimation of long-term uncertainty. At the confidence level β, the upper bound for each norm distance is defined in Eq. (4).

$$\alpha_1 = \frac{N_w}{2N_y} \ln \frac{2N_w}{1-\beta}; \quad \alpha_\infty = \frac{1}{2N_y} \ln \frac{2N_w}{1-\beta^j}, \quad \alpha_w = l_s \sqrt{\frac{2}{N_y} \cdot \log \frac{1}{1-\beta}} \tag{4}$$

Lower Level: In each long-term scenario, the HSRO method is applied to minimize the operation cost [14]. The uncertain factors, solar and wind energy, ambient temperature, electric demand, and hot water demand, are considered and modeled.

The robust method employs a "max-min" structure to address short-term fluctuations [15]. Specifically, the "max" identifies the worst-case scenario with the highest daily operational costs, and the "min" determines the scheduling results with minimal operation costs. In Eq. (5), $\boldsymbol{u}_w$ denotes the short-term uncertain variable set, including the electric demand, wind energy, solar radiation, ambient temperature, and hot water demand. A polyhedral set Eq. (6) is defined for each variable to control the uncertainty degree.

$$\boldsymbol{u}_w = \{\tilde{\boldsymbol{P}}_w^{\text{el}}, \tilde{\boldsymbol{P}}_w^{\text{wt}}, \tilde{\boldsymbol{S}}_w^{\text{rad}}, \tilde{\boldsymbol{T}}_w^{\text{out}}, \tilde{\boldsymbol{Q}}_w^{\text{hw}}\} \tag{5}$$

$$\begin{aligned}\mathcal{U}_w^{\text{el}} = \Big\{\tilde{\boldsymbol{P}}_w^{\text{el}} \in \mathbb{R}^{N_s \times N_t} \mid \tilde{P}_{wst}^{\text{el}} &= \hat{P}_{wst}^{\text{el}} + P_{wst}^{\text{el+}} \varsigma_{wst}^{\text{el+}} - P_{wst}^{\text{el}} \varsigma_{wst}^{\text{el-}},\\ \varsigma_{wst}^{\text{el+/-}} &= \{0, 1\},\\ \sum_{t=1}^{N_t} (\varsigma_{wst}^{\text{el+}} &+ \varsigma_{wst}^{\text{el-}}) \le \Gamma_s^{\text{el}}\Big\} \quad \forall w, s, t\end{aligned} \tag{6}$$

Besides, $\boldsymbol{y}_w$ denotes the continuous operating variable set. $\boldsymbol{P}$ denotes active power and hydrogen flow variables. $\boldsymbol{Q}$ denotes reactive power as well as thermal and cooling power variables. $\boldsymbol{Ra}$ denotes equipment ramping variables. $\boldsymbol{E}$ denotes energy storage variables. $\boldsymbol{Z}_w$ denotes the 0-1 operating variable set. $\boldsymbol{I}$ denotes equipment startup and shutdown variables. $\boldsymbol{\varepsilon}$ denotes equipment on and off status variables.

$$\boldsymbol{y}_w = \begin{bmatrix} \boldsymbol{P}_w^{\text{bs+/-}}, \boldsymbol{P}_w^{\text{chp,in/out}}, \boldsymbol{P}_w^{\text{ed,in/out}}, \boldsymbol{P}_w^{\text{grid+/-}}, \boldsymbol{P}_w^{\text{hp}}, \boldsymbol{P}_w^{\text{hs+/-}}, \boldsymbol{P}_w^{\text{shs+/-}}, \\ \boldsymbol{Q}_w^{\text{pv}}, \boldsymbol{Q}_w^{\text{wt}}, \boldsymbol{Q}_w^{\text{grid}}, \boldsymbol{Q}_w^{\text{chp,out}}, \boldsymbol{Q}_w^{\text{ach,in/out}}, \boldsymbol{Q}_w^{\text{ed}}, \boldsymbol{Q}_w^{\text{hp,c/h}}, \boldsymbol{Q}_w^{\text{st}}, \\ \boldsymbol{Q}_w^{\text{ts+/-}}, \boldsymbol{Ra}_w^{\text{chp}}, \boldsymbol{Ra}_w^{\text{ed}}, \boldsymbol{E}_w^{\text{bs}}, \boldsymbol{E}_w^{\text{hs}}, \boldsymbol{E}_w^{\text{hs}}, \boldsymbol{E}_w^{\text{shs}} \end{bmatrix} \tag{7}$$

$$z_w = \{\boldsymbol{I}_w^{\text{chp,on}}, \boldsymbol{I}_w^{\text{chp,off}}, \boldsymbol{I}_w^{\text{ed,on}}, \boldsymbol{I}_w^{\text{ed,off}}, \boldsymbol{\varepsilon}_w^{\text{chp}}, \boldsymbol{\varepsilon}_w^{\text{ed}}, \boldsymbol{\varepsilon}_w^{\text{grid+/-}}, \boldsymbol{\varepsilon}_w^{\text{shs+/-}}\} \tag{8}$$

3 Mathematical Formulation for Tri-Level Planning Model

In this section, the detailed formulation of the tri-level hybrid robust planning model is provided.

3.1 Objective Function

Detailed components in the objective function are given as follows. Equations (9)–(10) compute the investment cost encompassing all the energy equipment. In Eq. (11), the capital recovery factor ϕ is calculated. Equation (12) denotes the operation and maintenance (O&M) cost attributed to all the energy equipment. Equation (13) denotes the degradation cost for battery storage, μCHP, and electrolyzer device. In Eq. (14), the annual electricity trading cost is formulated.

$$C_{\psi}^{\text{inv}} = c_{\psi}^{\text{inv}} x_c^{\psi} \text{Cap}_c^{\psi} \phi \tag{9}$$

$$\begin{aligned} C^{\text{inv}} = (C_{\text{ach}}^{\text{inv}} + C_{\text{bs}}^{\text{inv}} + C_{\text{chp}}^{\text{inv}} + C_{\text{ed}}^{\text{inv}} + C_{\text{hp}}^{\text{inv}} + C_{\text{hs}}^{\text{inv}} + C_{\text{pv}}^{\text{inv}} + \\ C_{\text{shs}}^{\text{inv}} + C_{\text{st}}^{\text{inv}} + C_{\text{ts}}^{\text{inv}} + C_{\text{wt}}^{\text{inv}}) \end{aligned} \tag{10}$$

$$\phi = \frac{\sigma(1+\sigma)^{\text{Y}_\text{N}}}{(1+\sigma)^{\text{Y}_\text{N}} - 1} \tag{11}$$

$$C_w^{\text{om}} = \sum_s D_s \sum_t \left\{ \begin{array}{l} c_{\text{chp}}^{\text{on}} I_{wst}^{\text{chp}} + c_{\text{chp}}^{\text{off}} I_{wst}^{\text{chp}} + c_{\text{ed}}^{\text{on}} I_{wst}^{\text{ed}} + c_{\text{ed}}^{\text{off}} I_{wst}^{\text{ed}} + c_{\text{chp}}^{\text{om}} P_{wst}^{\text{chp,in}} + \\ c_{\text{bs}}^{\text{om}} (P_{wst}^{\text{bs+}} + P_{wst}^{\text{bs-}}) + c_{\text{ed}}^{\text{om}} P_{wst}^{\text{hp}} + c_{\text{hp}}^{\text{om}} P_{wst}^{\text{hp}} + c_{\text{pv}}^{\text{om}} P_{wst}^{\text{pv}} + \\ c_{\text{wt}}^{\text{om}} P_{wst}^{\text{wt}} + c_{\text{hs}}^{\text{om}} \left(P_{wst}^{\text{hs+}} + P_{wst}^{\text{hs-}}\right) + c_{\text{shs}}^{\text{om}} \left(P_{wst}^{\text{shs+}} + P_{wst}^{\text{shs-}}\right) + \\ c_{\text{ac}}^{\text{om}} Q_{wst}^{\text{ach}} + c_{\text{st}}^{\text{om}} Q_{wst}^{\text{st}} + c_{\text{ts}}^{\text{om}} \left(Q_{wst}^{\text{ts+}} + Q_{wst}^{\text{ts-}}\right) \end{array} \right\} \Delta t \tag{12}$$

$$C_w^{\text{deg}} = \sum_s D_s \sum_t \left(c_{\text{bs}}^{\text{deg}} \left(p_{wst}^{\text{bs+}} + p_{wst}^{\text{bs-}}\right) + c_{\text{chp}}^{\text{deg}} Ra_{wst}^{\text{chp}} + c_{\text{ed}}^{\text{deg}} Ra_{ws}^{\text{ed}} \right) \Delta t \tag{13}$$

$$C_w^{\text{grid}} = \sum_s D_s \sum_t \left(c_{st}^{\text{buy}} P_{wst}^{\text{grid+}} - c^{\text{sell}} P_{wst}^{\text{grid}-} \right) \Delta t \tag{14}$$

3.2 Constraints

Investment Constraints: In Eq. (15), the installed capacity for each equipment is restricted.

$$0 \le x_c^{\psi} \text{Cap}_c^{\psi} \le \text{Cap}^{\psi,\max} \tag{15}$$

Energy Balance Constraints: Equations (16) and (17) denote the active and reactive power balance constraints, respectively. The heating and cooling power balance are constrained in Eqs. (18) and (19), respectively.

$$\begin{aligned} \widetilde{P}_{wst}^{\text{pv}} + \widetilde{P}_{wst}^{\text{wt}} + P_{wst}^{\text{bs-}} - P_{wst}^{\text{bs+}} + P_{wst}^{\text{chp,out}} + P_{wst}^{\text{grid+}} - \\ P_{wst}^{\text{grid-}} - P_{wst}^{\text{hp}} - P_{wst}^{\text{ed,in}} = \widetilde{P}_{wst}^{\text{el}}; \forall w, s, t \end{aligned} \tag{16}$$

$$Q_{wst}^{\text{pv}} + Q_{wst}^{\text{wt}} + Q_{wst}^{\text{grid}} = \widetilde{P}_{wst}^{\text{el}} \tan(\cos^{-1}(\phi_{\text{el}})); \forall w, s, t \tag{17}$$

$$Q_{wst}^{\text{hp,h}} + Q_{wst}^{\text{chp,out}} + \widetilde{Q}_{wst}^{\text{st}} - Q_{wst}^{\text{ts+}} + Q_{wst}^{\text{ts-}} - Q_{wst}^{\text{ach,in}} = \widetilde{Q}_{wst}^{\text{hl}} + \widetilde{Q}_{wst}^{\text{hw}}; \forall w, s, t \tag{18}$$

$$Q_{wst}^{\text{ac,out}} + Q_{wst}^{\text{hp,c}} = \widetilde{Q}_{wst}^{\text{cl}}; \forall w, s, t \tag{19}$$

In Eq. (20), the hydrogen flow balance is formulated, where $P_{wst}^{\text{ed,out}}$ indicates the hydrogen power generated by ED, $P_{wst}^{\text{hs+}}/P_{wst}^{\text{hs-}}$ represent the charging and discharging power of HS, $P_{wst}^{\text{shs+}}/P_{wst}^{\text{shs-}}$ represent the charging and discharging power of SHS, $P_{wst}^{\text{chp,in}}$ represent the CHP input power.

$$P_{wst}^{\text{ed,out}} + P_{wst}^{\text{shs-}} - P_{wst}^{\text{shs+}} + P_{wst}^{\text{hs-}} - P_{wst}^{\text{hs+}} = P_{wst}^{\text{chp,in}}; \forall w, s, t \tag{20}$$

Electricity Trading Constraints: Equation (21) establishes the boundaries for power transactions with main grid. Equation (22) enforces that ZEBs cannot simultaneously buy and sell electricity. Equation (23) represents that the annual electricity injected into the grid cannot exceed the electricity buying from the grid. Equation (24) imposes a bound on the transaction reactive power.

$$0 \le P_{wst}^{\text{grid+}} \le \epsilon_{wst}^{\text{grid+}} P_{\max}^{\text{grid}}, 0 \le P_{wst}^{\text{grid-}} \le \epsilon_{wst}^{\text{grid-}} P_{\max}^{\text{grid}}; \forall w, s, t \tag{21}$$

$$\epsilon_{wst}^{\text{grid+}} + \epsilon_{wst}^{\text{grid-}} \le 1; \forall w, s, t \tag{22}$$

$$\sum_{s}\sum_{t} P_{wst}^{\text{grid+}} \Delta t - \sum_{s}\sum_{t} P_{wst}^{\text{grid-}} \Delta t \le 0; \forall w \tag{23}$$

$$-Q_{\max}^{\text{grid}} \le Q_{wst}^{\text{grid}} \le Q_{\max}^{\text{grid}}; \forall w, s, t \tag{24}$$

CHP and ED Operational Constraints: Equation (25) defines the specific startup and shutdown conditions for CHP and ED. The starting and turnoff status of them are limited in Eqs. (26)–(29). Equation (30) restricts the input power of these two equipment, and Eq. (31) restricts the ramping rates of them. In Eq. (32), the relationships between the input power and output power of these two equipment are formulated. In Eq. (33), the recovered thermal energy of CHP is formulated.

$$I_{wst}^{\{\cdot\},\text{on}} + I_{wst}^{\{\cdot\},\text{off}} \le 1, \{\cdot\} = \{\text{chp,ed}\}; \forall w, s, t \tag{25}$$

$$\sum_{t=k}^{k+N_{\min}^{\{\cdot\},\text{on}}-1} I_{wst}^{\{\cdot\},\text{on}} \le 1, \sum_{t=k}^{k+N_{\min}^{\{\cdot\},\text{on}}-1} \left(I_{wst}^{\{\cdot\},\text{on}} + I_{wst}^{\{\cdot\},\text{off}}\right) \le 1, \{\cdot\} = \{\text{chp, ed}\}; \\ \forall w, s, k \in [1, N_t - N_{\min}^{\{\cdot\},\text{on}} + 1] \tag{26}$$

$$\sum_{t=k}^{k+N_{\min}^{\{\cdot\},\text{off}}-1} I_{wst}^{\{\cdot\},\text{off}} \leq 1, \sum_{t=k}^{k+N_{\min}^{\{\cdot\},\text{off}}-1} \left(I_{wst}^{\{\cdot\},\text{on}} + I_{wst}^{\{\cdot\},\text{off}}\right) \leq 1, \{\cdot\} = \{\text{chp}, \text{ed}\}; \quad \forall w, s, k \in [1, N_t - N_{\min}^{\{\cdot\},\text{off}} + 1] \tag{27}$$

$$\sum_{t=k}^{k+N_{\max}^{\{\cdot\},\text{on}}} \epsilon_{wst}^{\{\cdot\}} \leq N_{\max}^{\{\cdot\},\text{on}}, \{\cdot\} = \{\text{chp}, \text{ed}\}; \forall w, s, k \in [1, N_t - N_{\max}^{\{\cdot\},\text{on}}] \tag{28}$$

$$I_{wst}^{\{\cdot\},\text{on}} - I_{wst}^{\{\cdot\},\text{off}} = \epsilon_{wst}^{\{\cdot\}} - \epsilon_{ws,t-1}^{\{\cdot\}}, \{\cdot\} = \{\text{chp}, \text{ed}\}; \forall w, s, t \tag{29}$$

$$\delta^{\{\cdot\}} \epsilon_{wst}^{\{\cdot\}} x_c^{\{\cdot\}} Cap_c^{\{\cdot\}} \leq P_{wst}^{\{\cdot\},\text{in}} \leq \epsilon_{wst}^{\{\cdot\}} x_c^{\{\cdot\}} Cap_c^{\{\cdot\}}, \{\cdot\} = \{\text{chp}, \text{ed}\}; \quad \forall w, s, t \tag{30}$$

$$P_{wst}^{\{\cdot\},\text{in}} - P_{ws,t-1}^{\{\cdot\},\text{in}} \leq Ra_{wst}^{\{\cdot\}} \leq Ra_{\max}^{\{\cdot\}}, \quad P_{ws,t-1}^{\{\cdot\},\text{in}} - P_{wst}^{\{\cdot\},\text{in}} \leq Ra_{wst}^{\{\cdot\}} \leq Ra_{\max}^{\{i\}}, \quad \{\cdot\} = \{\text{chp}, \text{ed}\}; \forall w, s, t \tag{31}$$

$$P_{wst}^{\{\cdot\},\text{out}} = \eta^{\{\cdot\}} P_{wst}^{\{\cdot\},\text{in}}, \{\cdot\} = \{\text{chp}, \text{ed}\}; \forall w, s, t \tag{32}$$

$$Q_{wst}^{\text{chp,out}} = \kappa^{\text{chp}} (P_{wst}^{\text{chp,in}} - P_{wst}^{\text{chp,out}}); \forall w, s, t \tag{33}$$

ACH and HP Operational Constraints: Equation (34) limits the ACH and HP input thermal power. In Eqs. (35) and (36), the energy conversion relationship of ACH and HP is formulated, respectively.

$$0 \leq Q_{wst}^{\text{ach,in}} \leq x_c^{\text{ach}} Cap_c^{\text{ach}}, 0 \leq P_{wst}^{\text{hp}} \leq x_c^{\text{hp}} Cap_c^{\text{hp}}; \forall w, s, t \tag{34}$$

$$Q_{wst}^{\text{ach,out}} = Q_{wst}^{\text{ach,in}} \eta^{\text{ach}}; \forall w, s, t \tag{35}$$

$$Q_{wst}^{\text{hp,h}} = \kappa^{\text{hp}} P_{wst}^{\text{hp}} \eta_{\text{h}}^{\text{hp}}, Q_{wst}^{\text{hp,c}} = (1 - \kappa^{\text{hp}}) P_{wst}^{\text{hp}} \eta_c^{\text{hp}}; \forall w, s, t \tag{36}$$

PV, ST, and WT Operational Constraints: The output power of PV, ST, and WT are formulated in Eqs. (37)–(39), respectively. Equation (40) defines the reactive output power boundaries for PV and WT.

$$0 \leq \widetilde{P}_{wst}^{\text{PV}} \leq \eta^{\text{PV}} x_c^{\text{PV}} Cap_c^{\text{PV}} \widetilde{S}_{wst}^{\text{rad}}; \forall w, s, t \tag{37}$$

$$0 \leq \widetilde{Q}_{wst}^{st} \leq \eta^{\text{st}} x_c^{\text{st}} Cap_c^{\text{st}} \widetilde{S}_{\text{rad}}^{wst}; \forall w, s, t \tag{38}$$

$$0 \leq \widetilde{P}_{wst}^{\mathrm{wt}} \leq \widetilde{\xi}_{wst}^{\mathrm{wt}} x_c^{\mathrm{wt}} Cap_c^{\mathrm{wt}}; \forall w, s, t \tag{39}$$

$$-\widetilde{P}_{wst} \tan(\cos^{-1}\phi_{\cdot}) \leq Q_{wst} \leq \widetilde{P}_{wst} \tan(\cos^{-1}\phi_{\cdot}), \quad \{\cdot\} = \{\mathrm{pv}, \mathrm{wt}\}; \forall w, s, t \tag{40}$$

Energy Storage Equipment Operational Constraints: Multi-type energy storage equipment is considered in the energy system. Equations (41)–(42) define the maximum charge and discharge power limits. Equations (43)–(44) denote that the charge and discharge state cannot simultaneously occur. Equation (45) defines the state of charge (SOC) boundary for each device. Equation (46) enforces uniformity in initial and terminal SOC to maintain consistency in each day. Equation (47) defines the SOC dynamics in daily operation. Eq. (48) defines the SOC status of SHS in the seasonal timescale. Equation (49) formulates the exclusive charging and discharging status. Equation (50) formulates the relationship between the charging/discharging power and equipment states.

$$0 \leq P_{wst}^{\{\cdot\},+/-} \leq x_c^{\{\cdot\}} Cap_c^{\{\cdot\}} \mu^{\{\cdot\}}, \{\cdot\} = \{\mathrm{bs}, \mathrm{hs}, \mathrm{shs}\}; \forall w, s, t \tag{41}$$

$$0 \leq Q_{wst}^{\mathrm{ts},+/-} \leq x_c^{ts} Cap_c^{\mathrm{ts}} \mu^{\mathrm{ts}}; \forall w, s, t \tag{42}$$

$$P_{wst}^{\{\cdot\}+} \cdot P_{wst}^{\{\cdot\}-} = 0, \{\cdot\} = \{\mathrm{bs}, \mathrm{hs}, \mathrm{shs}\}; \forall w, s, t \tag{43}$$

$$Q_{wst}^{\mathrm{ts}+} \cdot Q_{wst}^{\mathrm{ts}-} = 0; \forall w, s, t \tag{44}$$

$$x_c^{\{\cdot\}} Cap_c^{\{\cdot\}} \delta_{\min}^{\{\cdot\}} \leq E_{wst}^{\{\cdot\}} \leq x_c^{\{\cdot\}} Cap_c^{\{\cdot\}} \delta_{\max}^{\{\cdot\}}, \quad \{\cdot\} = \{\mathrm{bs}, \mathrm{hs}, \mathrm{shs}, \mathrm{ts}\}; \forall w, s, t \tag{45}$$

$$E_{ws,N_t}^{\{\cdot\}} = E_{ws,0}^{\{\cdot\}} = \left(\chi_c Cap_c^{\{\cdot\}}\right)/2, \quad \{\cdot\} = \{\mathrm{bs}, \mathrm{hs}, \mathrm{shs}, \mathrm{ts}\}; \forall w, s, t \tag{46}$$

$$E_{ws,t+1}^{\{\cdot\}} = E_{ws,t}^{\{\cdot\}}(1 - \eta^{\{\cdot\}}) + \left(P_{wst}^{\{\cdot\}+} \eta^{\{\cdot\}+} - P_{wst}^{\{\cdot\}-} \eta^{\{\cdot\}-}\right) \Delta t, \quad \{\cdot\} = \{\mathrm{bs}, \mathrm{hs}, \mathrm{shs}, \mathrm{ts}\}; \forall w, s, t \tag{47}$$

$$E_{w,s,0}^{\mathrm{shs}} = \left(1 - \eta^{\mathrm{shs}} D_{s-1}\right) \left(E_{w,s-1,0}^{\mathrm{shs}} + D_{s-1}\left(E_{w,s-1,N_t}^{\mathrm{shs}} - E_{w,s-1,0}^{\mathrm{shs}}\right)\right); \quad \forall w, s \in [2, N_s] \tag{48}$$

$$\epsilon_{ws}^{\mathrm{shs}+} + \epsilon_{ws}^{\mathrm{shs}-} \leq 1; \forall w, s \tag{49}$$

$$0 \leq P_{wst}^{\mathrm{shs}+} \leq \epsilon_{ws}^{\mathrm{shs}+} M, 0 \leq P_{wst}^{\mathrm{shs}-} \leq \epsilon_{ws}^{\mathrm{shs}-} M; \forall w, s, t \tag{50}$$

Indoor temperature constraint: Regarding the demand side, a Resistor-Capacitor (RC) thermal network model [16] is employed to model building's thermal dynamics, as depicted in Eqs. (51) and (52). In practice, the indoor temperature can fluctuate within a comfortable range, as is presented in Eq. (53).

$$C_1^{\text{room}} \frac{dT_{wst}^{\text{room},1}}{dt} = \sum_{j=2}^{5} \frac{(T_{wst}^{\text{wall}_1,j} - T_{wst}^{\text{room},1})}{R_{1,j}^{\text{wall}}} + \frac{(\widetilde{T}_{wst}^{\text{out}} - T_{wst}^{\text{room},1})}{R^{\text{win}}} + \theta^{\text{win}} A^{\text{win}} \widetilde{S}_{wst}^{\text{rad},1,3} + \widetilde{Q}_{wst}^{\text{hl},1} - \widetilde{Q}_{wst}^{\text{cl},1}; \forall w, s, t \tag{51}$$

$$C_{i,j}^{\text{wall}} \frac{dT_{wst}^{\text{wall},i,j}}{dt} = \frac{T_{wst}^{\text{room},i,j} - T_{wst}^{\text{wall},i,j}}{R_{i,j}^{\text{wall}}} + \frac{T_{wst}^{j} - T_{wst}^{\text{wall},i,j}}{R_{i,j}^{\text{wall}}} + \nu_{i,j}^{\text{rad}} \theta_{i,j}^{\text{wall}} A_{i,j}^{\text{wall}} \widetilde{S}_{wst}^{\text{rad},i,j}; \forall (i,j), w, s, t \tag{52}$$

$$T_{\min}^{\text{room}} \le T_{wst}^{\text{room},i} \le T_{\max}^{\text{room}}; \forall w, s, t, i \tag{53}$$

4 Solution Methodology

The proposed model shows a complicated tri-level coupling structure, and it cannot be effectively solved using commercial solvers. In this section, the methodology for solving the planning model is outlined.

4.1 Linearization of Non-convexity Constraints

The non-convexity in Eqs. (43), (44), (51), (52) makes the model computationally intractable, and the linearization methods are utilized to convert them.

Relaxation of Bilinear Term: The bilinear terms (43)–(44) can be relaxed directly without changing the optimality of the solution. A comprehensive proof is available in [17].

Differential Equation: The differential equations Eqs. (51)–(52) are converted into discrete difference state models, as follows.

$$C_1^{\text{room}} (T_{wst,t+1}^{\text{room}1} - T_{wst}^{\text{room},1}) = \Delta t \left(\sum_{j=2}^{5} \frac{(T_{wst}^{\text{wall},1,j} - T_{wst}^{\text{room},1})}{R_{1,j}^{\text{wall}}} + \frac{(\widetilde{T}_{wst}^{\text{out}} - T_{wst}^{\text{room},1})}{R^{\text{win}}} + \theta^{\text{win}} A^{\text{win}} \widetilde{S}_{wst}^{\text{rad},1,3} + \widetilde{Q}_{wst}^{\text{hl},1} - \widetilde{Q}_{wst}^{\text{cl},1} \right); \quad \forall w, s, t \tag{54}$$

$$C_{i,j}^{\text{wall}} T_{ws,t+1}^{\text{wall},i,j} = \left(C_{i,j}^{\text{wall}} - \frac{2\Delta t}{R_{i,j}^{\text{wall}}} \right) T_{wst}^{\text{wall},i,j} + \Delta t \left(\frac{T_{wst}^{\text{room}_i} + T_{wst}^{j}}{R_{i,j}^{\text{wall}}} + \nu_{i,j}^{\text{rad}} \theta_{i,j}^{\text{wall}} A_{i,j}^{\text{wall}} \widetilde{S}_{wst}^{\text{rad},i,j} \right); \forall w, s, t \tag{55}$$

4.2 Improved C&CG-AOP Algorithm

For brevity, the proposed tri-level model is expressed in a compact form.

$$\min_{x} \boldsymbol{a}^T \boldsymbol{x} + \sup_{\pi \in P} \mathbb{E}_{\pi_w} \left[\max_{u_w \in \mathcal{U}_w} \min_{y_w, z_w \in \Omega(x, u_w)} \boldsymbol{c}^T \boldsymbol{y}_w + \boldsymbol{d}^T \boldsymbol{z}_w \right] \tag{56a}$$

$$\text{s.t. } \boldsymbol{b}^T \boldsymbol{x} \le \boldsymbol{q}, \boldsymbol{x} \in \{0, 1\} \tag{56b}$$

$$\boldsymbol{k}^T (\boldsymbol{\pi} - \hat{\boldsymbol{\pi}}) = \boldsymbol{i}^T \boldsymbol{r} + \boldsymbol{j}^T \boldsymbol{v} \tag{56c}$$

$$\boldsymbol{m}^T \boldsymbol{r} + \boldsymbol{n}^T \boldsymbol{v} \le \boldsymbol{\alpha} \tag{56d}$$

$$\boldsymbol{e}^T \boldsymbol{y}_w + \boldsymbol{f}^T \boldsymbol{z}_w + \boldsymbol{g}^T \boldsymbol{u}_w \le \boldsymbol{l}_w - \boldsymbol{h}^T \boldsymbol{x}, \boldsymbol{z}_w \in \{0, 1\} \tag{56e}$$

where $\boldsymbol{\alpha}$, $\boldsymbol{q}$, $\boldsymbol{a}$, $\boldsymbol{b}$, $\boldsymbol{c}$, $\boldsymbol{d}$, $\boldsymbol{e}$, $\boldsymbol{f}$, $\boldsymbol{g}$, $\boldsymbol{h}$, $\boldsymbol{i}$, $\boldsymbol{j}$, $\boldsymbol{k}$, $\boldsymbol{m}$, $\boldsymbol{n}$ and $\hat{\boldsymbol{\pi}}$ denote the coefficient matrices. Equation (56b) denotes the investment constraints, corresponding to Eq. (15). Equations (56c)–(56d) represents the long-term uncertainty constraints Eq. (3). Equation (56e) denotes the operational constraints, corresponding to Eqs. (16)–(42), (45)–(50), and (53)–(55).

In general, the proposed tri-level model is computationally intractable and cannot be efficiently solved by existing commercial solvers. To address this, the original model is reformulated into a tractable form based on the inner coupling tri-level relationship. As depicted in Eq. (56), each long-term scenario w corresponds to a low-level subproblem (LLS) with "max-min" structure as in Eq. (57).

$$\text{(LLS)} \quad \max_{u_w \in \mathcal{U}_w} \min_{y_w, z_w} \boldsymbol{c}^T \boldsymbol{y}_w + \boldsymbol{d}^T \boldsymbol{z}_w \tag{57a}$$

$$\text{s.t. } \boldsymbol{e}^T \boldsymbol{y}_w + \boldsymbol{f}^T \boldsymbol{z}_w + \boldsymbol{g}^T \boldsymbol{u}_w \le \boldsymbol{l}_w - \boldsymbol{h}^T \boldsymbol{x}^*, \boldsymbol{z}_w \in \{0, 1\} \tag{57b}$$

Define a vector $\boldsymbol{LLS} = [LLS_1, \cdots, LLS_{N_w}]$, which denotes the objective set for all the low-level subproblems. Then, the original planning model is expressed as follows.

$$\min_{x} \boldsymbol{a}^T \boldsymbol{x} + \max_{\pi} \boldsymbol{LLS}^T \boldsymbol{\pi} \tag{58a}$$

$$\text{s.t. } \boldsymbol{b}^T \boldsymbol{x} \le \boldsymbol{q}, \boldsymbol{x} \in \{0, 1\} \tag{58b}$$

$$\boldsymbol{k}^T (\boldsymbol{\pi} - \hat{\boldsymbol{\pi}}) = \boldsymbol{i}^T \boldsymbol{r} + \boldsymbol{j}^T \boldsymbol{v} : \boldsymbol{\epsilon} \tag{58c}$$

$$\boldsymbol{m}^T \boldsymbol{r} + \boldsymbol{n}^T \boldsymbol{v} \le \boldsymbol{\alpha} : \boldsymbol{y} \tag{58d}$$

where ϵ and γ denote the dual variable associated with constraints Eqs. (58c) and (58d), respectively. As Eqs. (58c)–(58d) are linear constraints, the duality theory can be employed to transform Eq. (58c). Consequently, the original problem Eq. (58) can be transformed into an equivalent formulation Eq. (59).

$$\min_{x,\epsilon,\gamma\geq 0} \mathbf{a}^T\mathbf{x} + \hat{\pi}^T k\epsilon + \alpha^T\gamma \tag{59a}$$

$$\text{s.t. } b^T x \leq q, x \in \{0, 1\} \tag{59b}$$

$$m\gamma - i\epsilon \geq 0, n\gamma - j\epsilon \geq 0 \tag{59c}$$

$$k\epsilon \geq LLS \tag{59d}$$

The structure of the recast model Eq. (59) resembles the traditional two-stage RO model [18]. Consequently, it can be tackled by existing methods that are designed for the latter. To solve the reformulated model, an enhanced C&CG-AOP algorithm is proposed. Two iterative loops are contained. In the inner loop, the AOP approach is adopted in Algorithm 1 to solve the slave problem. In the outer loop, Algorithm 2 decomposes the reformulated model Eq. (59) into a master problem and a slave problem.

The Slave Problem: In the slave problem, the optimal x in the master problem is transferred to each low-level subproblem. To solve the slave problem, each low-level subproblem is further decomposed. To be specific, an u-fixed problem and a z-fixed problem are iteratively solved as follows.

$$\text{(UFP)} \quad \min_{y_w, z_w} c^T y_w + d^T z_w \tag{60a}$$

$$\text{s.t. } e^T y_w + f^T z_w + g^T u_w^{v*} \leq l_w - h^T x^*, z_w \in \{0, 1\} \tag{60b}$$

where u_w^v denotes the optimal of the z-fixed problem in the v-th iteration. Then the v-th optimal solution is fed back to the z-fixed problem:

$$\text{(ZFP)} \quad \max_{u_w, \lambda_w} -\lambda_w^T(l_w - h^T x^* - g^T u_w - f^T z_w^*) + d^T z_w^* \tag{61a}$$

$$\text{s.t.} \quad -\lambda_w^T e^T \geq c^T, \lambda_w^T \geq 0, u_w \in \mathcal{U}_w \tag{61b}$$

where λ_ω denotes the dual variable of constraint Eq. (60b). Using the big-M method, bilinear $\lambda_\omega^T u_\omega$ can be further reformulated into a tractable MILP problem [19].

The Master Problem: In the k-th iteration, optimal results u_w^{k*} in the slave problem are transferred to the master problem.

$$\text{MP} = \min_{x,\epsilon,\gamma\geq 0} a^T x + \hat{\pi}^T k\epsilon + \alpha^T\gamma \tag{62a}$$

$$\text{s.t. (8.59b), (8.59c)} \tag{62b}$$

$$\{k\epsilon\}_w \geq c^T y_w^m + d^T z_w^m, \forall w, 1 \leq m \leq k \tag{62c}$$

$$e^T y_w^m + f^T z_w^m + g^T u_w^* \leq l_w - h^T x, \quad \forall w, 1 \leq m \leq k \tag{62d}$$

where k represents the iteration number. $(\boldsymbol{y}_w^m, \boldsymbol{z}_w^m)$ denotes the newly introduced variable under $\boldsymbol{u}_w^{m^*}$. The comprehensive solution process is outlined as follows.

Algorithm 1: AOP algorithm for each LLS Eq. (57)

1 Initialize $v = 1$, relax z_ω as continuous, and solve LLS_w with $\boldsymbol{x}$, obtain $\boldsymbol{u}_\omega^v$;
2 **Repeat**;
3 Solve Eq. (60) with $(\boldsymbol{u}_w^{v^*})$, obtain $(\boldsymbol{y}_w^{v^*}, z_w^{v^*})$;
4 Solve Eq. (61) with $(z_w^{v^*}, \boldsymbol{x}^*)$, obtain $\boldsymbol{u}_w^{v+1^*}$;
5 $v = v + 1$;
6 **Until** $\boldsymbol{u}_w^v = \boldsymbol{u}_w^{v-1}$, and present the optimal solution;

Algorithm 2: AOP algorithm for each LLS Eq. (57)

1 Set a feasible $\boldsymbol{x}$ and convergence precision ϵ Set $k = 1$. Solve each LLSs with $\boldsymbol{x}^{0*}$, obtain $(\boldsymbol{y}_w^{k^*}, z_w^{k^*}, \boldsymbol{u}_w^{k^*})$. Set $LB = -\infty$, $UB = +\infty$;
2 **Repeat**;
3 Solve the master problem with $\boldsymbol{u}_\omega^{k*}$, obtain $\boldsymbol{x}^{k*}$, $\boldsymbol{\epsilon}^{k*}$, $\boldsymbol{\gamma}^{k*}$ and the objective V_k^{MP}, and update $LB = V_k^{\text{MP}}$;
4 Solve N_w LLSs in parallel with $\boldsymbol{x}^{k*}$. Output $(\mathbf{y}_w^{(k+1)^*}, \mathbf{z}_w^{(k+1)^*}, \mathbf{u}_w^{(k+1)^*})$ and $\boldsymbol{LLS}^{(k+1)^*}$, then update $UB = \boldsymbol{A}^\top \boldsymbol{x}^{k^*} + \hat{\pi}^T \boldsymbol{LLS}^{(k+1)*} + \alpha^T \boldsymbol{\gamma}^{k*}$;
5 $k = k + 1$;
6 **Until** $\frac{|UB-LB|}{UB} \leq \epsilon$, and output the solution;

---

5 Case Studies

In this section, a typical office building group is utilized for case studies. To be specific, it consists of 10 buildings, each of which comprises 20 floors. Each floor is divided into two zones with a similar structure. Each zone has four rooms, measuring 3 meters high, 8 meters wide, and 8 meters long. Nine typical days are selected to represent the seasonal operational conditions. All numerical simulations were implemented using Matlab-R2021a, with optimization solved by Gurobi 9.1.2. The optimality gap was set as 0.5%. The convergence threshold ϵ was chosen as 0.01, and the confidence level β was set to 0.9.

5.1 Comparative Analysis on Different Planning Methods

The proposed method is evaluated against four alternative methods to highlight its strengths. **M1**: This method does not consider long-term uncertainty and daily fluctuations, incorporating seasonal fluctuations using typical days. **M2**: This method models long-term uncertainty by the DRO approach, and utilizes typical days to model short-term uncertainty. Fluctuations in daily operation are not considered. **M3**: This method does not consider long-term uncertainty, and models seasonal-daily uncertainty by the HSRO approach. **M4**: The proposed method considers both long-term uncertainty and short-term uncertainty, while the hydrogen equipment is not considered.

The detailed planning results of **M1–M5** are presented in Tables 1 and 2. As is shown, compared to **M1**, **M2** and **M3**, the planning solution for **M4** and **M5** are more conservative. This stems from the consideration of both long-term uncertainty and short-term uncertainty, necessitating more equipment capacities to supply demand and complete the ZEBs' yearly zero energy objective. Moreover, even though the long-term uncertainties primarily impact ZEBs' heating/cooling demand, it is evident that energy equipment is altered after incorporating long-term uncertainty. Comparing **M5** to **M4**, the total cost is decreased. It denotes that hydrogen equipment can facilitate multi-energy complementation and enhance the planning economics.

To conduct a comprehensive evaluation of the five planning methods, a random scenario set of 800 scenarios is generated and simulated based on the obtained planning schemes. To be specific, long-term weather conditions are randomly selected,

Table 1 Annual costs of five planning methods (10^4\$)

	C^{inv}	C^{om}	C^{grid}	C^{deg}	C^{all}
M1	97.30	19.55	5.26	2.12	124.23
M2	100.99	21.09	5.74	2.36	130.18
M3	102.12	21.20	5.85	2.35	131.52
M4	107.18	24.09	6.97	3.05	141.29
M5	108.27	22.54	6.61	2.93	140.35

Table 2 Optimal schemes of five planning methods

	PV	WT	CHP	ED	ACH	HP	ST	BS	TS	HS	SHS
M1	13.0	1700	750	1000	350	900	1800	800	50	40	210
M2	13.5	1700	800	1100	400	900	1800	800	50	40	220
M3	13.5	1750	800	1100	350	900	1900	900	51	45	230
M4	17.5	2600	0	0	500	1100	2300	4850	55	0	0
M5	13.0	1850	800	1250	400	1000	2000	950	50	50	290

*Unit: 1000 m^2 for PV and ST; MW for ACH, CHP, ED, HP and WT; 100 kWh for energy storage equipment

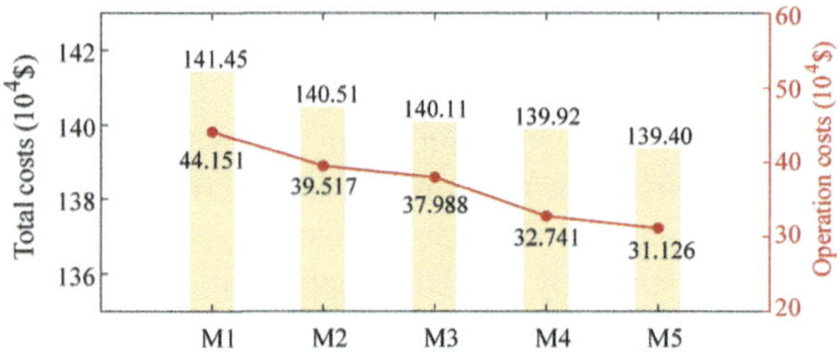

Fig. 4 Average total planning costs and operation costs of five planning methods

Table 3 Performance comparison of yearly zero energy objective

	M1	M2	M3	M4	M5
Infeasible rate	78.4%	38.3%	31.9%	0	0
Average E^{NTE} **(MWh)**	672.3	395.3	321.3	0	0

and source-demand uncertain scenarios are generated randomly within the polyhedral set. When the planning scheme report shows infeasible results, it indicates that the yearly zero energy objective cannot be completed. As a result, it requires buying more electricity. In such instances, the yearly zero energy objective will be relaxed, and the lower bound of additional imported electricity E^{NTE} is calculated. Two metrics are introduced to quantify the planning schemes' reliability with regard to the yearly zero energy objective. The first metric is the infeasible rate of testing scenarios. The second metric is the mean value of E^{NTE} among the test scenarios. For **M1–M5**, the planning costs are depicted in Fig. 4, and the evaluation results of these two metrics are presented in Table 3.

By analyzing the data from Fig. 4 and Table 3, several conclusions can be drawn: (1) the developed method M5 has superior performance in terms of total planning costs. Although **M5** incurs the highest investment cost, it achieves the lowest operation cost under both long-term and short-term uncertainties. As a result, its total planning cost is the lowest. (2) Taking diverse uncertainties into account proves in developing a more dependable planning strategy to complete the yearly zero energy objective. Both **M4** and **M5** adhere to the annual zero energy requirement, demonstrating the efficacy of the proposed methods in mitigating multiple timescale uncertainties. (3) Underestimating uncertainties can result in unsatisfactory planning results. In comparison to **M2** and **M3**, **M1** performs poorly as it overlooks both long-term and short-term uncertainties.

5.2 Operation Results Analysis

In this subsection, to verify the system dispatch results, the operation conditions in typical days are depicted in Fig. 5. As is shown, multiple energy equipment are coordinated to provide an energy supply of multi-energy demands. In terms of electric flow, wind and solar energy exhibit complementary characteristics; these two renewable energy forms primarily provide the electricity to supply electrical demand. Additionally, BS plays a crucial role in smoothing out wind and solar energy. During the summer season and transition season, excess electricity is exported to the main grid. Conversely, in winter, when the renewable energy is reduced, more electricity is purchased from the main grid. Consequently, the annual zero energy target can be completed.

From Fig. 5b and c, multiple energy conversion equipment are coordinated to meet the space heating/cooling demand. Despite outdoor temperature fluctuations, the occupants' comfort levels are maintained, as depicted in Fig. 5e. In terms of hydrogen flow, the electrolyzer equipment utilizes solar and wind energy to produce hydrogen energy. Then the CHP equipment can convert hydrogen energy into electrical and thermal energy. Additionally, the hydrogen storage equipment can further optimize the hydrogen dispatch to resist multiple timescale uncertainties.

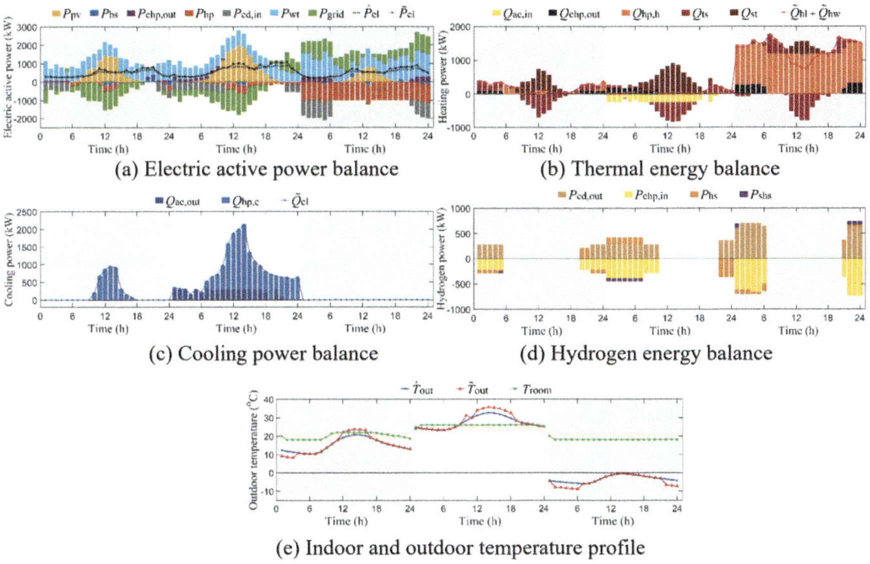

Fig. 5 Multi-energy operation condition in typical day 2, 5, 8

5.3 Sensitivity Analysis on System Uncertainties

In this subsection, the influence of long-term uncertainty and short-term uncertainty on the planning results is further evaluated by conducting a sensitivity analysis.

Short-term Uncertainty: For source-demand uncertain factors in the daily timescale, the sensitivity analysis is conducted in two aspects. Firstly, the collective influence of all the factors is evaluated. For wind energy, solar radiation, electric demand, and hot water demand, an increased interval of 5% (0.75°C) is set. With each incremental rise in the deviation ratios, the investment cost experiences an approximate increase of 1.86%, 1.76%, 1.60%, and 1.79% (2.58%, 2.11%, 2.28%, and 2.58%), respectively. Secondly, to evaluate the influence of each uncertain factor, the control variate method is utilized. As illustrated in Fig. 6b, the ambient temperature exerts the most significant influence. In comparison, the influence of the hot water demand is the least.

Long-term Uncertainty: DRO is employed to handle the long-term uncertain weather conditions. The ambiguity set ensures the inclusion of all potential probability distributions within a confidence level β. β directly influences the conservativeness of the planning results. The influence of confidence level is further evaluated and displayed in Fig. 7. Specifically, it compares the total planning cost in two settings under the DRO approach, i.e., with and without including the Wasserstein distance.

As shown in Fig. 7, a higher value of confidence level results in overestimated schemes to withstand weather condition uncertainty. Additionally, the DRO approach with Wasserstein distance leads to superior results compared to the approach without it. This is due to the stricter constraints imposed by the Wasserstein distance, especially noticeable when opting for a smaller confidence level. For ZEBs investors, selecting an appropriate confidence level is crucial to ensure the energy performance during the system life cycle.

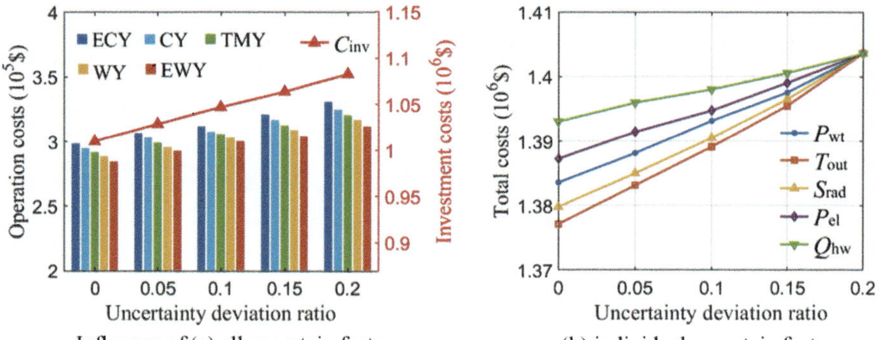

Influence of (a) all uncertain factors; (b) individual uncertain factor.

Fig. 6 Economic performance under different deviation ratios of uncertainties

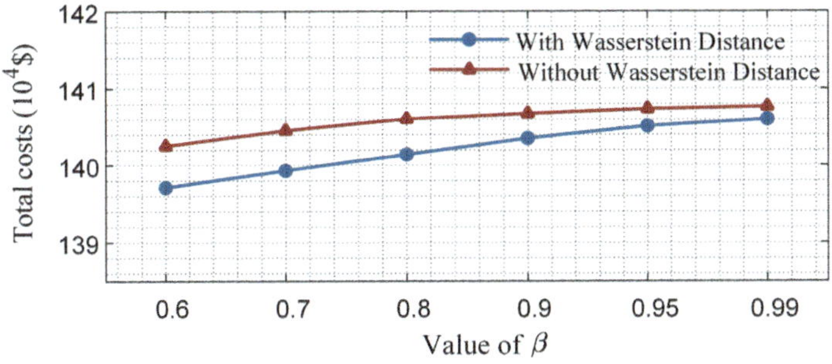

Fig. 7 System total costs under two settings with regarding to β

6 Conclusions

This chapter proposes a tri-level ZEBs planning approach considering both long-term uncertainty and short-term uncertainty. Long-term uncertainty related to future weather conditions is managed using a distributionally robust method with the forecast dataset. In the short-term timescale, a hybrid stochastic and robust method is employed to model the source and demand uncertainties in daily operations. The proposed model is reformulated and solved by an enhanced C&CG-AOP algorithm. Numerical results confirm the efficacy of the proposed planning approach. It can effectively coordinate multiple energy flows to enhance system efficiency and complete the yearly zero energy objective amidst uncertainties. A comprehensive sensitivity analysis further evaluates the influence of different uncertain factors on the planning schemes.

References

1. Taherahmadi J, Noorollahi Y, Panahi M (2021) Toward comprehensive zero energy building definitions: a literature review and recommendations. Int J Sustain Energy 40(2):120–148. (Taylor & Francis)
2. Peterson K, Torcellini P, Grant R et al (2015) A common definition for zero energy buildings. US Department of Energy, National Institute of Building Sciences, Washington, DC, USA
3. Avramidis II, Capitanescu F, Deconinck G (2021) From smart to sustainable to grid-friendly: a generic planning framework for enabling the transition between smart home archetypes. IEEE Trans Sustain Energy 12(3):1684–1694
4. Mehrjerdi H, Iqbal A, Rakhshani E et al (2019) Daily-seasonal operation in net-zero energy building powered by hybrid renewable energies and hydrogen storage systems. Energy Convers Manag 201:112156
5. Perera A, Nik VM, Chen D et al (2020) Quantifying the impacts of climate change and extreme climate events on energy systems. Nat Energy 5(2):150–159
6. Mehrjerdi H, Hemmati R, Shafiekhah M et al (2021) Zero energy building by multicarrier energy systems including hydro, wind, solar, and hydrogen. IEEE Trans Ind Inf 17(8):5474–5484

7. Robert A, Kummert M (2012) Designing net-zero energy buildings for the future climate, not for the past. Build Environ 55:150–158
8. Rey-Hernández JM, Yousif C, Gatt D et al (2018) Modelling the long-term effect of climate change on a zero energy and carbon dioxide building through energy efficiency and renewables. Energy Build 174:85–96
9. Moazami A, Nik VM, Carlucci S et al (2019) Impacts of future weather data typology on building energy performance-Investigating long-term patterns of climate change and extreme weather conditions. Appl Energy 238:696–720
10. Mehrtash M, Capitanescu F, Heiselberg PK et al (2021) A new bi-objective approach for optimal sizing of electrical and thermal devices in zero energy buildings considering environmental impacts. IEEE Trans Sustain Energy 12(2):886–896
11. Mehrtash M, Mozafari G, Hua K et al (2022) Stochastic optimal device sizing model for zero energy buildings: A parallel computing solution. IEEE Trans Ind Appl 58(3):3275–3284
12. Pan G, Gu W, Hu Q et al (2021) Cost and low-carbon competitiveness of electrolytic hydrogen in China. Energy Environ Sci 14(9):4868–4881
13. Yan J, Hu B, Xie K et al (2020) Data-driven transmission defense planning against extreme weather events. IEEE Trans Smart Grid 11(3):2257–2270
14. Pan G, Gu W, Lu Y et al (2020) Optimal planning for electricity-hydrogen integrated energy system considering power to hydrogen and heat and seasonal storage. IEEE Trans Sustain Energy 11(4):2662–2676
15. Sharma S, Verma A, Xu Y et al (2021) Robustly coordinated bi-level energy management of a multi-energy building under multiple uncertainties. IEEE Trans Sustain Energy 12(1):3–13
16. Lu Y, Yu X, Jin X et al (2021) Bi-level optimization framework for buildings to heating grid integration in integrated community energy systems. IEEE Trans Sustain Energy 12(2):860–873
17. Li Z, Wu L, Xu Y et al (2021) Stochastic-weighted robust optimization based bi-layer operation of a multi-energy home considering practical thermal loads and battery degradation. IEEE Trans Sustain Energy 13(2):668–682
18. Zhang C, Xu Y, Dong ZY (2020) Robustly coordinated operation of a multi-energy micro-grid in grid-connected and islanded modes under uncertainties. IEEE Trans Sustain Energy 11(2):640–651
19. Qiu H, Long H, Gu W et al (2021) Recourse-cost constrained robust optimization for microgrid dispatch with correlated uncertainties. IEEE Trans Ind Electron 68(3):2266–2278

5G and Beyond Networks in Local Energy Systems

Jiaying Guo and Shen Wang

Abstract This chapter explores the transformative impact of 5G and beyond networks on local energy systems, emphasizing the potential benefits and addressing key challenges. It begins with an overview of the evolution of mobile networks, highlighting the technical requirements and services enabled by 5G technology. The role of communication systems in advancing local energy systems, facilitating real-time data exchange, efficient energy distribution, and enhancing system reliability is discussed. The chapter then examines how 5G-enabled technologies and applications can support localized energy systems, focusing on areas such as smart metering, dynamic pricing, energy theft detection, and blackout prevention. However, the integration of 5G networks introduces significant research challenges, particularly concerning cybersecurity, consumer privacy, and service scalability. To address these challenges, the chapter proposes potential solutions, including the use of advanced encryption techniques, federated learning, edge computing, and network slicing. By leveraging these innovations, 5G and beyond networks can support the development of secure, scalable, and efficient local energy systems, promoting sustainable energy management and resilience.

1 5G and Beyond Networks

1.1 Introduction to the Evolution of Mobile Networks

Mobile networks have experienced several generations of evolution in performance and functionality, and now we have entered the 5G era, the fifth generation of wireless network technology. In the early 1980s, the first generation (1G) was launched as the original cellular mobile network and introduced by an analog radio signal.

J. Guo · S. Wang (✉)
School of Computer Science, University College Dublin, Belfield, Dublin, Ireland
e-mail: shen.wang@ucd.ie

J. Guo
e-mail: jiaying.guo@ucdconnect.ie

W. Hua et al. (eds.), *Digitalisation of Local Energy Systems*, Green Energy and Technology, https://doi.org/10.1007/978-3-031-77833-9_9

1G mobile systems were designed only for basic voice calls and implemented with simple infrastructure, resulting in many neglected issues. The signal coverage and channel capacity are limited, while the voice quality is low. Additionally, information security was not considered in these early systems. The next generation of mobile networks, 2G, took a big leap toward digital technology in the 1990s, enabling better voice quality and increased capacity. Digital encryption was first introduced to secure data transmission. The improved network performance also supported basic data services, such as Short Message Service (SMS) and Multimedia Messaging Service (MMS), and provided limited internet access. Subsequently, mobile broadband was introduced with 3G in the early 2000s, paving the way for the emergence of smartphones and various mobile applications. Data bandwidth and throughput experienced exponential growth, greatly enhancing customer services. Around 2010, 4G was rolled out to meet the continued need for faster data speeds, greater network efficiency, and support for a broader range of applications. 4G users can relish advanced services, such as High-Definition (HD) multimedia streaming and 3D gaming.

1.2 Technical Requirements and Services of 5G Networks

With the evolution of previous generations of cellular mobile networks, network performance has continuously improved, as shown in Table 1. Although the vision of the Internet of Everything has been long anticipated, 4G technology remains insufficient to support it. The emergence of new services has increased user demands, often leading to congestion in mobile communications. Internet of Things (IoT) applications in industry still have stringent requirements for throughput, latency, and reliability. Therefore, 5G, as the next-generation technology, must provide the capabilities needed to meet these requirements and support practical operations.

International Mobile Telecommunications 2020 (IMT-2020), developed by the International Telecommunication Union (ITU), defines usage scenarios and technical requirements for 5G mobile networks [1]. The 5G architecture is designed to provide three primary services: enhanced Mobile Broadband (eMBB), Ultra-Reliable Low-Latency Communication (URLLC), and massive Machine Type Communication (mMTC).

eMBB This focuses on human-centric applications by increasing system capacity and user-experienced data rates. The improved performance and Quality of Service (QoS) allow users to access multimedia content, services, and data across a wide coverage area. eMBB supports data-intensive services such as 3D video, Ultra-High-Definition (UHD) TV, and virtual/augmented reality, providing a more immersive user experience under the 5G framework. It is also applicable in high-mobility scenarios, ensuring seamless connectivity and high data rates for users moving at high speeds, such as in fast-moving vehicles or trains. Accordingly, the key technical requirements most relevant to eMBB are data rate and mobility. Data rate refers to the amount of data transmitted over a communication link per unit of time. In 5G networks, the peak data rate can reach up to 20 Gbps, while user-experienced data

Table 1 Summary and comparison of mobile network technology generations

	1G	2G	3G	4G	5G
Deployment	1980s	1990s	2000s	2010s	2020s
Data rate	2.4 Kbps	64 Kbps	2 Mbps	100 Mbps ~ 1 Gbps	1–20 Gbps
Latency	>1 s	300~400 ms	100~200 ms	30~50 ms	1~5 ms
Number of connections	Low	Improved to 1G	Moderate	High	Very High, 1M devices/km^2
Frequency	800 ~ 900 MHz	850~1900 MHz	800 MHz ~ 2.1 GHz	600 MHz ~ 5.9 GHz	600 MHz ~ 71 GHz
Standard	AMPS, NMT, TACS	GSM, CDMA	UMTS, CDMA2000	LTE, WiMAX	5G NR
Technology	Analog cellular	Digital cellular	Wideband CDMA, IP technology	All-IP architecture, seamless integration with LAN, WAN, PAN, WLAN	mmWave, Massive MIMO, beamforming, network slicing
Multiplexing	Frequency-Division Multiple Access (FDMA)	Time Division Multiple Access (TDMA), Code Division Multiple Access (CDMA)	CDMA	Orthogonal Frequency-Division Multiple Access (OFDMA), for downlink (from tower to phone) , and Single-Carrier Frequency Division Multiple Access (SC-FDMA) for uplink	Orthogonal Frequency-Division Multiplexing (OFDM), Non-Orthogonal Multiple Access (NOMA)
Switching	Circuit switching	Circuit switching	Circuit switching for voice and packet switching for data transmission	Packet switching	Packet switching
Service	Voice	Digital voice, SMS, MMS, limited internet access	Mobile video streaming, video calls, improved internet access	HD multimedia streaming, online gaming, wearable devices	IoT, cloud computing, devices with AI capabilities

rates are expected to reach 100 Mbps. Spectrum efficiency is improved by increasing the information rate transmitted over a given bandwidth, which supports higher data rates and area traffic capacity. Area traffic capacity, which indicates the total data rate of users per-unit area, is targeted at 10 Mbps/m^2. Enhanced mobility in eMBB allows a maximum speed of 500 km/h with acceptable QoS.

URLLC This use case of 5G emphasizes the reliability and availability of network capabilities. 5G provides the possibility to realize mission-critical and tactile Internet applications that involve machine-to-machine and human-to-machine interaction in the IoT, such as distribution automation in smart grids, autonomous driving, remote surgery, etc. Ultra-low latency network communications facilitate real-time control in smart grids and intelligent transportation systems. The critical technical requirements for enabling URLLC are latency and reliability. End-to-end latency is ideally limited to 1 millisecond, ensuring near-instantaneous data transmission. Reliability is also a crucial factor, with tested scenarios achieving a failure rate as low as 10^{-9} for sub-millisecond radio transmission in factory automation [2].

mMTC This supports high connection density for massive device deployment. A typical scenario is an enormous number of devices connected to a base station but with a relatively small amount of non-latency-sensitive data being transferred [3]. This is characteristic of applications and services in smart cities, involving smart homes, smart grids, and smart retail. The primary technical requirements for mMTC are connection density and network energy efficiency. 5G can support connection density of up to 10^6 devices/km^2. Improved energy efficiency aims to support efficient data transmission under load or power constraints, and it also reduces energy consumption when no data is being transmitted. Additional challenges for mMTC include congestion in accessing mobile networks, huge signaling overhead, and QoS provisioning [4].

With dramatic improvements in ultra-fast data speeds, low latency, and massive device connectivity, 5G technology holds great promise for transformative applications across industries. It unlocks new possibilities in the digital age. One particularly impactful area for 5G is in local energy systems, where advanced communication systems are essential for efficient operation and management. By enabling real-time data exchange and precise control, 5G can significantly enhance the reliability and efficiency of local energy grids. This supports the deployment of smart grid technologies and facilitates the integration of renewable energy sources, paving the way for more sustainable and responsive energy systems.

1.3 The Role of Communication Systems in Local Energy Systems

Energy systems are a striking example of where integration with 5G communications will be revolutionary. Traditional energy systems are centralized and operate unidirectionally, with limited information flow and control. In these systems, energy is generated by large power plants and delivered to users through high-voltage transmission lines, substations, and distribution networks. However, with the integration of 5G communication, energy systems are undergoing a paradigm shift towards more localized and interactive models.

Local energy systems leverage 5G-enabled communication networks to provide a more flexible and controllable energy configuration, including power generation from renewable sources (e.g., solar, wind, hydroelectric), energy storage solutions, and smart control systems. Intelligent devices such as sensors and smart meters, which form the Advanced Metering Infrastructure (AMI), rely on 5G for real-time data exchange and decision-making. 5G enables these devices to collect and analyze data continuously, allowing for efficient monitoring, control, and optimization for local energy generation, distribution, and consumption. 5G's high-speed, low-latency communication capabilities enable remote control systems to manage automated and distributed energy-delivery networks, making dynamic electricity flow from local electricity generators to customers. The two-way flows of electricity and information, facilitated by 5G, bring more interaction between utilities and customers, creating mutual benefits and supporting load balancing and demand response initiatives. In addition, 5G supports the use of Unmanned Aerial Vehicles (UAVs) and robots, offering cost-efficient and safe solutions for infrastructure maintenance and inspection, thus replacing human operators. These robots can also be remotely controlled to perform complex repair operations, improving safety and efficiency. Moreover, 5G can handle massive device connectivity, which is crucial to integrating renewable energy sources into local energy systems. This reduces reliance on traditional energy sources. Virtual power plants, which aggregate diverse renewable energy, rely on robust communication networks to function effectively. By enabling these advancements, 5G communication systems not only enhance the efficient and intelligent functionality of local energy systems but also contribute to the broader goals of sustainability.

2 How 5G and Beyond Networks Will Benefit the Local Energy Systems

2.1 5G-Enabled Technologies for Localized Energy Systems

Within local energy systems, various innovative technologies harness the capabilities of 5G communication to optimize operations and enhance efficiency. These technologies contribute to improved reliability, resilience, and sustainability in local energy systems, as demonstrated in the following examples [5]:

- **Advanced metering infrastructure (AMI)**: AMI is a technical framework composed of systems and networks that use smart meters to collect data and provide intelligent management [6]. A smart meter is a digital device that measures and records real-time electricity consumption while transmitting additional information, including peak electricity use, voltage, and other power characteristics. The data is generally transmitted into a central utility database for analysis and billing purposes. AMI requires two-way communication between utilities and consumers at regular intervals (e.g., 15 minutes or hourly) [7]. With the integration of 5G communication, AMI can achieve faster and more reliable data transmission. Specifically, utilities can monitor the quality of power delivery and consumption in real time, detect abnormal measurements to prevent energy theft, and send immediate pricing signals or requests to distributed meters to manage peak energy usage. Conversely, AMI allows consumers to receive timely information and commands from utility companies and then control their energy consumption more effectively. Leveraging 5G's low latency and massive connectivity features, AMI is a milestone in transforming traditional energy systems where devices and infrastructure at all levels can be universally connected and controlled [8].

- **Demand response**: Demand response refers to rapid and accurate adjustments in electricity consumption in response to grid conditions. Traditionally, demand response programs have operated on timescales of minutes or even hours. However, advanced demand response integrated with 5G can achieve millisecond-level precision, significantly improving grid stability and resilience. Utilities can balance supply and demand in real-time, preventing an avalanche effect of high loads, from a small-scale power outage or a spike to a large-scale blackout. Common demand response programs are divided into two categories: pricing-based and incentive-based [9]. Incentive-based programs offer users fixed or variable incentives through payments or contracts to reduce consumption during peak times. Examples include interruptible load programs, where penalties on bill credits are imposed if response criteria are not met; direct load control, which allows utilities to remotely turn off smart-home appliances; frequency-controlled load curtailment; and capacity market programs. Pricing-based methods, like dynamic price, provide different electricity rates throughout the day to encourage users to shift their demand away from peak periods. This includes strategies like time-of-use pricing and critical peak pricing, which provide higher rates during peak demand times and lower rates during off-peak hours. By leveraging the ultra-fast, low-latency communication of 5G, advanced demand response systems enhance grid stability and efficiency, empowering utilities and consumers to actively maintain load balance.
- **Distribution automation**: Distribution automation is a mechanism designed to improve the reliability of electricity supplies and reduce operation and maintenance costs. Power distribution systems consist of various devices that monitor and control power distribution, including voltage regulators, capacitor bank controllers, fault detectors, switches, and transformers [10]. These in-use devices can be upgraded with communication and control components, enabling more advanced functionality. Advanced distribution automation systems can quickly detect and respond to faults or outages, thereby reducing downtime and minimizing the impact of disruptions [9]. Communication and remote control in distribution automation can isolate affected areas and reroute power, enabling self-healing

capabilities in the grid. Effective decision-making and execution in distributed automation rely heavily on real-time data collection and configuration adjustments provided by AMI. Meanwhile, demand response systems deliver demand information to distributed automation systems, which automatically analyze customer demands and schedule resources to quickly meet user requirements and ensure the stability of the energy system. Thus, we integrate AMI and demand response into distribution automation in Fig. 1, further illustrating the technical requirements of 5G to enable these technologies in local energy systems.
- **Infrastructure maintenance and inspection**: Traditional grid inspections rely on ground staff or low-flying helicopters, which poses significant risks of accidents and injuries to inspectors, especially in remote or hazardous areas, and incur high equipment and labor costs [5]. UAVs offer a cost-efficient alternative, providing ubiquitous coverage in hard-to-reach areas and flexible mobility for rapid response in emergencies [4]. Equipped with high-resolution cameras and sensors, UAVs can conduct visual inspections of power lines, towers, and substations, while thermal cameras can detect overheating components, helping to avoid equipment failures. Robots are also increasingly utilized in grid maintenance, enhancing safety and efficiency. These robots can autonomously inspect grid equipment and perform certain maintenance tasks, such as replacing or repairing equipment components. By integrating 5G communication, both UAVs and robots can transmit real-time data and be operated remotely. The high-speed, reliable connectivity provided by 5G enables these devices to communicate continuously with control centers, allowing for real-time monitoring, diagnostics, and immediate response to potential issues.
- **Virtual power plants**: A virtual power plant is a cloud-based, software-driven system that aggregates a variety of distributed energy resources and participates in the energy market as a single entity, including solar panels, wind turbines, energy storage systems, and more flexible loads like electric vehicles. While performing the basic functions of a conventional power plant, a virtual power plant optimizes energy generation and dispatch, facilitates the integration of renewable energy sources, and further helps balance supply and demand on the grid. Enabled by advanced mobile networks, particularly 5G, users owning distributed energy resources can dynamically act as both load and generator, responding to real-time scheduling operations of virtual power plants [11].

Further development of the above-mentioned technologies in local energy systems relies on 5G communication networks, each with its specific characteristics and requirements. Figure 1 provides a summary of their specific requirements for leveraging 5G to achieve the expected improvements. Three of these technologies, AMI, demand response, and distribution automation, rely heavily on the widespread adoption of smart devices to achieve efficient and reliable grid control and are categorized together for analysis. A large number of distributed meters necessitates high connection density, while the load balancing of local energy systems has stringent requirements on low latency of data transmissions. For infrastructure maintenance and inspection, UAVs and robots offer the highest mobility, requiring high data rates to transmit high-quality images and videos. In contrast, independent and distributed

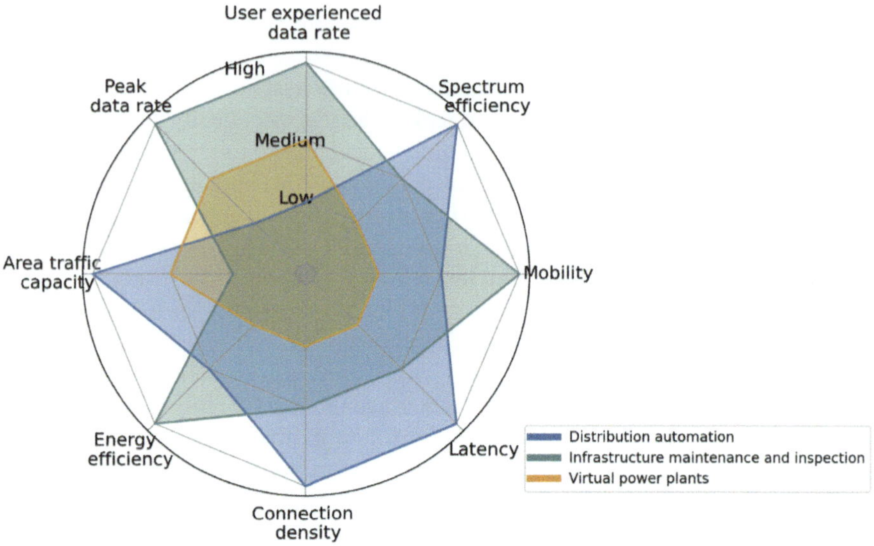

Fig. 1 Key features of 5G-enabled local energy system technologies. (It is noted that distribution automation covers AMI and demand response as it interacts with the consumption and demand of energy customers)

renewable energy resources for virtual power plants are currently not deployed on a large scale. For example, only a small fraction of households have installed solar panels. However, utilizing green and clean energy is a sustainable solution that continues to be strongly advocated.

2.2 5G-Enabled Applications for Localized Energy Systems

Traditional electric utilities face high costs associated with meter readings, customer services, field services, and theft management [7]. Local energy systems integrated with advanced 5G communication technologies are expected to significantly reduce these costs. The Electric Power Research Institute (EPRI) has estimated the costs and benefits of smart grid implementation [12]: a fully developed smart grid could result in savings between 1.3 and 2.0 trillion dollars, with deployment expenditures ranging from 338 to 476 billion dollars over approximately 20 years. Modern energy infrastructures improve energy generation and distribution through extensive data acquisition and analysis. However, the deployment of smart grids is uneven around the world. While developed nations are heavily investing in local energy systems to integrate with promising 5G networks, developing countries still suffer relatively high operation and maintenance costs due to traditional infrastructures [13]. In this section, we discuss three dominant applications of local energy systems. Advanced 5G has brought multiple innovative approaches to optimize these practical applications.

2.2.1 Energy Theft Detection and Prevention

Electrical losses are inevitable at any stage of local energy systems. While losses during generation are relatively easy to define, losses in transmission and distribution are more challenging to quantify [6]. These losses can be categorized into technical and non-technical. Technical loss occurs naturally due to power dissipation in transmission lines and components, including corona discharges, leakage currents, dielectric losses, open circuit losses, impedance, and other variable losses [13]. These losses are inherent to the system and account for about 20% of the generated energy capacity. In contrast, non-technical losses, which account for 10$\sim$40%, are mainly caused by energy theft, metering inaccuracies, billing errors, and unmetered supply. Among them, energy theft is the predominant contributor. In the United States, industry consensus is that the theft loss ranges from 0.5 to 3.5% of annual gross revenues. Energy theft not only negatively affects the quality of energy supply but also increases the generation load and the risk of power outages, deteriorates utility operations, and imposes excessive costs for consumers.

Malicious actors can engage in energy theft by tampering with meters or bypassing them altogether to tap directly into power lines, ensuring unmetered consumption. In traditional power systems, electromechanical meters are easy to manipulate and lack built-in security mechanisms. Common methods to disrupt the operation of these meters include attaching magnets to interrupt the rotating disk of traditional electricity meters or inserting viscous fluid to lower usage readings [6, 13]. Besides, the pressure coil of the meter can be damaged by switching the incoming and outgoing terminals.

The implementation of smart meters can essentially eliminate simple and rampant energy theft by physical manipulations. Smart meters, as part of AMI, enable two-way communication, allowing utilities to detect zero readings and promptly receive alerts about potential tampering. AMI systems, embedded with digital communication infrastructure in each household, provide granular data on energy consumption at the end-user level. Individual data collected by smart meters is transmitted through gateways and aggregated for analysis at a regional or city level, enhancing the detection of anomalies in power distribution that could indicate energy theft. This data transmission largely relies on implementing 5G in the smart grid, allowing a massive number of smart meters to access mobile networks.

With smart meters and advanced communication networks, data-driven detection mechanisms on energy consumption measurements are expected to mitigate energy theft effectively [14]. Reliable and precise end-user consumption collected in smart meters and transmitted to utilities by communication networks is the premise of detecting energy theft using data-driven methods. These methods use statistical techniques to characterize normal consumption patterns, identify unusual energy utilization, and accurately localize energy theft. Machine learning techniques, such as support vector machines, are widely explored to classify customer load for detecting energy theft [8, 13]. A deep learning model proposed in [15] uses a temporal convolutional network on a real energy consumption dataset to identify theft. In partially deployed smart grid infrastructures (e.g., public residential areas with hundreds of households), detection methods such as those in [14] show the potential by combining data-driven schemes with smart meters/AMIs. Specifically, collected energy measurements are analyzed using Fast Fourier Transform (FFT) to profile consumption into frequency components, identifying normal or abnormal behaviors for each consumer according to the spectral entropy of signals. The k-means clustering model is also used to detect similar energy usage patterns among households, which enhances theft detection at a regional level. It is noted that a finer granularity of energy consumption information that considers more practical factors (e.g., temperature and time of the year) would lead to more accurate detection and identification of an observed anomalous pattern. The advanced 5G mobile networks, combined with a variety of interconnected sensors in smart meters, are expected to increase the granularity and accuracy of energy measurements.

Energy theft detection often occurs after the illegal activity has taken place, so post-detection approaches that rely on energy consumption data cannot restore substantial financial losses that have already been caused. Recent studies have started exploring the infrastructure upgrade in AMI to introduce direct electricity theft prevention mechanisms. Indicators of energy theft derived from observed abnormal data should be summarized and modeled, such as false signatures on data in smart meters, incorrect pricing, and flagged meter status, to trigger preventative actions. A rule-based prevention model proposed in [16] operates across several sub-zones of the AMI network in a distributed way. By monitoring consumers in each sub-zone, the model selects four indicative parameters of energy theft to define targeted detection rules. A fuzzy inference system then implements prevention actions based on these rules. In this model, the sub-zone division facilitates tracking fraudulent consumers. Each consumer's smart meter must be connected to a central meter that records and monitors its zone.

Smart meters, equipped with advanced sensors and supported by 5G communication networks, represent a significant advancement in addressing energy theft. By enabling real-time monitoring, rapid data transmission, and the implementation of data-driven detection and prevention mechanisms, 5G technology enhances the accuracy, reliability, and responsiveness of theft detection systems. As energy systems become increasingly digitalized and interconnected, the role of 5G in safeguarding against energy theft will continue to grow, ensuring that localized energy systems remain secure, efficient, and sustainable.

2.2.2 Blackout Detection and Isolation

Conventional power systems are spread over large geographic regions, with high capacity and long transmission distances to meet the ever-increasing demand for electricity. When any type of abnormal events or faults occur within the power system, for example, line tripping or voltage collapse, it can be complex to restore instantly. The chain reaction accidents will cause devastating consequences of large-scale power blackouts. Nearly 90% of power outages are caused by inappropriate operations and adjustments in distribution networks [8]. As modern society is heavily dependent on electricity, power outages can result in enormous economic losses and daily life disorders, e.g., water supply interruption, traffic paralysis, Internet service loss, etc. Therefore, control and protection techniques are necessary in power systems to avoid or mitigate power outages.

To maintain a continuous and stable power supply, the frequency and voltage of the power network must remain within permissible ranges. Overloaded transmission lines lead to frequency or voltage drops. If the transmission system lacks spare capacity or contains an electromagnetic loop network, the transmission line will be overloaded, triggering successive overload protection [17]. Overload protection quickly isolates the overloaded section to prevent further damage, but this can lead to load distribution problems and cascading faults throughout the system. Under-frequency load shedding (UFLS) and under-voltage load shedding (UVLS) techniques are used to maintain the frequency and voltage within a safe operating range by shedding or disconnecting certain loads in the case of a power disturbance before overload protection. Whereas, these techniques cannot always prevent power blackouts. For instance, Bangladesh's power system experienced a system collapse on November 1, 2014, which started from an unplanned station outage. The UFLS was activated, but the load shedding was not enough to deal with the power disturbance, resulting in a blackout lasting about 24 h [18]. Other protection methods, including fault isolation and condition monitoring diagnosis, are used along with load shedding [19]. However, power systems are still at risk of blackout and outage due to control maloperation and protection malfunction.

Power system blackouts can be caused by various factors, the most intuitive of which is equipment failure, including line tripping or overloading, voltage collapse, and poor maintenance on outdated equipment. Natural disasters and extreme weather also pose challenges to power grid operations. The conventional power grid structure is vertically integrated and complex, with many links requiring coordination, which delays prompt response times during outages. Decentralized distribution automation, supported by 5G technology, offers a promising solution to mitigate power blackouts and improve power system reliability. A smaller area of power blackout brings less adverse impact and a shorter restoration interval. Decentralized structures replace the centralized control centers with local distribution terminals that use peer-to-peer (P2P) communication to identify fault locations, isolate them, and initiate restoration processes. The self-healing capabilities of these distribution automation systems are expected to automatically handle faults without manual intervention and minimize processing time. It replaces the hours-long process of crews patrolling fault areas looking for fault indications and ensuring protection or restoration [20]. The ultra-low latency and high reliability of 5G networks support the possibility of achieving millisecond-level grid control and protection operations, enhancing the resilience of energy systems.

With more equipment automation, multi-agent systems are increasingly explored to coordinate and control distribution automation in a decentralized mode for blackout detection and isolation. Grid components can be regarded as intelligent agents, such as transformers, switches, lines, buses, and generators [19]. These agents communicate and interact with neighboring agents to achieve autonomous operation and fault management. In the multi-agent system proposed in [20], three types of agents, line agent, load agent, and decentralized production agent, collaborate to maintain grid stability. Line agents collect the electrical current by phasor measurement units (PMUs) and compare it with the threshold value for fault detection. When any abnormal operation is detected, line agents send a trip signal to circuit breakers connected at the two ends of the distribution line for fault isolation. Load agents request the load supply from decentralized production agents for fast network restoration. This multi-agent system effectively localizes faults and restores power through agent communication and coordination, as demonstrated in a 20 kV power distribution system simulation for Kenitra City in Morocco. Other artificial intelligence techniques have been applied for blackout detection in distribution systems, including artificial neural networks (ANN), fuzzy logic, and genetic algorithms. Fuzzy logic, for instance, is integrated into multi-agent systems for decentralized voltage control of power systems [21]. The proposed method contains two types of control agents, i.e., master and local. They are designed in two steps for fault detection and restoration, and optimization of the power system, respectively. Upon fault detection, spectral graph partitioning divides power systems into subsystems according to reactive power flow. Local control agents use the Sugeno fuzzy method to adjust the reference voltage of generators, switch capacitors, or load shedding for voltage control. Master control agents adopt teaching-learning-based optimization to restore the grid to its optimal state. Compared with the centralized control method, such as model predictive control, multi-agent systems are more computationally efficient and more resilient to single-point failures.

The effectiveness of these advanced detection and isolation techniques depends on real-time measurement data and reliable communication within distribution networks. The number of real-time measurement devices and the amount of data they collect significantly impact fault detection and localization accuracy. Modelling approaches, such as the Quasi-Steady-State (QSS) direct current power flow model, provide insights to assess the power system's vulnerability and analyze the potential cascading effect [18]. These models facilitate accurate decision-making for voltage control and grid restoration, although they require comprehensive grid information. The high-speed, two-way communication in power systems supported by 5G mobile networks is promising to provide real-time accurate data for modeling approaches and optimize control approaches for addressing power faults and cascading outages. Digital twins would be a possible solution that utilizes real-time data collected to create a digital replica. Potential outages can be predicted, and preventive actions taken in advance[19]. During an outage, digital twins can also assist in optimizing the restoration process by simulating different strategies for the digital replica.

Wide-Area Monitoring, Protection, and Control (WAMPAC) is an emerging technique coupled with a low-latency mobile communication network to enhance power system reliability [18]. WAMPAC utilizes real-time data collected by PMUs to monitor power system conditions. Global positioning system (GPS) deployed in WAMPAC allows a common time reference with synchronized timestamps [22].

WAMPAC technology facilitates fault location and finding the cause of system outages over large geographic areas. It also helps the UFLS method to deal with the under-frequency problem in grid protection [23].

Natural disasters and extreme weather put significant stress on energy transmission and distribution, potentially leading to unexpected power outages [17]. Leveraging real-time meteorological data and 5G technology enhances the capabilities of power systems to anticipate and mitigate potential issues. Power grid operators can proactively identify areas at risk of power outages by continuously monitoring weather conditions such as severe storms, high temperatures, or lightning activity. With 5G technology enabling high-speed data transmission and low-latency communication, meteorological data can be collected and analyzed in real time. Power utilities can develop effective protection plans and emergency response strategies, enabling a more robust and resilient power grid.

Additionally, microgrid technology is a decentralized approach to energy supply in localized areas. Integration of local renewable energy sources can be used as auxiliary power supplies for municipal services in emergencies and power blackouts [17]. Microgrids can operate either connected to the primary grid or in an isolated mode. Distributed backup power and energy storage technology also enhances energy resilience where grid infrastructure is inadequate during power blackouts. The control and management of micro-grids involve advanced technologies such as smart meters, sensors, optimization algorithms, and 5G communication networks. Microgrid technology plays a crucial role in enabling reliable energy solutions and promoting power protection and isolation schemes at a local level.

The transition from traditional, centralized power grids to decentralized, automated energy systems underscores the importance of advanced communication technologies. As energy systems become more complex and distributed, the role of 5G in enabling real-time monitoring, rapid response, and autonomous control is critical to preventing blackouts and ensuring a reliable power supply. By facilitating seamless data exchange and supporting innovative technologies like digital twins, multi-agent systems, and WAMPAC, 5G strengthens the resilience and stability of modern power systems.

2.2.3 Dynamic Pricing

Dynamic pricing is a practical method that allows customers to adjust their electricity usage patterns according to various electricity prices, thereby reducing costs without significantly curtailing regular energy consumption. During peak time (e.g., 5–7 pm), energy prices get higher due to higher energy demand. The goal of dynamic pricing by utilities and energy companies is to balance load between peak and off-peak periods, which can reduce power generation and distribution costs. Customers can lower their bills or even profit by adjusting their energy usage behavior. Dynamic pricing, as a demand response scheme, benefits from the interaction between users and electricity companies through smart meters. Bidirectional communication flow in the smart grid allows utilities to monitor the energy usage of customers meanwhile helping customers manage their consumption. Here are some common dynamic price schemes [9, 24]:

- Time-Of-Use (TOU) price: Also known as the time-of-day dynamic price scheme, TOU sets different price rates throughout the day. For example, energy companies may offer lower night rates (nearly half of daytime electricity prices). Typically, TOU pricing has two or three tiers: a day rate (8 am to 5 pm and 7 pm to 11 pm), a night rate (11 pm to 8 am), and a special peak rate (5 pm to 7 pm). This scheme is widely adopted for residential consumers to schedule appliance usage during lower-cost periods and save on energy bills compared to flat-rate pricing.
- Critical Peak Price (CPP): CPP is a similar dynamic price scheme to TOU but with distinguished rates for peak and off-peak periods in a day. The price difference in CPP shows a stronger signal for peak load shaving. The peak period price could be five times higher than the off-peak price. CPP is usually rolled out a few days each year when utilities are under high demand pressures, such as during sweltering summers or cold winters. In this case, the excess cost of generation and distribution is unusually high relative to daily electricity prices in the TOU. Besides, a peak-time rebate is an incentive approach as part of the CPP scheme. Customers can earn credits and redeem them later if they consume energy below a certain level during critical peak periods.
- Inclining Block Rate (IBR): IBP pricing is determined by a customer's total electricity usage over a certain time interval (hour/day/month). The scheme contains multiple price tiers, with the per-unit rate increasing as energy consumption exceeds predefined thresholds. This encourages consumers to shift their load schedules or cut energy usage to avoid higher rates.
- Real-Time Price (RTP): RTP is the most intuitive form of dynamic pricing, as electricity prices change more frequently than the above pricing schemes to reflect real-time market conditions. Price adjustments are typically announced a day or hours in advance, or even within minutes. Hence, the RTP scheme is challenging in practice due to its extreme reliance on advanced communication technology to handle high data collection and transmission rates. The low-latency advantages of 5G cellular networks are particularly suited for RTP applications.

Dynamic pricing aims to reduce peak energy demand by incentivizing changes in high-consumption behavior. Since electricity consumption in buildings accounts for a significant portion of total demand, especially in those with central air-conditioning systems, shifting load during peak times plays a vital role in effective demand response. To be specific, customers, on the demand side, must be informed of dynamic prices to impact their usage behaviors. They can obtain bill savings while helping the grid keep load balance and improve reliability. In turn, the electricity utilities collect consumers' behavior, and then analyze the demand fluctuation and peak load, thereby lowering power generation costs and maintaining healthy grid operations. Dynamic pricing is optimized a day in advance, or at least a few hours in advance, for anticipated power management. The interaction between electricity companies and customers is expected to obtain economic profits and achieve a win-win situation. Among the existing approaches to investigate dynamic pricing, game-theoretical methods are an effective method for the win-win bidirectional interaction in the demand response of local energy systems. A study using the Stackelberg game model with on-site data from buildings on a university campus in Hong Kong showed an about 40% reduction in the grid demand fluctuations and a 2.5–8.3% decrease in electricity bills for campus buildings [25].

When focusing on concrete building energy consumption, household appliances are the main cause of energy consumption. If smart plugs can control their operating period, load shifting and saving bills can be realized according to dynamic pricing. Among household electrical appliances, some are controllable and interruptible, such as washing and drying machines, which can shift the load to an off-peak demand period. Smart plugs connected through networks can control appliances with optimal switching cycles and minimum costs. Existing studies have validated this approach. One of the common evolutionary algorithms, particle swarm optimization, has been used to address the load shift problem in controlled appliances [26]. Such methods significantly reduce average electricity bills under a two- or three-tier TOU price system.

On the other hand, dynamic pricing can also be used to generate profits for customers rather than just reducing billing. By incorporating renewable energy resources into local energy systems, dynamic pricing can benefit residential customers with independent energy production, such as those equipped with photovoltaic systems and battery storage. Selling surplus renewable energy to the electricity market allows customers to engage in arbitrage while benefiting electric utilities by minimizing production and storage costs and reducing carbon emissions. A day-ahead dynamic pricing mechanism introduced in [27] aims to maximize the social welfare of residential customers and the profits of electric companies while promoting load balance in the local grid. The optimal conditional decomposition technique is used to handle this nonlinear programming problem, which decomposes the objective functions between the different sides, i.e., customers and utilities. This decision-making of day-ahead dynamic pricing involves addressing non-convex NP-hard constraint optimization, considering factors like market prices of selling and purchasing electricity, maintenance and operation costs, customer demand, photovoltaic generation, and reactions to price changes. A Monte Carlo scenario-based technique is applied to model the uncertainty of photovoltaic generation and customers' demands.

In addition, dynamic pricing models are widely applied for electric vehicle (EV) charging. The charging stations in the hotspot areas experience high load during afternoon peak traffic hours (e.g., 2 pm and 6 pm). The load demand for plug-in EVs overlaps with residential peak loads, risking network instability and potential power outages. Dynamic pricing can incentivize EV load-shifting to off-peak periods and encourage car owners to drive toward underutilized charging stations for load balance. Addressing the peak load-shaving problem in EV charging is increasingly gaining attention. For instance, a rule-based heuristic solution for updating electricity price and charging strategy proposed in [28] aims to minimize the overlapped time between EV charging load and daily residential peak load. The queue occupancy, waiting time, and capacity of a charging station are considered in the model. The demand estimation in charging stations affects charging prices during various timeslots. Additionally, energy scheduling in office buildings equipped with photovoltaic systems and EV charging is an ongoing research focus. Power companies have developed special EV charging plans, such as offering reduced rates from 2 am to 6 am.

Existing research on dynamic pricing has examined the impact of various factors. Still, prediction uncertainty continues to pose challenges in determining electricity prices in advance and optimizing load control. Advanced adaptive control strategies, such as model predictive control, offer real-time optimization capabilities for dynamic pricing and power demand management, enhancing flexibility and robustness. Furthermore, machine learning algorithms enable the analysis of customer behavior and precise forecasting of renewable power generation, improving the overall effectiveness of dynamic pricing. By utilizing extensive data from smart devices, including household demand collected from 5G communication networks, dynamic prices can be determined with greater reliability and accuracy.

In conclusion, Fig. 2 summarizes the technical requirements of 5G networks to enable the above three applications in local energy systems. Dynamic pricing requires the energy consumption of customers to be collected and transmitted at special intervals. High data rates of 5G support a large amount of data aggregated from diverse smart meters. EVs, with their high mobility and significant impact on peak energy demand, also influence dynamic pricing schemes as they are both energy consumers and potential storage solutions. Compared to dynamic pricing, energy theft detection and prevention mechanisms necessitate a higher connection density within 5G networks, as they involve monitoring customer usage meters and connected transmission

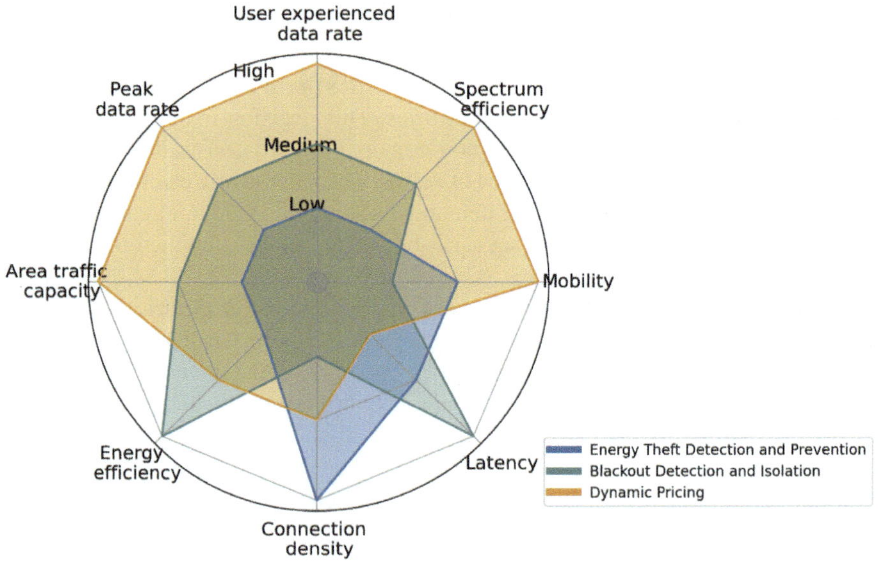

Fig. 2 Key features of 5G-enabled local energy system applications

lines across the electricity utilization network. Blackout detection and isolation methods work in the distribution network, which includes a variety of connected devices for energy distribution and monitoring. Although these devices are fewer in number compared to the extensive metering infrastructure, they still demand ultra-low latency from 5G networks to ensure rapid fault detection, isolation, and restoration, thereby avoiding power outages and severe societal consequences. While solutions for these applications continue to advance, their practical implementation remains dependent on the improved 5G network performance. The ongoing development and deployment of robust 5G infrastructure are essential to fully realize the potential of these smart applications, ensuring reliable, efficient, and sustainable local energy systems.

3 Research Challenges and Possible Solutions

This section discusses the primary research challenges currently faced by local energy systems that integrate communication networks and explores potential solutions to these issues. One of the most critical issues is the pervasive communication requirement for managing and controlling local energy systems, which makes them vulnerable to cyberattacks. Such attacks pose a serious threat to the reliability and availability of energy system operations and services. Another significant concern is the potential leakage of customer data privacy, which could hinder the widespread adoption of smart devices. In addition, the scalability of services is the key challenge that must be addressed to achieve broad coverage of intelligent local energy systems in the future, rather than being limited to core areas during the current development phase. A concise summary of the primary research challenges and the corresponding proposed solutions can be seen in Table 2, providing a framework for the discussions that follow. Investment in infrastructure upgrades is crucial to promote the advancements of techniques and practical applications, but it is not investigated in this section.

3.1 Cyber Security

As more grid equipment is connected and controlled remotely through faster data transmission speeds in 5G networks, concerns about cyber threats and attacks have increased significantly. Customer data collected by deployed smart meters may be forwarded through multiple communication nodes, about four to five times, before reaching the utility company [7]. Communication in the distribution network occurs all the time to achieve real-time load balancing, increasing the risk of security threats with each communication link. The Federal Bureau of Investigation (FBI) has reported that the attackers can hack into a smart meter using only moderate computer knowledge and low-cost tools and software readily available on the Internet [31]. Then, the attacker can access sensitive customer data collected by the smart

Table 2 Summary of Research Challenges and Proposed Solutions in Local Energy Systems

Challenges	Description	Proposed solutions
Cyber security	As energy systems become more interconnected through 5G networks, they are increasingly vulnerable to cyber threats such as data breaches, DoS attacks, and false data injection	• Implement digital credentials for smart meters [22] • Utilize blockchain for secure data transmission and verification • Employ robust encryption techniques • Develop fault detection schemes for cyber threats [29]
Consumer privacy	The frequent collection of detailed energy usage data by smart meters can expose sensitive information about consumers, leading to privacy concerns	• Aggregate data and reduce transmission frequency [6] • Use federated learning for decentralized model training [15] • Apply local differential privacy measures • Implement P2P computing algorithms to ensure privacy in data sharing [13]
Service scalability	Managing the complexity of local energy systems with numerous interconnected devices and ensuring real-time control and coordination presents scalability challenges	• Adopt decentralized smart control systems (e.g., multi-agent systems) [20] • Leverage edge computing to reduce latency and enhance local decision-making [19] • Implement network slicing to allocate resources based on application needs [30]

meter, leading to data breaches and privacy concerns. In addition, smart meters can serve as a portal for malicious access to the broader energy system, launching further attacks on other connected devices or compromising other smart appliances. Power service disruptions cause inconvenience to customers and incur losses for utility providers. The consequence of grid load balance and stability is more severe and unpredictable due to security threats.

Cyber attacks, such as data tampering, can modify data at various stages, during collection, storage in smart meters, or during transmission over the network to utility companies [6]. In particular, data integrity is critical both for accurate customer consumption analysis and for the reliable execution of control commands from utility companies to meters. Denial of Service (DoS) attacks are another common threat to cause interruption of energy services. Generating a high volume of network traffic floods the target service and connection, exhausting the communication resources. In some cases, DoS attacks exploit system vulnerabilities by sending malformed packets, resulting in service unavailability. Additionally, WAMPAC systems are vulnerable to time-based attacks [22]. These systems cover a wide geographical area and rely on synchronized timestamps for fast fault location. Slowing down packets or altering their timestamps can compromise the functionality and reliability of WAMPAC systems.

To prevent cyber attacks and protect the integrity of energy systems, smart meters should be equipped with digital credentials to ensure authorized data access and secure communication with other energy devices. Blockchain technology shows promise in enhancing energy system security [22]. Blockchain-based encryption mechanisms can maintain the confidentiality and integrity of transmitted data, e.g., the electrical current measurements collected by PMUs, thereby preventing unauthorized access and tampering. Establishing secure, direct P2P communication channels between smart meters and energy utilities can eliminate potential vulnerabilities in energy network middleware. Furthermore, blockchain can be used to verify the digital credentials of smart meters, which facilitates accountability in billing schemes on energy consumption data in AMI.

Leveraging communication technology, cyber attacks have even become a more achievable and convenient energy-stealing method than conventional physical attacks. Techniques such as spoofing attacks and man-in-the-middle attacks can be used to manipulate meter readings [8]. Cyber attacks can not only deceive actual energy consumption but even generate fraudulent energy production data to solicit illegal payments [32]. In some cases, attackers may distribute a portion of their appliance usage costs to similar appliances in neighboring legitimate households, thus reducing their energy costs [14]. Research in [32] utilizes a Self-Organizing Map (SOM) for clustering consumers based on similar energy consumption patterns, and a multilayer perceptron ANN to distinguish between legitimate and fraudulent customers within each cluster. This approach considers energy theft arising from both cyber and physical attacks.

Given the enormous economic losses and serious social impacts caused by power outages, the threat of cyber attacks puts great pressure on fault detection and protection. For instance, malicious cyber attacks caused the Ukrainian blackout in 2015. To counter such threats, a fault detection and isolation scheme has been proposed in [29] against false data injection attacks. This method utilizes an unknown input interval observer to prevent data tampering on the running state of the generator or generator terminal bus. Furthermore, substation security challenges, including malicious fault injection attacks (FIA) and hardware Trojans, are analyzed in [33]. Compared with FIA, hardware Trojans are more challenging to detect. Therefore, implementing robust cybersecurity protection measures is critical to mitigate the attack risk and protect the integrity and reliability of local energy systems.

3.2 Consumer Privacy

Energy usage patterns of customers can potentially expose sensitive information about their personal lives. Analyzing energy consumption data can reveal details about the number of people in a household and their daily routines, including when they leave and return home, and the frequency of appliance usage such as washing machines, dishwashers, and air conditioners [6]. Smart meters automatically send energy consumption data to utility companies at 15-minute intervals [7], which can provide a detailed picture of a household's activities. This data could also disclose information about health conditions, appliance malfunctions, or even the status of home alarm systems. Consequently, consumers may resist the deployment of AMI if they fear their data could be disclosed and used without their consent [13].

Protecting the privacy of energy users is crucial, as consumers need assurance that their data will remain confidential. However, there must be a trade-off between maintaining privacy and preserving the utility of the data. One approach to enhancing privacy is to aggregate data and transmit it at less frequent intervals [6]. This reduces the amount of data sent and lowers the risk of revealing detailed personal habits, but it can also decrease the accuracy and timeliness of the data. In cases where precise data is not critical, sampling or estimating data rather than collecting continuous real-time information can help mitigate privacy risks while still meeting the operational needs of energy systems. Only data that provides sufficient utility and is protected by adequate privacy protection measures should be transmitted, meanwhile ensuring compliance with relevant regulations.

Federated learning offers a promising solution for maintaining data privacy in smart grids. This approach enables the use of consumers' energy consumption data without transferring it to a centralized server, thus preserving privacy. Federated learning models are trained across multiple distributed servers, using local data to customize the model. Only the model updates, not the raw data, are sent over the network, which is also efficient in bandwidth usage and computation demand. Fed-Detect, proposed in [15], is a federated learning framework specifically designed for energy theft detection. In this framework, each detection station only observes and processes data from local consumers under a local differential privacy scheme. Additionally, a secure protocol encrypts training parameters exchanged between the data/control center and detection station, supporting aggregated model updates. P2P computing algorithms are also utilized to identify energy theft while safeguarding customer privacy [13]. These algorithms facilitate efficient communication, resource sharing, and coordination between grid components in a decentralized manner.

As distributed renewable energy sources become more integrated into local energy systems, it should draw attention to safeguarding data privacy on the billing mechanisms between utilities and customers with renewable energy production and storage. For example, household customers who have photovoltaic systems or EVs can generate electricity independently from the grid or serve as standalone energy storage to avoid sudden blackouts and outages. If this on-site generation unit produces more electricity than is consumed, the excess can be fed back into the grid and purchased by the utilities. This process involves multiple billing mechanisms to purchase or offset, such as net metering, net purchase and sale, and feed-in tariff. The privacy issue of residential customers during these transactions should not be disregarded in future studies. Existing optimization algorithms for dynamic pricing, such as the optimal conditional decomposition technique [27], are designed with different objective functions on two sides of energy transactions to guarantee privacy during data exchanges.

With 5G networks connecting a vast array of devices and sensors within local energy systems, these networks generate substantial volumes of real-time data. The processing and analysis of this data necessitate robust data management systems to ensure that privacy is maintained while still leveraging the benefits of real-time information flow.

3.3 Service Scalability

Traditional energy systems typically generate electricity at centralized power plants and transmit it through high-voltage grid networks to utilities and communities. Substations convert high-voltage, large-capacity power from transmission lines into low-voltage, small-capacity power and deliver it to thousands of households. This vertically integrated energy system is unidirectional, with limited information flow and control. In contrast, local energy systems, supported by advancements in communication technology, provide a more flexible and controllable energy model. Numerous devices and sensors are deployed to monitor and manage energy use, creating a bidirectional flow of information and achieving more efficient energy distribution. However, the introduction of pervasive communication infrastructure into energy systems presents challenges for traditional centralized structures, especially when managing complex grid configurations with numerous connections. Coordinating diverse elements and ensuring precise, real-time control is a significant challenge for scalable and efficient energy service.

Decentralized smart control in local energy systems provides a flexible and scalable approach to grid management and coordination. The multi-agent fault protection system architecture described in [20] involves multiple intelligent electronic devices as agents, i.e., line agent, load agent, and decentralized production agent, and coordinates them to achieve more accurate fault detection and outage protection. Multi-agent systems have the advantages of high computation efficiency and enhanced robustness over a centralized approach. They do not require aggregation of information from the entire energy system while promising to avoid single points of failure and improve overall system reliability.

Besides, a scalable local energy system not only promotes decentralized smart control, but it also includes distributed energy resources (e.g., solar panels) and energy storage devices (e.g., electric vehicles) with the need for energy transition. Leveraging these diverse renewable energy resources, local energy systems can reduce reliance on traditional energy sources and achieve a more sustainable energy supply.

Various applications within local energy systems require 5G networks, such as smart metering, fault detection, and load balancing. Local energy systems must maintain a reliable, stable, and continuous energy supply. Scalable 5G networks must be capable of managing the numerous connections between devices and infrastructures in local energy systems. 5G communication networks are expected to ensure the network has no latency, failures, or interruptions that could adversely impact energy production and distribution. Thus, prompt measures for 5G networks are also important to deal with potential network congestion. Edge computing brings computation tasks closer to the relevant devices, which reduces the need to transmit all data to a centralized center and enables faster data processing. Integration of edge computing to 5G networks can enhance energy system efficiency and support faster responses by local decision-making [19]. Additionally, network slicing within 5G networks can allocate virtual network instances for diverse grid services or applications [30]. This guarantees critical services have the necessary network resources to meet low-latency and reliable requirements. Customizing network slices allows for specific characteristics and dedicated network portions to support different applications effectively.

4 Conclusion

This chapter highlights the significant impact of 5G networks on enhancing the efficiency, reliability, and scalability of local energy systems. By enabling real-time monitoring, data exchange, and control, 5G supports the integration of advanced technologies such as smart metering, dynamic pricing, and energy theft detection, which are essential for modernizing energy management and ensuring grid stability. The chapter also discusses key challenges, including cybersecurity, consumer privacy, and service scalability, and proposes solutions such as blockchain technology, federated learning, and edge computing to mitigate these issues.

Looking beyond 5G, future communication technologies such as 6G are expected to offer even greater capabilities, including higher data rates, lower latency, and advanced AI integration. These advancements will further support the development of local energy systems, enable more effective demand response strategies, and enhance the ability to manage distributed energy resources. The evolution towards these next-generation networks will continue to drive innovations in local energy systems, realizing more resilient, sustainable, and intelligent energy infrastructures.

References

1. Series M (2015) IMT vision–framework and overall objectives of the future development of IMT for 2020 and beyond. Recomm ITU 2083(0)
2. Yilmaz ONC, Wang Y-PE, Johansson NA, Brahmi N, Ashraf SA, Sachs J (2015) Analysis of ultra-reliable and low-latency 5G communication for a factory automation use case. In: 2015 IEEE international conference on communication workshop (ICCW), pp 1190–1195
3. Navarro-Ortiz J, Romero-Diaz P, Sendra S, Ameigeiras P, Ramos-Munoz JJ, Lopez-Soler JM (2020) A survey on 5G usage scenarios and traffic models. IEEE Commun Surv Tutor 22(2):905–929

4. Shehab MJ, Kassem I, Kutty AA, Kucukvar M, Onat N, Khattab T (2021) 5G networks towards smart and sustainable cities: a review of recent developments, applications and future perspectives. IEEE Access 10:2987–3006
5. Zhang L, Mei C, Li J, Liang Y, Song J et al (2019) A survey on 5G network slicing enabling the smart grid. In: 2019 IEEE 25th international conference on parallel and distributed systems (ICPADS), pp 911–916
6. Ghosal A, Conti M (2019) Key management systems for smart grid advanced metering infrastructure: a survey. IEEE Commun Surv Tutor 21(3):2831–2848
7. Jackson J (2014) Smart grids: an optimised electric power system. In: Future energy, pp 633–651
8. Xia X, Xiao Y, Liang W, Cui J (2022) Detection methods in smart meters for electricity thefts: a survey. Proc IEEE 110(2):273–319
9. Makhadmeh SN, Khader AT, Al-Betar MA, Naim S, Abasi AK, Alyasseri ZAA (2019) Optimization methods for power scheduling problems in smart home: survey. Renew Sustain Energy Rev 115:109362
10. Kabalci E, Kabalci Y (2019) From smart grid to internet of energy. Academic Press
11. Liu C, Yang RJ, Yu X, Sun C, Wong PSP, Zhao H (2021) Virtual power plants for a sustainable urban future. Sustain Cities Soc 65:102640
12. Gellings C et al (2011) Estimating the costs and benefits of the smart grid. Electric Power Research Institute USA
13. Prasad J, Samikannu R et al (2017) Overview, issues and prevention of energy theft in smart grids and virtual power plants in Indian context. Energy Policy 110:365–374
14. Jindal A, Schaeffer-Filho A, Marnerides AK, Smith P, Mauthe A, Granville L (2020) Tackling energy theft in smart grids through data-driven analysis. In: 2020 International conference on computing, networking and communications (ICNC), pp 410–414
15. Wen M, Xie R, Kejie L, Wang L, Zhang K (2021) Feddetect: a novel privacy-preserving federated learning framework for energy theft detection in smart grid. IEEE Internet Things J 9(8):6069–6080
16. Otuoze AO, Mustafa MW, Abioye AE, Sultana U, Usman AM, Ibrahim O, Omeiza IOA, Abu-Saeed A (2022) A rule-based model for electricity theft prevention in advanced metering infrastructure. J Electric Syst Inf Technol 9(1):1–17
17. Zeng B, Ouyang S, Zhang J, Hui S, Geng W, Ming Z (2015) An analysis of previous blackouts in the world: lessons for China's power industry. Renew Sustain Energy Rev 42:1151–1163
18. Haes Alhelou H, Hamedani-Golshan ME, Njenda TC, Siano P (2019) A survey on power system blackout and cascading events: research motivations and challenges. Energies 12(4):682
19. Fan D, Ren Y, Feng Q, Liu Y, Wang Z, Lin J (2021) Restoration of smart grids: current status, challenges, and opportunities. Renew Sustain Energy Rev 143:110909
20. Azeroual M, Boujoudar Y, Bhagat K, El Iysaouy L, Aljarbouh A, Knyazkov A, Fayaz M, Qureshi MS, Rabbi F, EL Markhi H (2022) Fault location and detection techniques in power distribution systems with distributed generation: Kenitra City (Morocco) as a case study. Electric Power Syst Res 209:108026
21. Shahbazi H, Karbalaei F (2020) Decentralized voltage control of power systems using multi-agent systems. J Modern Power Syst Clean Energy 8(2):249–259
22. Vahidi S, Ghafouri M, Au M, Kassouf M, Mohammadi A, Debbabi M (2023) Security of wide-area monitoring, protection, and control (WAMPAC) systems of the smart grid: a survey on challenges and opportunities. IEEE Commun Surv Tutor
23. Haes Alhelou H, Hamedani-Golshan ME, Njenda TC, Siano P (2019) Wide-area measurement system-based optimal multi-stage under-frequency load-shedding in interconnected smart power systems using evolutionary computing techniques. Appl Sci 9(3):508
24. Dutta G, Mitra K (2017) A literature review on dynamic pricing of electricity. J Oper Res Soc 68(10):1131–1145
25. Tang R, Wang S, Li H (2019) Game theory based interactive demand side management responding to dynamic pricing in price-based demand response of smart grids. Appl Energy 250:118–130

26. Chang J-C, Wu T-H (2020) Demand side management of power for time-of-use pricing based on particle swarm optimization. In: 2020 3rd IEEE international conference on knowledge innovation and invention (ICKII), pp 317–320
27. Parandeh K, Bagheri A, Jadid S (2023) Optimal day-ahead dynamic pricing of grid-connected residential renewable energy resources under different metering mechanisms. J Modern Power Syst Clean Energy 11(1):168–178
28. Moghaddam Z, Ahmad I, Habibi D, Masoum MAS (2019) A coordinated dynamic pricing model for electric vehicle charging stations. IEEE Trans Transp Electrif 5(1):226–238
29. Wang X, Luo X, Zhang M, Jiang Z, Guan X (2020) Detection and isolation of false data injection attacks in smart grid via unknown input interval observer. IEEE Internet Things J 7(4):3214–3229
30. Wijethilaka S, Liyanage M (2021) Survey on network slicing for Internet of Things realization in 5G networks. IEEE Commun Surv Tutor 23(2):957–994
31. Krebs B (2012) FBI: smart meter hacks likely to spread. Krebs on Security
32. de Souza MA, Pereira JLR, de Alves GO, de Oliveira BC, Melo ID, Garcia PAN (2020) Detection and identification of energy theft in advanced metering infrastructures. Electric Power Syst Res 182:106258
33. Chattopadhyay A, Ukil A, Jap D, Bhasin S (2017) Toward threat of implementation attacks on substation security: case study on fault detection and isolation. IEEE Trans Ind Inf 14(6):2442–2451

Traffic-Informed Electric Vehicle Coordination in Integrated Local Energy and Transportation Systems

Yujian Ye and Quan Yuan

Abstract As a part of the global decarbonization agenda, the electrification of the transport sector involving the large-scale integration of electric vehicles (EV) constitutes one of the key initiatives. However, the upstream generation of carbon emissions associated with EV charging demand is still accountable which should not be overlooked. In this context, efficient coordination of EV flows and their charging demand in the coupled transportation and power networks promises significant decarbonization potential. Nevertheless, such potential can only be realized under adequate incentive mechanisms. In this chapter, a novel low-carbon EV charging coordination approach in a coupled transportation and power network is proposed and formulated as a bi-level optimization. In the upper level, an AC optimal power flow problem is formulated and solved to determine the optimal operation for the power system. Then, the carbon emission flow tracing is performed to compute rational locational-differentiated price signals. Given price-based incentives, the lower level employs traffic user equilibrium to describe the distribution of EV path flow and charging demands, taking into account the uncertain traffic conditions, and the aggregate charging demands are fed back to the upper level. The bi-level problem is solved iteratively through a modified particle swarm algorithm with enhanced convergence properties. Case studies demonstrate the effectiveness of the proposed coordination method in effectively mitigating the global carbon emission of the coupled networks.

1 Introduction

Environmental and energy security concerns have paved the way for a remarkable transformation in our energy systems. Imagine a future where renewable energy sources and electric transportation seamlessly merge to tackle climate change,

Y. Ye (✉) · Q. Yuan
School of Electrical Engineering, Southeast University, Nanjing, Jiangsu, China
e-mail: yeyujian@seu.edu.cn

Q. Yuan
e-mail: yuanquan_seu@seu.edu.cn

W. Hua et al. (eds.), *Digitalisation of Local Energy Systems*, Green Energy and Technology, https://doi.org/10.1007/978-3-031-77833-9_10

which is not confined to China but extending globally, driven by the urgent need to decarbonize our energy infrastructure [1, 2]

A quarter of global greenhouse gas emissions stem from electricity and heat production [3]. The burning of fossil fuels like coal, natural gas, and oil for generating power and heat stands as the single largest contributor to these emissions. Another challenge is that the transportation sector is responsible for 14% of global greenhouse gas emissions, primarily from fossil fuel-powered vehicles that dominate our streets, skies, and seas. Shockingly, a whopping 95% of transportation energy relies on petroleum-based fuels, with gasoline and diesel leading the way. In this context, transportation electrification serves as a significant approach to reduce carbon emission in the transportation sector, and thus increase the electric vehicle stock and penetration [4, 5].

While the introduction of electric vehicles (EVs) promises zero on-road emissions, their overall impact stretches beyond just driving. The emissions tied to their manufacturing and charging should also be considered [6]. As electric vehicles intertwine transportation and electric networks, their movements and charging behaviors set off a chain reaction, impacting the flow of electricity and the flow of traffic. Imagine each EV as a puzzle piece, its choices influencing the bigger picture. The energy consumed and the carbon emitted from these choices are intricately woven into the fabric of the Integrated Local Energy and Transportation Systems (ILETS). To master this puzzle, managing the way EVs are charged becomes a crucial step toward an economy that thrives on both low carbon and affordability.

In the realm of managing EV charging within ILETS, scholars' explorations have revealed that EVs serve a threefold purpose: First, they hold the power to ease traffic congestion through strategic travel and charging decisions [7–9]. Second, they offer a chance for smarter energy consumption, with slow and fast charging patterns playing their roles. Slow charging helps shift loads to off-peak times, smoothing out energy demand peaks [10–13]. Fast charging, on the other hand, involves higher energy requirements and speaks to the spatial distribution of charging stations and travel routes. This challenge invites us to devise charging strategies that respond to variable traffic conditions [14–16]. Last, EVs, when coordinated, contribute to the overall social welfare by reducing costs, both monetary and temporal [17–19]. However, the true potential of these coordinated charging behaviors to mitigate carbon emissions is an area still in its infancy.

Addressing vehicular carbon emissions involves intricate models that bridge the gap between transportation and emissions. Some studies have delved into assigning traffic based on fuel emission models for traditional vehicles [20]. And when it comes to electrified transportation, solutions have been sought through quantifying emissions and integrating them into planning [21, 22]. But what if we told you there's more to this puzzle? What if we dive into the specifics of each energy load, their temporal and spatial variations, tracing carbon emissions, and then using these insights to incentivize better energy consumption choices? This is the heart of what we're about to explore.

In the following pages, we will unveil a groundbreaking approach that transforms EVs into not just vehicles but into dynamic contributors to a cleaner, greener future. Section 2 outlines the proposed ILETS framework and the traffic user equilibrium (TUE) model of EV fleets. Section 3 details the modeling of charging loads integrating into the distributed power grid along with the carbon flow tracing model to decide the low-carbon incentives for EV charging. Section 4 formulates the bi-level operation of EV charging coordination in the local energy system, and describes the solving algorithm. Section 5 presents the case studies and illustrative results demonstrating the value of the proposed method.

2 Traffic-Informed Behavior Modeling of Electric Vehicle Fleets

The interaction between the transportation network, the power grid, and the EV charging loads is shown in Fig. 1. In the coupled transportation and power networks, the path or charging station selection of each EV is influenced by the traffic information, e.g., the traffic flow, the waiting time, and the charging price at different

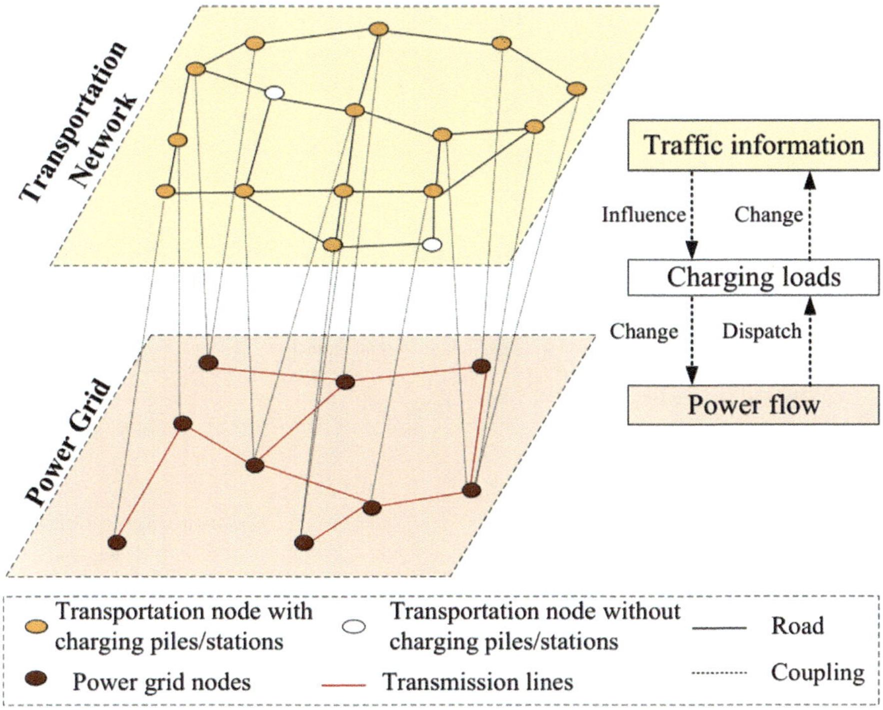

Fig. 1 Interaction between transportation network, power grid, and charging loads of electric vehicles

fast charging stations (FCS), etc. In addition, the charging decisions of the EV users are related to the spatial-temporal distribution of the aggregate charging demands. From the grid operator's perspective, the traffic information should be available to accurately evaluate and coordinate the charging loads.

Meanwhile, after the power grid operator dispatches the charging loads, the traveling and charging decision of EVs, especially those with fast charging demand, will conversely result in a change in the traffic conditions. The traffic network can be abstractly described by a connected graph $G = [N, A]$. Each EV traveling in G departs from an origin and arrives at a destination, forming an O-D pair. A path is composed of multiple end-to-end connected links, while a link can be shared by multiple different paths. Each O-D pair can be connected by different paths, where the EV flows will be distributed.

The TUE problem can be formulated as follows:

$$\min \sum_{a \in A} \int_0^{x_a} W t_a(n) dn + \sum\nolimits_{z \in Z} W \int_0^{x_z} t_z(n) dn + \lambda_z P_t(x_z^0) \tag{1}$$

$$t_a(x_a) = t_a^0 \left(1 + 0.15 \left(\frac{x_a}{c_a}\right)^4\right), \forall a \tag{2}$$

$$t_z(x_z) = t_z^0 \left(1 + \mu_z \frac{x_z}{(c_z - x_z)}\right), \forall z \tag{3}$$

$$x_a = \sum_{j \in J} \sum_{i \in I_j} f_{i,j} \delta_{i,j,a}, \forall a \tag{4}$$

$$x_z = \sum_{j \in J} \sum_{i \in I_j} f_{i,j} \Pi_{i,j,z}, \forall z \tag{5}$$

$$\sum_{i \in I_j} f_{i,j} = d_j, \forall j \tag{6}$$

$$f_{i,j} \geq 0, \forall i, \forall j \tag{7}$$

$$P_z^{EV} = \sum_{j \in J} \sum_{i \in I_j} f_{i,j} \Pi_{i,j,z}, \forall z \tag{8}$$

The objective function Eq. 1 is composed of three cost components: (i) total travel time cost on the links, (ii) total waiting time cost at FCS, and (iii) total electricity purchasing cost from the local energy system at FCS. a denotes the value of time. Constraints Eqs. 2 and 3 describe the relationships between the path flow fi,j and flows on link a and at FCS z. Driven by the congestion effect, the traveling time of EVs on link a is represented as a BPR latency function of the EV flows, following the commonly adopted practice in [17, 18]. Besides, FCSs are sparsely distributed in the ILETS, so the binary variable *ai,j,z* is employed to make sure each feasible path contains at least one FCS. Essentially, constraint Eq. 4 suggests that the flow on link *a* is not strictly bounded by its capacity *ca*, congestion will be penalized through a sharp increase in the traveling time. The *Davidson function* [17] based on the *queuing theory* is employed to model the queuing time of EV flows at FCS *z*, as expressed in constraint Eq. 5. Constraints Eq. 6 defines the charging price at FCS *z*, including energy price and extra carbon emission price. Constraints Eq. 7 and 8 ensure that all path flows are non-negative, and the summation of flows over all connecting paths for an O-D pair *j* equals to the EV traveling demand of the same O-D pair.

The sources of nonlinearities in the TUE formulation include the nonlinear BPR latency function and the Davidson function. These nonlinear functions can be approximated by piece-wise linear (PWL) functions and augmented in TUE as linear constraints with integer variables. In this article, the PWL approximation is performed by introducing *type 2 special ordered set variables* (SOS2), which promises more efficient branching strategies, and is desirable for branch-and-bound solvers [23]. To perform SOS2 linearization on constraint Eq. 4, for example, we partition the feasible interval of x_a is into smaller segments. In each segment, $t_a(x_a)$ is approximated by a linear function, and constraint Eq. 4 can be re-expressed as:

$$x_a = \sum_m x_a^m \eta_a^m, \forall a \tag{9}$$

$$t_a = \sum_m t_a\left(x_a^m\right) \eta_a^m, \forall a \tag{10}$$

$$\sum_m \eta_a^m = 1, \forall a \tag{11}$$

$$0 \leq \left\{\eta_a^m, \forall m\right\} \in SOS2, \forall a \tag{12}$$

where x_a^m denotes the break point of each segment, and η_a^m denotes continuous weights of each segment. The SOS2 linearization of Eq. 5 can be carried out in the same logic.

3 Low-Carbon Incentives for Electric Vehicle Charging Coordination

The FCS in the traffic network is served by a power distribution network (PDN), which is usually a radial network, and can be represented by a tree topology. Given the distribution of EV charging demands through the TUE model in Sect. 2, the PDN is modeled by ACOPF. In addition to the beneficial impact promised by mobile flexible resources such as EVs (such as directing EV flows to FCS connected by abundant wind resources) in decarbonization of the distribution system, we further explore the beneficial impact from non-mobile resources such as carbon capture storage (CCS) in reducing carbon emissions of co-located distributed generators (DG). The general ACOPF can be formulated as:

$$\min \sum_{k} \left[a_k (P_k^G)^2 + b_k P_k^G\right] \tag{13}$$

$$P_{ky}^{l'} + P_y^{NG} - r_{ky}^{l'} k_y^{l'} = \sum_{r \in K_y} P_{yr}^{l'} + P_y^D, \forall l \tag{14}$$

$$Q_{ky}^{l'} + Q_y^G - x_{ky}^{l'} k_y^{l'} = \sum_{r \in K_y} Q_{yr}^{l'} + Q_y^D, \forall l \tag{15}$$

$$U_y = U_k - 2(r_{ky}^{l'} P_{ky}^{l'} + x_{ky}^{l'} Q_{ky}^{l'}) + (z_{ky}^{l'})^2, \forall l \tag{16}$$

$$i_{ky}^{l'} U_k \geq (P_{ky}^{l'})^2 + (Q_{ky}^{l'})^2, \forall l \tag{17}$$

$$i_{ky}^{l'} \leq I^{l'}, \forall l \tag{18}$$

$$P_{ky}^{l'} \geq 0, \forall l \tag{19}$$

$$Q_{ky}^{l'} \geq 0, \forall l \tag{20}$$

$$P_{G_k}^{min} \leq P_k^G \leq P_{G_k}^{max}, \forall k \tag{21}$$

$$Q_k^{min,G} \leq Q_k^G \leq Q_k^{max,G}, \forall k \tag{22}$$

$$U^{min} \leq U_k \leq U^{max}, \forall k \tag{23}$$

$$CE_k = \beta P_k^G - CC_k, \forall k \in K^C \tag{24}$$

$$CE_k = \beta P_k^G, \forall k \in K \backslash K^C \tag{25}$$

$$CC_k = \epsilon P_k^C, \forall k \in K^C \tag{26}$$

$$P_k^{NG} = P_k^G - P_k^C, \forall k \in K^C \tag{27}$$

where a_k and b_k are respectively the quadratic and linear term of the cost function of generator k, P_k^G, Q_k^G, and P_k^N denote the active generation power, reactive generation power, net generation power of generator k, $P_{ky}^{'}$, $Q_{ky}^{'}$ are the active and reactive transmission power of line l between node k and node y, while $i_{ky}^{'}$ denotes the line's quadratic current. $r \in K_y$ is defined as the subset node r belonging to the set of node y.

$r'_{ky}, x'_{ky}, z'_{ky}$ are the resistance, reactance, and impedance of line l. P_y^D, Q_y^D represent the in-shiftable loads of node y, here representing the power demands except electric vehicle charging demands. U_k is the quadratic voltage of node k. $\overline{I}_l$ is the united quadratic current of line l. $P_{k,\min}^G$ and $P_{k,\max}^G$ denote the upper and lower bound of the active output generation, while $Q_k^{G,\min}$ and $Q_k^{G,\max}$ are the reactive upper and lower bound. $U^{\min}$, $U^{\max}$ denote the minimum and maximum value of the quadratic voltage. CE_k and CC_k are the carbon emission and carbon captured in node k. $k \in K^C$ denote the node set K with carbon capture systems (CCS), featuring positive carbon captured value, where $k \in K \backslash K^C$ are the nodes without CCS, featuring zero carbon capture value. P_k^C is the power consumed by the capture system, while β and ε are respectively the carbon emission intensive index and the carbon capture efficiency.

In the formulation, objective Eq. 13 is to minimize the generation cost of the local energy system. Constraints Eqs. 14–15 express the nodal active and reactive power balance of the PDN. For example, $P_N G$ involve output of DGs, CCS, and photovoltaic (PV), and $P_L$involve EV charging demands at FCS, namely $P_E V$, and electric loads. Constraints [13–17, 24–26] together form the power flow constraints of PDN after second-order cone programming (SOCP). Notably, the original formulation of the ACOPF problem is non-convex and NP-hard, to pursue non-convexity and foster higher computational efficiency, the SOCP convex relaxation technique to employed. Convex relaxations increase the feasible space to include the non-convex feasible space. Qian et al. [18], Yuan and Tang [19], Wang et al. [20], Fang et al. [21] are the constraints of the CCS.

Since carbon flow corresponds to the energy consumption, carbon flow tracing relies on the power flow allocation. The carbon flow tracing clarifies where the energy supply of each sink comes from, and how much percentage of power comes from each generator. Therefore, the essence of carbon flow tracing is to clarify the power flow mapping from sources to sinks. The sources refer to generators, and the sinks represent the load connected to each node, namely, each node could have multiple sinks.

The mapping between sources and sinks can be expressed as:

$$P_d = F P_c \tag{28}$$

where P_d, P_c are the output of sources and input of sinks. $\boldsymbol{F}$ is the flow mapping matrix representing the contribution of sources to sinks. For any given node x in the distribution power network, its relationship between sources and sinks can be expressed as:

$$P_c = C P_x \tag{29}$$

$$P_d = D P_x \tag{30}$$

where P_x denotes the nodal gross input flow. C, D are the contributions from sources to x and the input power flow of sinks origin from x, respectively. The elements of C and D are defined as:

$$\alpha_{mn} = \begin{cases} 1 & m = n \\ -\frac{P_{mn}}{P_{xn}} & n \in m^+ \\ 0 & n \neq m, n \notin m^+ \end{cases} \tag{31}$$

$$\beta_{mn} = \begin{cases} \frac{P_{dm}}{P_{xn}} & n \in m^+ \\ 0 & n \notin m^+ \end{cases} \tag{32}$$

where $\boldsymbol{m+}$ is the set of nodes belonging to the input branches of node m. $P_n m$ denotes power flow from m to n. P_{dm} denotes the power of Z_{th} sinks related to m. $P_{x,n}$ is the P_{th} row of matrix P_x. In this article, the net loss is transferred to sinks, and the equations for loss transfer are as follows:

$$\Delta P_d = D\Delta P_x \tag{33}$$

$$\Delta P_{xm} = \sum_{n\in m^+} \Delta P_{mn} \tag{34}$$

Where ΔP_d is the allocated net loss to sinks. ΔP_x is the nodal input loss and $\Delta P_{x,m}$ is its elements. Define a loss allocation matrix L, in which:

$$\tau_{mn} = \begin{cases} \alpha_{mn}^{(-1)} \frac{\Delta P_{mn}}{P_{xn}} & n \in m^+ \\ 0 & n \notin m^+ \end{cases} \tag{35}$$

where $\alpha_{mn(-1)}$ is the row m column n element of A_{-1}. Now we have the final mapping:

$$\Delta P_d = DC^{-1}LP_c \tag{36}$$

$$P_d = FP_c = DC^{-1}P_c \tag{37}$$

$$P_d' = (F - DC^{-1}L)P_c \tag{38}$$

Noting the P_d is the power of sinks, including the net load, and P_d is the actual input power of sinks. Consequently, the amount of carbon emission of each load/sink can be expressed as:

$$E_d^{ce} = \varepsilon^G FP_c \tag{39}$$

For a local energy system with transportation-electricity coupling, it is still possible to utilize the decarbonization potential of EV demands by transferring the charging load of EV to nodes with lower carbon emission intensity, such as nodes with higher renewable generation penetration. As talked above in Sect. 1, the EV charging demands can be influenced based on differentiated price signals. Therefore, to reduce the carbon emission, the price incentive could be designed to vary with the carbon emission, and the specific strategy is introduced in the next chapter.

4 Optimal Operation of ILETS Considering Electric Vehicle Fleets

Mathematically, the price-based optimal operation of a local energy system considering EV charging coordination is a bi-level optimization problem. The upper level problem simulates the energy system operation and decides the optimal charging prices of FCS, which is implemented by the ACOPF model and the carbon emission flow tracing model introduced in Sect. 3. At the lower level of the proposed framework, TUE is solved according to the augmented prices to minimize the total charging cost of EV users. The charging demands for EV in this sub-problem are then transferred to the upper level. This procedure iterates until an equilibrium is reached.

Hence, the carbon price (CP) is introduced under the carbon tax policy. The contribution of per MWh electricity to the total carbon emissions can vary widely for different nodes depending on their locations in the PDN. Therefore, the money paid by EV users is actually used as incentives for them to charge at low-carbon FCS, whose electricity originates from clean renewable energy resources such as PV and wind. The loads can be charged according to their responsible carbon emissions, as generation-based incentives are less sensitive and explainable for EV users. Given the carbon emission flow tracing model, the carbon prices for loads can be obtained by identifying their responsible carbon emissions.

The locational CP per MWh of electricity demand equals the product of the constant carbon tax γ and the corresponding nodal average carbon emission intensity, as:

$$\lambda_z^C = \gamma \frac{CE_z}{P_z^{EV}}, \quad \forall z \tag{40}$$

For clarity, the unit of λ_z^C, γ, CE_z, P_z^{EV} are CNY/MWh, CNY/ton, ton/h, and MW, respectively.

Consequently, the locational electricity-carbon augmented prices can be obtained by combining the electricity price λ_z^e and the carbon price λ_z^C as Eq. 6.

In the proposed model, the two levels are iteratively optimized to reach an equilibrium. At the upper level, both the electricity prices and carbon prices are decided with fixed electric demands. At the lower level, the augmented prices are used as parameters of the TUE sub-problem to update the electric demands. These demands are then transferred to the upper level for the next iteration. Equilibrium is achieved when the electric demands of each FCS at each time period are close enough between two iterations, i.e.:

$$\left| \frac{P_z^{EV}(k) - P_z^{EV}(k-1)}{P_z^{EV}(k)} \right| \leq \zeta, \forall z \tag{41}$$

where ζ denotes the tolerance and k is the iteration index. The main procedures of the solving method are presented as Algorithm 1. In this article, the particle is set as λ_z^e in z-dimension to find the maximum fitness function value. Noting that the fitness function of PSO should be non-negative, so the fitness function is set as the reciprocal of Eq. 12.

Algorithm 1: The Adopted Modified PSO Algorithm.

Input: $k_{\max}$, initial acceleration constants c_1, c_2, random searching function r_1, r_2, inertial weight Ω.

Output: Particle having the best fitness value $F_{d,g}^k$.

Initialization: $k = 0$
Set $k_{\max}$, initial acceleration constants c_1, c_2, random searching function r_1, r_2, inertial weight Ω
Set $\lambda_c^z = 0, \forall z$
for *each particle i* **do**
  for *each dimension d* **do**
    Initialize position $x_{i,d}$ and velocity $v_{i,d}$ randomly within permissible range
  end
end
for $k \leftarrow 1$ **to** $k_{\max}$ **do**
  for *each particle i* **do**
    Solve TUE Eqs. 1–9 to obtain p_z^{EV}
    Solve ACOPF Eq. 23 to obtain the fitness value for each particle.
    Solve carbon flow tracing Eqs. 24–38 to update λ_z^c
    if *the fitness value is better than* $F_{i,d,p}^k$ *in history* **then**
      replace $F_{i,d,p}$ with current fitness value.
    end
  end
  Choose the particle having the best fitness value as $F_{d,g}^k$.
  for *each particle i* **do**
    for *each dimension d* **do**
      Calculate the velocity according to

$$v_{i,d}^{k+1} = \Omega v_{i,d}^k + c_1 r_1 (F_{i,d,p}^k - x_{i,d}^k) + c_2 r_2 (F_{d,g}^k - x_{i,d}^k) \tag{10.42}$$

      Update particle position according to

$$x_{i,d}^{k+1} = x_{i,d}^k + v_{i,d}^{k+1} \tag{10.43}$$

      Update acceleration constants according to

$$c_1 = c_1 (1 - \frac{k}{k_{\max}}), c_2 = c_2 \frac{k}{k_{\max}} \tag{10.44}$$

      ;
    end
  end
  if *condition Eq. 41 holds or* $k > k_{\max}$ **then**
    Break.
  end
end

The PSO algorithm is modified by adjusting the acceleration constant $c1$ and $c2$ (as shown in step 21 in this Algorithm). As the constant $c1$, $c2$ reflect how the particles find the next position, while $c1$ concerns each individual particle's information and $c2$ concerns global information. In this context, $c1$ should decrease and $c2$ should increase as the iteration goes on, in order to explore as much of the whole feasible space as possible, while avoiding falling into global optimization. The constant $c1$ and $c2$ could be updated every iteration according to 10.42. Such algorithm modification has also been proved effective in [27–29], exhibiting faster convergence speed with respect to the conventional PSO algorithm.

By applying the modified PSO algorithm, the overall problem can converge to the TUE and optimal augmented pricing equilibrium after a finite iterations. In the next section, the two case studies verify the effectiveness of the proposed local energy system coordination method.

5 Example Research

In this section, we first validate the effectiveness of the proposed coordination method on a simple coupled 6-node traffic network (TN) and 6-node PDN networks. Then, we consider a real-world test system in Nanjing, China, involving a coupled 20-node TN, 21-node PDN networks, and corroborate the beneficial impact of the proposed low-carbon EV charging coordination. In both test cases 1 and 2, the aggregate EV charging behavior will be coordinated by electricity prices only and the augmented electricity-carbon price, respectively, to validate the effectiveness of the proposed low-carbon coordination method. The numerical data of both scenarios are optimal results obtained from the modified PSO algorithm convergence.

5.1 Test Case 1: 6-Node TN and 6-Node PDN System

The topology of the 6-node ILETS is shown in Fig. 2. TN contains 9 links with 400 veh/h road capacity. Three FCS with 600 veh/h capacity and 100 kW charging power are located at nodes T2, T4, and T5. The unity power factor for EV charging demand is assumed. The parameter of Davids on function is set as $\mu = 0.05$. The O-D demand is set to $1,000$ veh/h from node T1 to node T6. The value of unit time ω corresponds to a normalized distribution with mean 20 CNY/h and deviation 5 CNY/h. DGs are located at nodes E1, E2, and E4, while PV generators are located at nodes E2 and E4. DGs share the sameset of cost parameters: $a = 1,000$ CNY/(MWh)$_2$ and $b = 65,000$ CNY/MWh.

The bi-level optimization problem is iteratively solved until an equilibrium is reached. Figure 3 illustrates the TUE coordination under LMP pricing and the augmented electricity-carbon pricing, as well as the corresponding charging price at each FCS.

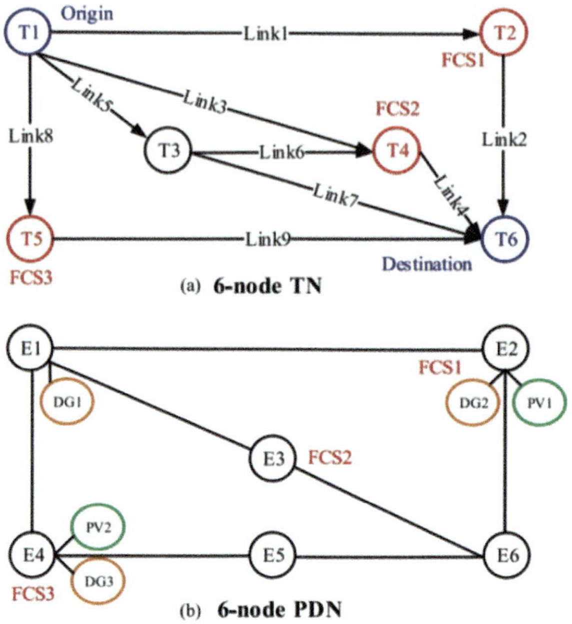

Fig. 2 Topology of the 6-node ILETS

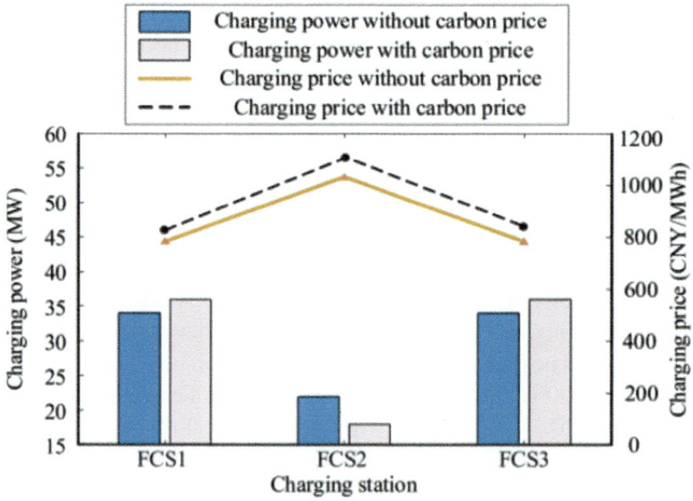

Fig. 3 Charging demands and charging price at each FCS in 6-node ILETS

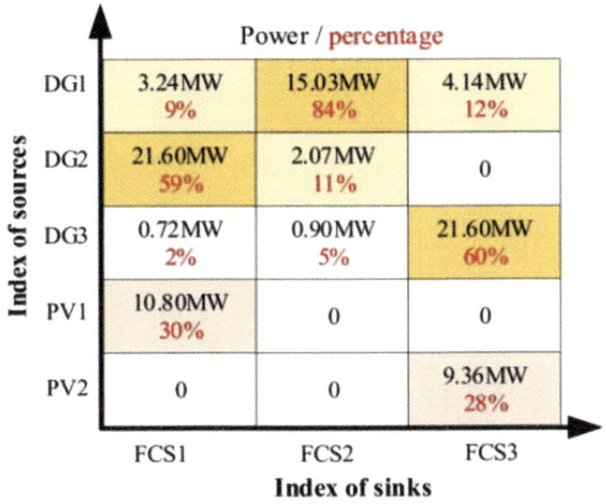

Fig. 4 Carbon emission flow mapping of the 6-node ILETS

As shown in Fig. 3, the augmented price increases the charging price at each FCS while enlarging the price difference between them. FCS 1 and 3 are equipped with PV, resulting in less per MWh carbon emission and consequently lower augmented charging price. As the carbon emission flow tracing results indicate different obligations of GHGE at each node, the proposed coordination method actually encourages EV users to charge at FCS with more cleaner and low-carbon energy supply.

Figure 4 shows the carbon emission flow mapping from sources to sinks at convergence. According to the carbon tracing results, DGs supply the major energy consumption of each FCS and incur inevitable GHGE, which contributes to the non-negative carbon price. More specifically, only FCS 1 and 3 have PV supply, and FCS 2 only has DG power supply, so the carbon price at FCS 2 will be higher than that in FCS 1 and 3, even though FCS 2 has less charging power.

The carbon emission of DGs and PVs w/wo carbon price (CP) are presented in Table 1. According to Table 1, the coordination increased the local consumption of PV, namely at FCS 1 and 3. The output of DG1 significantly decreased. Additionally,

Table 1 Carbon emission of each generator with/without augmented carbon price

Generator	Carbon intensity (ton/MWh)	Emission without CP (ton)	Emission with CP (ton)
DG1	1.205	32.468	27.690
DG2	0.900	23.309	21.303
DG3	1.015	23.265	23.568
PV1	0.025	0.4253	0.4500
PV2	0.022	0.3742	0.3960

Table 2 UE and ACOPF operational state with/without augmented carbon price

Item	Without CP	With CP	Percentage (%)
UE cost (CNY)	40,250	42,986	+6.36
Operational cost of PDN (CNY)	75,209	71,430	–5.02
DG output (MW)	75.764	69.300	–8.53
PV consumption (MW)	24.962	30.176	+20.89
Net loss (MW)	6.115	4.876	–20.41
Carbon emission (ton)	76.842	73.407	–4.47

Table 2 shows the UE and OPF results, as well as the overall carbon emission of the ILETS w./w.o. carbon price at convergence. The UE cost inevitably increased due to the employment of a carbon price, and the extra charging cost is borne by EV users in the total UE cost. The reason for the OPF cost reduction and the overall carbon emission reduction is threefold: decreasing of DG output, reduction of net loss, and increase in PV consumption. The DG output decreased by 6.464 MW by 8.53%, and the net loss reduced by 1.248 MW by 20.41%, and the PV consumption increased by 5.216 MW by 20.89%. The global GHGE consequently reduced by 4.47%. These results validate that the proposed method leads to the redistribution of EV charging loads, and therefore effectively increases renewable energy consumption and reduces the global GHGE of the ILETS.

The convergence curve of the modified PSO algorithm is shown as Fig. 5. Both w./w.o. CP curve convergence at 7th iteration, and the scenario with CP can converge to a lower OPF value due to the spatial redistribution of EV charging loads. The adopted algorithm has stable convergence performance, and seldom falls into oscillating or local optimum. Notably, the SOS2 linearization mentioned in Eq. 10 is adopted in the lower level TUE problem to speed up the bi-level problem solving and converging.

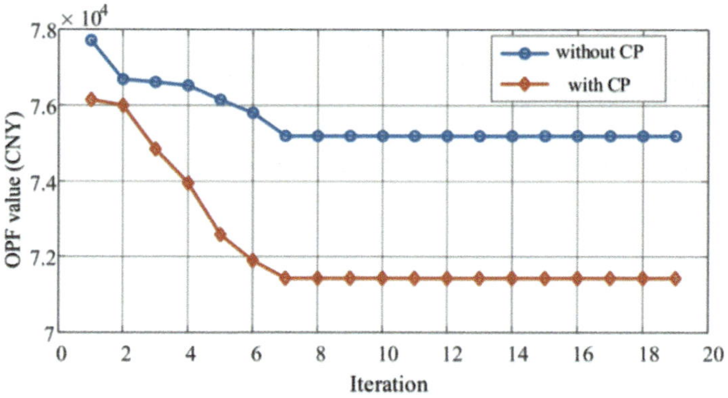

Fig. 5 The convergence curve of the modified PSO algorithm of 6-node ILETS

5.2 Test Case 2: 20-Node TN and 21-Node PDN System

The topology of the 20-node TN and 21-node PDN is simplified from the real-world data of Nanjing city as shown in Fig. 6.

The TN involves 54 traffic links and 6 FCS located at node T4, T7, T10, T12, T13, T18 (referred to as FCS 1-6), of which FCS 1, 2, 4, and 5 are characterized by PV-assisted stations. The connected location of FCS in the PDN is highlighted in Fig. 6b in red font next to the relevant PDN node. The charging power and capacity of each FCS are 100 kW and 2,800 veh/h, respectively, and the capacity of each link is 1,200 veh/h. The O-D demand is set to 10,000 veh/h. In order to reflect the diversity of EV users in O-D choice, we partition the TN into three different regions highlighted with different colors in Fig. 6a. Nodes at each region correspond to different probabilities of being sampled as an EV user's origin or destination. In this case, the probability corresponding to the orange/green/blue region is 0.6 /0.25 /0.15, resembling a case where the city center is the most popular choice for EV traveling demand. For more parameters related to the SOCP implementation of the PDN, associated with DG, fixed demand, PV, and CCS, please refer to [30].

Following a similar solving process in test case 1, the EV charging coordination result is presented in Fig. 7. Carbon emission flow mapping from sources to sinks at convergence is illustrated as Fig. 8.

According to Figs. 7 and 8, after applying the augmented charging price, the FCS with large-scale charging demand and intensive carbon emission will be charged for more carbon tax. This result corresponds with the motivation of integrating carbon emission obligation into charging price, so that EV users will be encouraged to charge at FCS with cleaner and low-carbon energy origins.

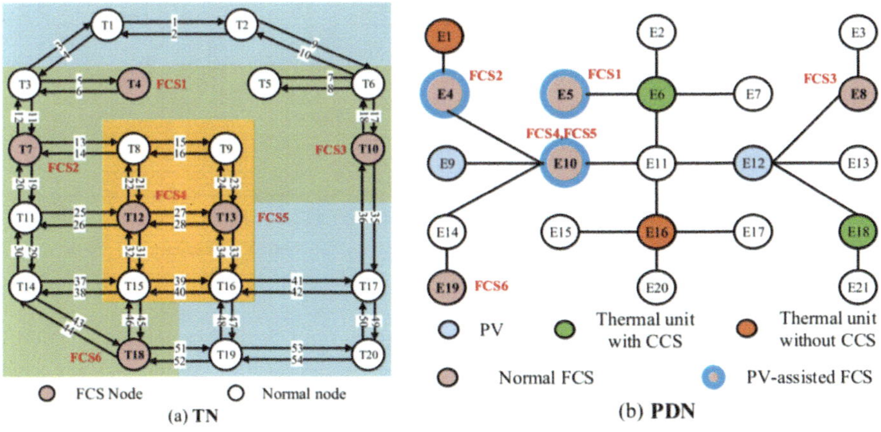

Fig. 6 The convergence curve of the modified PSO algorithm of 6-node ILETS

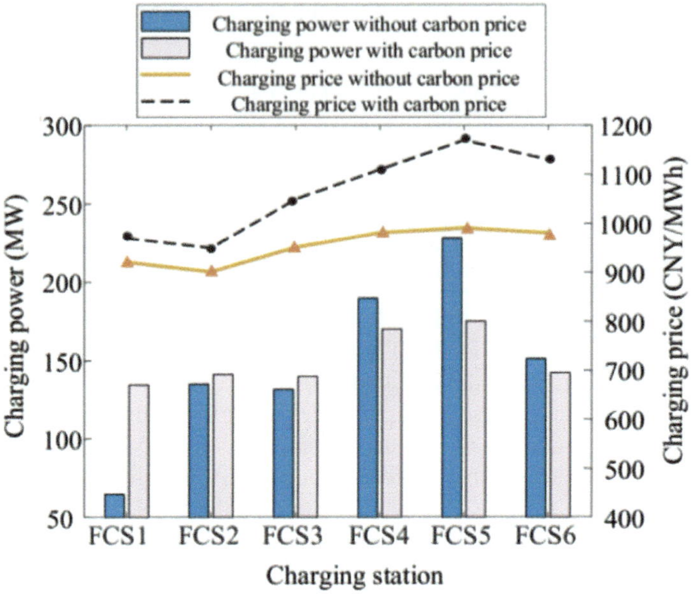

Fig. 7 Charging demands and charging price of each FCS in the real-world scale ILETS

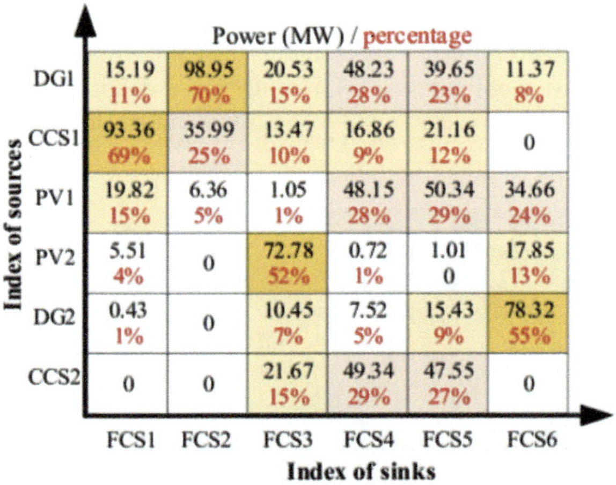

Fig. 8 Carbon emission flow mapping in the real-world scale ILETS

Consequently, the aggregated EV charging demand is distributed more evenly across all FCS. More specifically, FCS1 receives more charging demands after coordination due to its suitable location, i.e., close to the central region of TN, while avoiding heavy traffic flow. The energy supply of FCS1 mainly comes from CCS1, which also contributes to the carbon emission reduction.

The link traffic flow and FCS queueing time studies are also conducted, as shown in Figs. 9 and 10, respectively. The scenario with CP has better traffic congestion

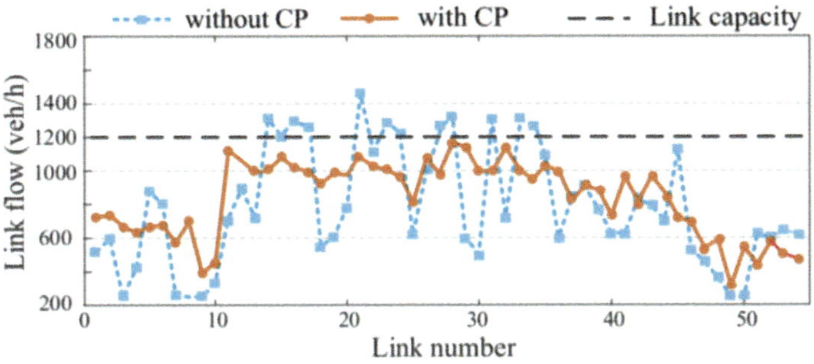

Fig. 9 Link flow w./w.o. carbon price in the real-world scale ILETS

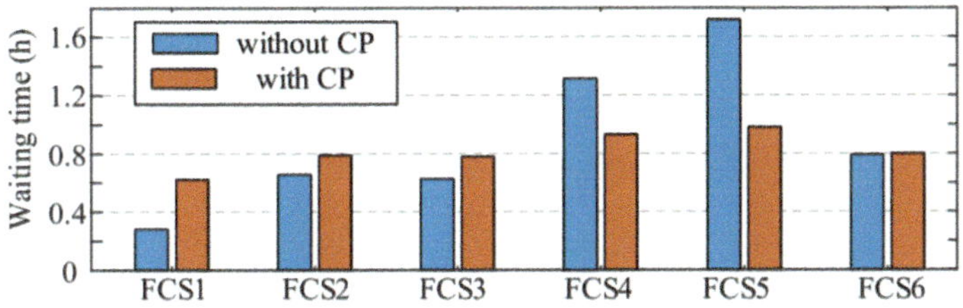

Fig. 10 Carbon emission flow mapping in the real-world scale ILETS

Table 3 Operational state with/without carbon price under real-world scale coupled networks

Item	Without CP	With CP	Percentage (%)
UE cost (CNY)	15.056	15.303	+1.64
Operational cost of PDN (CNY)	175.00	167.00	-4.57
DG output (MW)	678.43	640.88	-5.53
PV consumption (MW)	228.10	258.25	+13.22
Net loss (MW)	29.49	22.09	-25.09
Carbon capture (ton)	138.22	153.84	+11.30
Carbon emission (ton)	645.47	598.67	-7.25

management performance than the scenario without CP, because the EV transferred to FCS 1–3 simultaneously avoids the heavy traffic congestion in the city central regions, and consequently reduces the link flow of those overloaded links. The variation of queueing time at each FCS in Fig. 10 corresponds to their charging loads in Fig. 7.

Some numerical results are also obtained and presented in Table 3. Although the charging cost of EV users is increased due to the integration of carbon price, a significant improvement in traffic congestion is witnessed in TN (Fig. 9). Hence, the UE cost is slightly increased when these two factors interact. Due to the redistribution of the EV charging demands, PV absorption increased, and the overall DG output and net loss decreased, which leads to the reduction of OPF cost and GHGE. Notably, the EV coordination along with the CCS captures carbon emissions contributes to an overall reduction of 7.25% in GHGE. Therefore, the effective coordination of EV path flow and charging demands under adequate incentives contributes to the reduction of global carbon emissions of ILETS. The distribution of spatial-temporal EV charging demands can exploit its decarbonization potential while satisfying the charging requirements.

The convergence curve of the modified PSO algorithm in real-world scale ILETS is shown as Fig. 11. According to Fig. 11, the scenario with carbon price has a quicker converging speed than the scenario without carbon price. In a more sophisticated test system, the augmented electricity-carbon price will enlarge the locational price difference, assisting the particle swarm more efficiently in finding the global optimal fitness function. Also, the curve with the CP scenario converges to a lower OPF value, indicating the merit of applying a carbon price.

To conclude, Efficient coordination of EV flows and their charging demands in ILETS plays a vital role toward the low-carbon operation of the coupled networks, and the decarbonization potential can only be realized under adequate incentive mechanisms. In this context, this article proposes a bi-level EV charging coordination approach to mitigate the global carbon emission of the ILETS. After solving the ACOPF problem, the augmented electricity-carbon pricing method based on carbon emission flow tracing is obtained in the upper level, and the lower level solves the traffic assignment TUE problem to optimize EV path flow and charging demands. Further, the proposed bi-level optimization model is iteratively solved by a modified PSO algorithm until convergence. Case studies are carried out on a 6-node test case and a real-world scale system involving an ILETS in Nanjing city to validate the effectiveness of the proposed coordination method in alleviating the global carbon emission of the coupled networks.

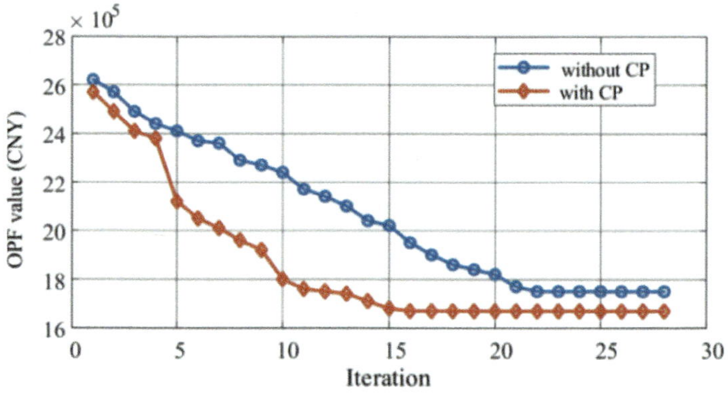

Fig. 11 The convergence curve of the modified PSO algorithm of real-world scale ILETS

References

1. The State Council of PRC (2022) China to develop high-quality new energy in new era. https://english.www.gov.cn/policies/latestreleases
2. Teng F, Ding Z, Hu Z, Sarikprueck P (2020) Technical review on advanced approaches for electric vehicle charging demand management, part I: applications in electric power market and renewable energy integration. IEEE Trans Ind Appl 56(5):5684–5694
3. United States Environmental Protection Agency (2022) Global greenhouse gas emissions data. https://www.epa.gov/ghgemissions/global-greenhouse-gas-emissions-data
4. Wang H, Ye Y, Wang Q, Tang Y, Strbac G (2023) An efficient LP-based approach for spatial-temporal coordination of electric vehicles in the electricity-transportation nexus. IEEE Trans Power Syst 38(3):2914–2925 May
5. Isik M, Dodder R, Kaplan PO (2021) Transportation emissions scenarios for New York city under different carbon intensities of electricity and electric vehicle adoption rates. Nat Energy 6:92–104 Jan
6. Yuan Q, Ye Y, Tang Y, Liu X, Tian Q (2023) Low carbon electric vehicle charging coordination in coupled transportation and power networks. IEEE Trans Ind Appl 59(2):2162–2172. (Mar–Apr 2023)
7. Zhong R, Xu R, Sumalee A, Ou S, Chen Z (2021) Pricing environmental externality in traffic networks mixed with fuel vehicles and electric vehicles. IEEE Trans Intell Transp Syst 22(9):5535–5554 Sep
8. Liu X (2020) Dynamic response characteristics of fast charging station-EVs on interaction of multiple vehicles. IEEE Access 8:42404–42421
9. Sarrafan K, Muttaqi KM, Sutanto D (2020) Real-time estimation of model parameters and state-of-charge of LI-ion batteries in electric vehicles using a new mixed estimation model. IEEE Trans Ind Appl 56(5):5417–5428
10. Lu X, Chan KW, Xia S, Zhang X, Wang G, Li F (2019) A model to mitigate forecast uncertainties in distribution systems using the temporal flexibility of EVAs. IEEE Trans Power Syst 35(3):2212–2221 May
11. Li Y et al (2023) Dense skip attention-based deep learning for day-ahead electricity price forecasting. IEEE Trans Power Syst 38(5):4308–4327 Sept
12. Alharbi B, Jayaweera D (2022) Impact assessment of plug-in electric vehicle charging locations on power systems with integrated wind farms incorporating dynamic thermal limits. IEEE J Mod Power Syst Clean Energy 10(3):710–718 May
13. Wang B, Dehghanian P, Wang S, Mitolo M (2019) Electrical safety considerations in large-scale electric vehicle charging stations. IEEE Trans Ind Appl 55(6):6603–6612
14. Morlock F, Rolle B, Bauer M, Sawodny O (2020) Forecasts of electric vehicle energy consumption based on characteristic speed profiles andrealtime traffic data. IEEE Trans Veh Technol 69(2):1404–1418 Feb
15. Gupta V, Konda SR, Kumar R, Panigrahi BK (2020) Electric vehicle driver response evaluation in multiaggregator charging management with EV routing. IEEE Trans Ind Appl 56(6):6914–6924
16. Sun X, Qiu J (2021) Hierarchical voltage control strategy in distribution networks considering customized charging navigation of electric vehicles. IEEE Trans Smart Grid 12(6):4752–4764 Nov
17. Wei W, Mei S, Wu L, Shahidehpour M, Fang Y (2017) Optimal trafficpower flow in urban electrified transportation networks. IEEE Trans Smart Grid 8(1):84–95 Jan
18. Qian T, Shao C, Li X, Wang X, Shahidehpour M (2020) Enhanced coordinated operations of electric power and transportation networks via EV charging services. IEEE Trans Smart Grid 11(4):3019–3030 Jul
19. Yuan Q, Tang Y (2021) Electric vehicle demand response technology based on traffic-grid coupling networks. Proc CSEE 41(5):1627–1636 Mar

20. Wang Y, Szeto W, Han K, Friesz TL (2018) Dynamic traffic assignment: a review of the methodological advances for environmentally sustainable road transportation applications. Transp Res Part B Methodol 111:370–394 May
21. Fang S, Xu Y, Li Z, Ding Z, Liu L, Wang H (2019) Optimal sizing of shipboard carbon capture system for maritime greenhouse emission control. IEEE Trans Ind Appl 55(6):5543–5553
22. Wei X, Zhang X, Sun Y, Qiu J (2022) Carbon emission flow oriented tri-level planning of integrated electricity-hydrogen-gas system with hydrogen vehicles. IEEE Trans Ind Appl 58(2):2607–2618
23. Vielma JP (2015) Mixed integer linear programming formulation techniques. SIAM Rev 57(1):3–57 Feb
24. Das S, Acharjee P, Bhattacharya A (2021) Charging scheduling of electric vehicle incorporating grid-to vehicle and vehicle-to-grid technology considering in smart grid. IEEE Trans Ind Appl 57(2):1688–1702
25. Huang Z, Fang B, Deng J (2020) Multi-objective optimization strategy for distribution network consideringV2G-enabled electric vehicles in building integrated energy system. Prot Control Mod Power Syst 5(7):1–8 Jan
26. Qiu D, Ye Y, Papadaskalopoulos D, Strbac G (2020) A deep reinforcement learning method for pricing electric vehicles with discrete charging levels. IEEE Trans Ind Appl 56(5):5901–5912. (Sept/Oct 2020)
27. Chang H-H, Lin L-S, Chen N, Lee W-J (2013) Particle-swarmoptimization-based nonintrusive demand monitoring and load identification in smartmeters. IEEE Trans Ind Appl 49(5):2229–2236
28. Upadhyay P, Marriboina V, Kumar S, Kumar S, Shah MA (2022) An enhanced hybrid glowworm swarm optimization algorithm for traffic-aware vehicular networks. IEEE Access 10:110136–110148
29. Zhou Q et al (2021) Modified particle swarm optimization with chaotic attraction strategy for modular design of hybrid powertrains. IEEE Trans Transport Electrific 7(2):616–625 Aug
30. Yuan Q, Ye Y, Tang Y, Liu Y, Goran S (2022) A novel deep-learning based surrogate modeling of stochastic electric vehicle traffic user equilibrium in low-carbon electricity-transportation nexus. Appl Energy 315, Art no 118961. (Apr 2022)

Safe Deep Reinforcement Learning for Renewable Energy Integrated Power System

Mengfan Zhang and Qianwen Xu

Abstract Deep reinforcement learning (DRL) is a promising solution for the coordination of power systems with high penetration of inverter-based renewable energy sources (RESs). Yet, when adopting the DRL-based control method, the safe and optimal operation of the system cannot be guaranteed at the same time, as the conventional DRL agent is not designed to solve the hard constraint problem. To address this challenge, a deep neural network (DNN) assisted projection-based DRL method for safe control of distribution grids is proposed in this section. First, a finite iteration projection algorithm is proposed to guarantee hard constraints by converting a non-convex optimization problem into a finite iteration problem. Next, a DNN-assisted projection method is proposed to accelerate the calculation of projection and achieve the practical implementation of hard constraints in the DRL problem. Finally, a DNN Projection embedded twin-delayed deep deterministic policy gradient (DPe-TD3) method is proposed to achieve optimal operation of distribution grids with guaranteed 100% safety of the distribution grid. The safety of the DRL training is guaranteed via the embedded Projection DNN in TD3 with participation in the gradient return process, which could smoothly and effectively project the DRL agent actions into the feasible area, thus guaranteeing the safety of data-driven control and the optimal operation at the same time. The case studies and comparisons are conducted in the IEEE 33 bus system to show the effectiveness of the proposed method.

M. Zhang · Q. Xu (✉)
Department of Electrical Engineering, KTH Royal Institute of Technology, Stockholm, Sweden
e-mail: qianwenx@kth.se

M. Zhang
e-mail: mezhang@kth.se

W. Hua et al. (eds.), *Digitalisation of Local Energy Systems*, Green Energy and Technology, https://doi.org/10.1007/978-3-031-77833-9_11

1 Introduction

The increasing integration of renewable energy sources (RESs) into the grid makes the modern power system more sustainable and effective [1]. Yet, the inherent characteristics of wind power and solar energy make the power supply more intermittent, thus, have introducing a significant challenge to the voltage security of power systems [2].

Traditionally, the extra devices, e.g., static Var compensator and static Var generators, are adopted to address the voltage stability issues, which directly adjust the reactive power of the system [3, 4]. However, they are not adequate to mitigate the fast voltage violations caused by the rapid renewable sources. Meanwhile, the additional devices will introduce extra cost to the system operation. Smart inverters, which serve as the interface of RESs to be connected to the grid, have, in theory, full controllability and can be controlled to provide extra reactive power supply for voltage support. Hence, it is promising to control inverter-based RESs in distribution grids to guarantee the grid voltage security.

To coordinate inverter-based RESs in distribution grids, conventional methods are mainly model based. In [5], a two-stage stochastic method is proposed to coordinate devices for voltage control in distribution grids. In [6], a three-stage robust optimization method is proposed to coordinate inverter-based RESs for voltage regulation. However, these model-based methods require an accurate physical model for the distribution grid and an accurate stochastic model to manage the uncertainty, which poses high stress on domain knowledge and human effort. Improper models or inaccurate parameters will lead to impractical results or even security violations. Moreover, they may need a high computation time that is not suitable for real-time operation.

To address the above-mentioned issues in model-based methods, deep reinforcement learning (DRL) based methods are proposed in recent works [7–11]. The agent of the DRL algorithm takes actions to interact with the environment to seek the optimal solutions of the voltage control [12]. The DRL-based method, which is trained with the feeding data, could achieve the satisfied voltage control performance both in the transmission system and the distribution system [13–16].

However, a significant challenge of existing DRL-based methods in power system applications is that they cannot guarantee safe operation and achieve optimal performance at the same time. During the training, the agent of the DRL algorithm explores the environment and takes every possible action to minimize the given reward function. The exploration of the agent cannot be hard constrained; thus, the generated actions cannot be constrained, which will eventually damage the safe operation of power systems. There are multiple works combining DRL and mathematical programming to address the safety constraints in DRL training. Reference [17] applies DRL combined with binary linear programming to develop a near-optimal solution of the scheduling problem of large-scale shared on-demand EV fleet, where the constraint of demand-supply gap is formulated by an explicit formula, and a constraint-based method is proposed for vehicle rebalancing to reduce the exploration space in

DRL training process. Yet, this method is not suitable for the safe control of distribution grids, as the voltage security constraints of the distribution grids cannot be represented by an explicit formula during the training. Meanwhile, Ref. [18] adds the Lagrangian relaxation procedure in the cost function to guarantee the safety of the DRL training. Yet, the relaxation cannot strictly achieve 100% safe operation of the system and is not suitable for the safe control of distribution grids. To address the safety issue in distribution system operation using the DRL method, several methods are proposed [19, 20], which can be classified into two groups. One group of methods is to directly regulate the voltage deviation of each bus in the system. For example, Refs. [21, 22] summarize the negative voltage deviation as the rewards, where the agents interact with the environment to achieve the optimal actions of the system. However, this method ignores the transmission loss of the system when designing the algorithm, which will degrade the system's operational performance. The other group of methods tries to transform the hard constraint optimization issue into a soft-constraint optimization problem, where the effects of physical constraints on the system are represented with penalty functions. The penalty functions are added to the reward function to force the agent to act in the feasible area. As illustrated in [23], the penalty of voltage violation is added to the reward function of the DRL algorithm, which could regulate the voltage of each bus and reduce the network loss to some extent. But the coefficient of penalty needs to be adjusted manually for good performance, which is usually hard to achieve. To avoid the extra adjustment of the hyperparameter, a Lagrange multiplier-based soft-constraint DRL method is proposed in [24]. The Lagrange multiplier is used to replace the coefficient of penalty, which participates in the DRL training for optimal performance. However, this group of methods still cannot strictly guarantee the 100% safety of the system operation, especially when two objectives of the reward function have conflicts [25].

In this section, we propose a groundbreaking safe DRL method to achieve optimal control with guaranteed safety to fill this research gap. Section 2 formulates the grid control problem as a deep reinforcement learning problem. Section 3 presents the proposed projection algorithm, and the proposed DNN-assisted projection algorithm to guarantee the hard constraint. The proposed DPe-TD3 method is presented in Sect. 4 to achieve optimal operation with guaranteed safe operation of distribution grids. The case studies and comparisons with existing methods are presented in Sect. 5 to validate the effectiveness of the proposed method.

2 DRL Problem Formulation

The voltage control problem in distribution grids can be formulated as a hard-constrained optimization problem. The optimization problem is formulated in Sect. 2.1, and the DRL problem is preliminarily formulated in Sect. 2.2.

2.1 System Description

For an n-bus distribution system, the power flow equations are depicted as follows:

$$\begin{aligned} p_{ij} &= g_{ij}\left(v_i^2 - v_i v_j \cos\delta_{ij}\right) - b_{ij} v_i v_j \sin\delta_{ij} \\ q_{ij} &= b_{ij}\left(-v_i^2 + v_i v_j \cos\delta_{ij}\right) - g_{ij} v_i v_j \sin\delta_{ij} \end{aligned} \tag{1}$$

where the complex power flowing from bus i to bus j is represented by $s_{ij} = p_{ij} + jq_{ij}$, $i, j \in N(N = 1, 2 \ldots, n)$, p_{ij} and q_{ij} are the active power and reactive power, respectively; $Y_{ij} = g_{ij} + jb_{ij}$ are the transmission line admittance; v_i and δ_i are the voltage magnitude and angle of bus i, respectively; and the voltage angle difference between bus i and bus j is $\delta_{ij} = \delta_i - \delta_j$.

For a distribution grid with high penetration of inverter-based RESs, there are voltage security issues caused by RESs, and it is promising to control the interface inverters of RESs to guarantee the grid voltage security. To maximize the utilization of RESs, the active power of inverters is fixed at the maximum available power of RESs, and the reactive power of inverters is adjustable and is to be controlled. Thus, the optimization problem of the distribution grid using the inverter interface RESs is modeled as follows:

$$\begin{aligned} &\min_{q_c} P_{loss} \\ &s.t. \quad \begin{cases} u_{lower} \leq |v_i| \leq u_{upper} \\ \text{constraints} \quad (1) \end{cases} \end{aligned} \tag{2}$$

where v_i denotes the voltage of bus i, q_c denotes the reactive power of controllable power electronics converters. u_{lower} and u_{upper} are the lower bound and upper bound of the bus voltage. The P_{loss} is the transmission loss of the distribution system, which can be written as follows:

$$P_{loss} = \sum_{i \in N} r_i i_i^2 \tag{3}$$

Due to the fluctuations of RESs, the above model-based optimization needs to be recalculated under the changing of RESs, which has a high computation time that is not suitable for real-time operation, and may lead to impractical results with the model uncertainties. Thus, the DRL-based method is adopted.

2.2 DRL Problem Formulation

The distribution grid optimization problem can be modeled as a Markov Decision Process and solved by DRL-based methods. The preliminaries of DRL problem formulation are presented in this subsection.

For DRL methods, in each time step $t = 0, 1, 2, 3, \ldots, T$, the agents interact with the environment by observing the state $s_t \in S$ and choosing the action $a_t \in A$ to find the optimal policy o through the whole operation period. The optimization process is to maximize the reward $J(\pi)$. In DRL methods, the policy π is represented with a deep neural network; thus, the policy is determined by the parameters of the neural network θ. And the reward function is written as $J(\theta)$. Combined with the problem described in the Sect. 2.1, the action and state, policy and reward are introduced as follows:

Action and State: As the inverter-based RESs work on the maximum power point tracking (MPPT) mode, the controllable variable is the reactive power of each inverter. Thus, the actions of the DRL agents are the reactive power of inverter-interfaced RESs. After the agents take action, the voltages at each bus will change. The voltage, active power, and reactive power of each bus are selected as the states. For an n-bus system that contains m renewables integrated inverters, the action of each inverter should obey the maximal rated power constraints, so the feasible action space A_f could be represented as follows:

$$\begin{aligned} A_{ci} &= \{q_{ci} : {p_{ci}}^2 + {q_{ci}}^2 < s_{ci,\max}^2\} \\ A_f &= A_{c1} \times \ldots \times A_{cm} \end{aligned} \tag{4}$$

where A_{ci} represents the action space of the i-th controllable inverter. The state space contains the voltage, active power, and reactive power of each bus, and the feasible state action S_f is represented as follows:

$$\begin{aligned} S_i &= \{p_i, q_i, v_i : {p_{ci}}^2 + {q_{ci}}^2 < s_{ci,\max}^2, u_{lower} \le |v_i| \le u_{upper}\} \\ S_f &= S_1 \times \ldots \times S_n \end{aligned} \tag{5}$$

Meanwhile, the feasible state action Sf should obey the power flow constraints.

Reward: The objective of voltage control is to regulate the voltage magnitude of each bus with minimal transmission loss of the system, where the transmission loss P_{loss} has been introduced in the previous section. As the normal DRL method cannot deal with the constrained optimization problem, a penalty term is added to the reward function to transform the constrained optimization problem into a non-constrained optimization problem. Thus, the reward is written as follows:

$$r = -P_{loss} - k \cdot f(v_i) \tag{6}$$

where $f(v_i)$ is the function to count the voltage violation.

However, the above-mentioned method could not strictly guarantee the safety of the training, especially when two objectives conflict. Meanwhile, if more constraints are added to the primary problem, the convergence cannot be guaranteed. Only if the state space S and the action space A are regulated with the hard constraints can the safety of system operation in the distribution grid be guaranteed.

3 Deep Neural Network For Projection

To address the challenge that normal DRL methods cannot guarantee the hard constraints, a novel finite iteration projection algorithm is proposed to guarantee hard constraints, and a DNN-assisted projection algorithm is proposed to accelerate the calculation of projection and achieve the practical implementation.

3.1 Proposed Finite Iteration Projection Algorithm

The core idea of our proposed method is to project the infeasible actions into the feasible action space to regulate actions with hard constraints.

The L2-norm projection can be adopted to achieve optimal actions with guaranteed safe operation. Consider the $P_F : R^n \rightarrow F$ project from the action space $a \in R^n$ to the closest point $\hat{a}$ in the feasible space $F \in R^n$. Then the problem is formulated as follows:

$$P_F := \arg\min_{\hat{a} \in F} \left\| a - \hat{a} \right\|_2^2 \tag{7}$$

As the projected actions $\hat{a}$ must be subject to the power flow constraints, which is non-convex, so that the feasible space F is not convex. Thus, it is challenging to achieve an optimal solution using the traditional methods (e.g., interior/exterior point method). To address this challenge, a finite iteration projection algorithm is proposed by converting the non-convex optimization problem into a finite iteration problem.

Figure 1 illustrates the principle of the proposed projection algorithm. Consider the problem that projects action a to the non-convex feasible region F with the smallest L2-norm. Then there exists $\hat{a} \in F$, such that the $\left\| \hat{a} - a \right\|_2^2$ is smallest. Sample a random point $a_0 \in F$, then the distance between a_0 and a can be calculated and marked as d_{upper}, which represents the largest distance with the projected action a in this sample process. Obviously, the shortest distance with the a is d_{lower} is 0. Then we can calculate the point a_k with the distance d_k by solving the following problem:

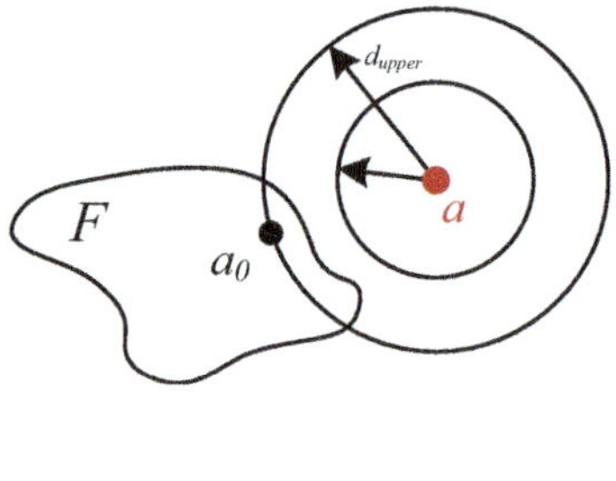

Fig. 1 Diagram of the finite iteration projection algorithm

$$\begin{cases} \|a - a_k\|_2^2 = d_k \\ \text{constraints} \quad (1) \\ u_{lower} \le |v_i| \le u_{upper} \end{cases} \tag{8}$$

where a_k can be easily solved. If the a_k is feasible, then update the a_{k+1} by solving the problem in (8) by changing the distance to $d_{k-1} - \frac{1}{2} * d_{upper}$. And if the constraints are not feasible, then update the ak+1 by solving the optimization problem in (8) by changing the distance to $d_{k-1} + \frac{1}{2} * d_{upper}$. After k times iteration, the error is limited to $1/2^k * (d_{upper} - d_{lower})$. Then with the proposed algorithm, the non-convex optimization problem Eq. (7) is transferred into a feasible iteration problem with linear rate of convergence. The detailed finite iteration projection algorithm is presented in Algorithm 1.

Algorithm 1: Finite iteration projection

Input: Infeasible action a, feasible space F.
Output: Projected action $\hat{a}$.
1 **initialize** ε as the acceptable tolerance of the algorithm
2 **sample** random action a_0 in F
3 **calculate**distance $d_{upper} = \|a - a_0\|_2^2$, $d_{lower} = 0$
4 **initialize** $d_0 = (d_{upper} - -d_{lower})/2$
5 **for** k = 1, ..., K, **do**
6 **calculate** a_k with (8)
7 **if** $1/2^k(d_{upper} - d_{lower}) < \varepsilon$ **end for**
8 **else do**
9 **if** a_k in Space F
10 **update** d_k in (8) with $d_k = d_{k-1} - \frac{1}{2^k} \times d_{upper}$
11 **else**
12 **update** d_k in (8) with $d_k = d_{k-1} + \frac{1}{2^k} \times d_{upper}$
13 **end for**

Proposition 1 *A feasible solution $\hat{a}$ can be found by Algorithm 1 in finite iterations.*

Proof. Let Ω, i.e., $a_1, a_2, \ldots, a_k, \ldots, a_K$, denote the set of actions found by Algorithm 1. For $k = 1$, if Problem (8) is infeasible as shown in Line 11 of Algorithm 1, d_1 is set to d_{upper}, as shown on Line 12 of Algorithm 1. When it comes to $k = 2$, with given d_1 as d_{upper}, a_0 is one feasible solution of Problem (8), indicating $a_2 = a_0$. If Problem (8) is feasible in the first iteration, a1 exists. It indicates Ω is nonempty. Due to the existence of power flow constraints (1) in Problem (8), these actions within Ω are feasible regarding the security constraints. Based on the linear decrease of searching diameter in Line 7 of Algorithm 1, it terminates in at most $log_2(d_{upper}/\varepsilon)$ iterations. As $\hat{a} \in \Omega$ is nonempty and Algorithm 1 terminates in $log_2(d_{upper}/\varepsilon)$ iterations, it finishes the proof.

3.2 DNN Assisted Projection for Practical Implementation

As the proposed finite iteration projection algorithm in Sect. 3.1 needs numerous iterations, it is time-consuming to add this algorithm in every time step of DRL training and algorithm implementation in real applications. To reduce the calculation burden and achieve practical implementation, a DNN-assisted projection method is proposed. The idea is to use a DNN to mimic the relationship between states s and actions a to the projected action â to accelerate the training and implementation. Meanwhile, as the DNN has the capability to return the gradient, it can be directly connected to the action network in the DRL algorithm.

The implementation of the proposed DNN-assisted projection method is presented as follows. Before the DNN training, the dataset should be prepared. As the states are decided by the actions in the distribution system, the average sampling is used to sample actions within the possible operating range. The states can be calculated by the power flow shown in Eq. (1). Then Algorithm 1 is adapted to generate the projection actions with corresponding states; thus, the dataset contains pairs of states and projection actions. After the training dataset is established, the data will be fed into the neural network training. A three-layer DNN as shown in Fig. 2 is trained to mimic Algorithm 1. The input of the neural network is the states of the system, i.e., voltage, active/reactive power; the outputs are the projection actions, i.e., adjustable reactive power of controllable power electronics inverters. The ReLU activation function is selected as the activation function of the first and second layers, and the tanh function is adapted in the output layer to regulate the output of the DNN within the operating range.

Fig. 2 DNN structure

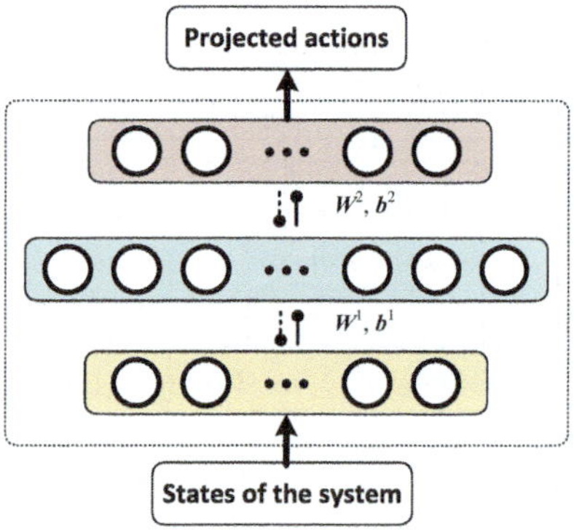

4 Proposed DNN Projection Embedded TD3

Based on the proposed projection algorithm, we further propose a DPe-TD3 algorithm for safe and optimal voltage control of distribution grids. The key idea is to embed the trained projection DNN into the actor networks in the TD3 algorithm, where the projection DNN participates in the gradient return during the DRL training. The overall algorithm is shown in Fig. 3 and illustrated below.

The structure of the proposed DPe-TD3 algorithm contains actor network π_θ with parameter θ_π, target actor network π_θ^T with parameter θ_π^T, critic network Q_1 with parameter $\theta_{Q,1}$ and critic network Q_2 with parameter $\theta_{Q,2}$, and target critic networks Q_1^T, Q_2^T with parameters $\theta_{Q,1}^T$, $\theta_{Q,2}^T$, and the projection DNN with trained parameters derived in projection process. For actor networks, we use a four-layer neural network, where the input of the actor network is the states of the distribution system (i.e., voltage, reactive power and active power of each bus), each hidden layer contains 128 neurons, and the output is the action (i.e., reactive power of smart converters). The ReLU is adopted as the activation function of the hidden layer, while the tanh function is adopted as the activation function for the output layer. For critic networks, we also use a four-layer neural network. The input is the states of the distribution system and the actions of each agent; each hidden layer contains 128 neurons. The ReLU is adopted as the activation function of the hidden layer, while there is no activation function for the output layer.

As shown in Fig. 3, the projection process (including the proposed finite iteration projection algorithm, Projection DNN training, and embedding) is highlighted in

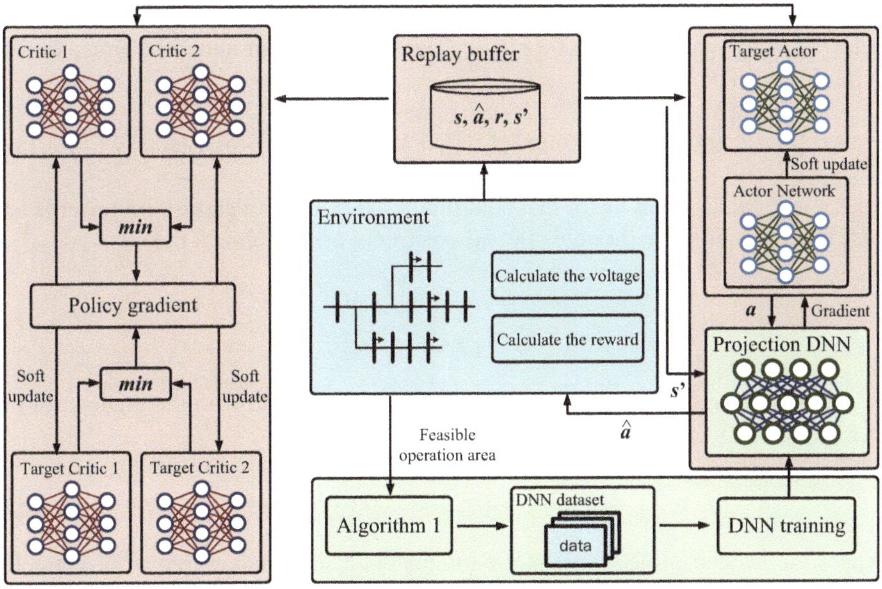

Fig. 3 Proposed DPe-TD3 algorithm for safe control of distribution grids.

green color in the figure. First, the actor network and critic network are initialized with random parameters, and the states are initialized. The actor network is used to find the optimal voltage control policy π, which could generate the optimal action based on the current states. At time step t, the states s generated by the environment, which consists of active/reactive power and voltage of each bus, are sent to the actor network, then the action $\pi_\theta(s)$ can be generated. To enable the preferable exploration of the agents, the Gaussian noises are added into the original actions, $a = \pi_\theta(s) + \varepsilon$. The active power and loads are from historic or predictive data, which are fixed during the time step t. And the generated actions are projected into the feasible region through a trained projection DNN, where the input of the DNN is the states, which consist of the active power, voltage, and generated reactive power. The outputs of the projection DNN are the feasible actions $\hat{a}$. Then the actions $\hat{a}$ are excluded in the environment to calculate the next step states s' using the power flow Eqs. (1) and the rewards using the following equation:

$$r = -P_{loss} \tag{9}$$

The feasible action $\hat{a}$, states s, s', and reward r are sent to the replay buffer to guarantee the independence of each sample [26]. If the replay buffer is fully filled, the oldest pair $(\hat{a}_t, r, s, s')$ is replaced with the newest one. During the training, the mini-batch is randomly sampled from the replay buffer.

In the critical networks, the clipped double Q learning technique is adapted to avoid the Q-value overestimation, where two Q-value functions $Q_1(s', \hat{a}')$ and $Q_2(s', \hat{a}')$ are generated by next step states and actions using the corresponding critical networks, respectively. Here, $\hat{a}'$ is generated with s' and added Gaussian noise, i.e., $\hat{a}' = \pi_\theta(s') + \varepsilon$. Then the minimum value between $Q_1(s', \hat{a}')$ and $Q_2(s', \hat{a}')$ is selected as Q-value instead of a single Q-value generation network. Thus, the TD error $\delta_i(t)$ can be written as:

$$\delta_i(t) = r_t + \gamma \cdot \min\left[Q_1\left(s', \hat{a}'\right), Q_2\left(s', \hat{a}'\right)\right] - Q_i\left(s, \hat{a}\right) \tag{10}$$

where γ is the discount factor that determines the priority of short-term reward. The loss functions of the actor network and critic network are shown below:

$$\begin{aligned} J_{c,i}\big|_{i=1,2} &= \mathbb{E}\left[\tfrac{1}{2}\delta_i(t)^2\right] \\ J_a &= \mathbb{E}\left[-Q_1\left(s, \hat{a}\right)\right] \end{aligned} \tag{11}$$

Thus, the parameters θ_π of the actor network are trained with the gradient of the performance function:

$$\nabla_{\theta_\pi} J_a\left(\theta_\pi\right) = \mathbb{E}\left[-\nabla_{\hat{a}} Q_1\left(s, \hat{a}\right) \nabla_{\theta_\pi} \pi_\theta\left(s\right)\right] \tag{12}$$

where the trained projection DNN participates in the gradient return of the actor network training with fixed parameters (i.e., the weights w and bias b of the DNN remain unchanged During the forward pass, the actions and states are fed into the projection DNN then generates the feasible actions. In the backward pass, the gradient is calculated with the same invariant weights w and bias b. Throughout the parameters updating process, the parameters of the projection DNN are kept fixed. And the critic networks are trained by minimizing the loss function:

$$\begin{aligned}\nabla_{\theta_{Q,i}} J_{c,i}\left(\theta_{Q,i}\right)\big|_{i=1,2} &= \mathbb{E}\left[-\delta_i\left(t\right)\nabla_{\theta_{Q,i}} Q_i\left(s,\hat{a}\right)\right] \\ &\approx \frac{1}{B}\sum_{k=1}^{B} \nabla_{\theta_{Q,i}} Q_i\left(s,\hat{a}\right)\cdot\left[r_k - Q_i\left(s_k,\hat{a}_k\right) + \gamma\cdot\min\left[Q_1\left(s_k{}',\hat{a}_k'\right), Q_2\left(s_k{}',\hat{a}_k'\right)\right]\right]\end{aligned} \tag{13}$$

where B is the size of the mini-batch, which is sampled randomly to reduce the correlation among the data. To guarantee training stability, the target actor network and target critic network with the lower updating rate are introduced [27]. The parameters of target networks are updated through:

$$\theta_\pi^T \leftarrow \tau\cdot\theta_\pi + (1-\tau)\cdot\theta_\pi^T \tag{14}$$

$$\theta_{Q,i}^T\big|_{i=1,2} \leftarrow \tau\cdot\theta_{Q_i} + (1-\tau)\cdot\theta_{Q,i}^T \tag{15}$$

where τ is a soft update coefficient, which is much lower than 1 to improve the stability of the learning. The pseudo-code of the proposed DPe-TD3 algorithm is shown in Algorithm 2.

Algorithm 2: DPe-TD3

Input: Parameters of network, Projection DNN D.
Output: Neural network agents.

initialize actor network π_θ and critic network Q_1, Q_2 with the corresponding parameters $\theta_{Q,1}$ and $\theta_{Q,2}$.
initialize target actor network $\pi_\theta^T(s)$ and target critic networks Q_1^T, Q_2^T with the corresponding parameters θ_π^T and $\theta_{Q,1}^T$, $\theta_{Q,1}^T$.
initialize replay buffer N
initialize batch size B
for episode = 1, ..., m, **do**
    for $t = 1, \ldots, T$ **do**
        select action a_t with exploration noise $a_t \sim \pi_\theta(s_t) + \varepsilon$
        project a_t with projection DNN $\hat{a}_t = D(s_t)$
        execute action $\hat{a}_t$ with (1)
        observe reward r_t and new state s_t'
        store state $(\hat{a}_t, r_t, s_t, s_t')$ in replay buffer N
        sample mini-batch of $B \sim$ replay buffer N randomly
        update the critic network through (13)
        update the actor network through (12)
        update target critic network through (15)
        update target actor network through (14)
    end for
end for

---

5 Case Study and Validation

The modified IEEE 33-bus distribution system is adopted to validate the proposed DPe-TD3 algorithm. The system diagram is shown in Fig. 4, where 6 solar PV systems are integrated into the system through their interface inverters, and eight variable loads are connected to the different buses. The one-week solar irradiance data are obtained from the NREL historical database, where the resolution is 1 min, the data are visualized as shown in Fig. 5. Assuming the PV panels are installed in the same area, the maximum active powers of different PV inverters are dependent on the installed PV panel area, which is shown as:

$$P_{i,\max} = k_i \cdot A_i \cdot l_{ir} \tag{16}$$

where k_i is the solar power conversion efficiency, A_i is the installed PV panel area, l_{ir} is the solar irradiance. As the interface PV inverters connected to the distribution system are controlled in the MPPT mode to fully utilize the solar, the active power produced by the PV inverter is equal to the maximum power that the PV panel can provide, i.e., $P_i = P_{i,max}$. Meanwhile, 5 typical loads data are acquired from the NREL database as well, which contains scenarios of a hospital, hotel, restaurant, supermarket, and warehouse. The loads data are visualized as shown in Fig. 6. The data sampling resolution of the loads is 1 minute.

The grid-connected PV panel area of each bus and loads distribution are shown in Table 1 . To validate the effectiveness of the proposed method, several experiments are conducted, and the comparison results are shown in this section.

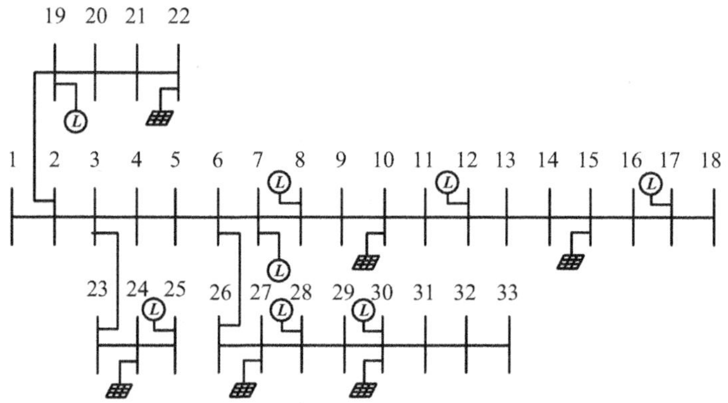

Fig. 4 IEEE 33 bus distribution system with PV inverters and variable loads

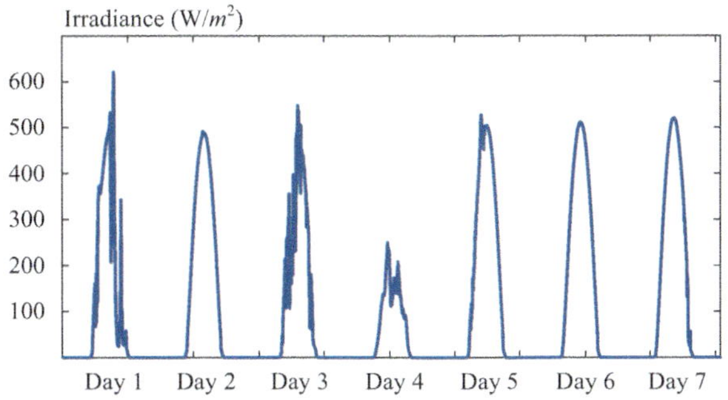

Fig. 5 One-week Irradiance profile

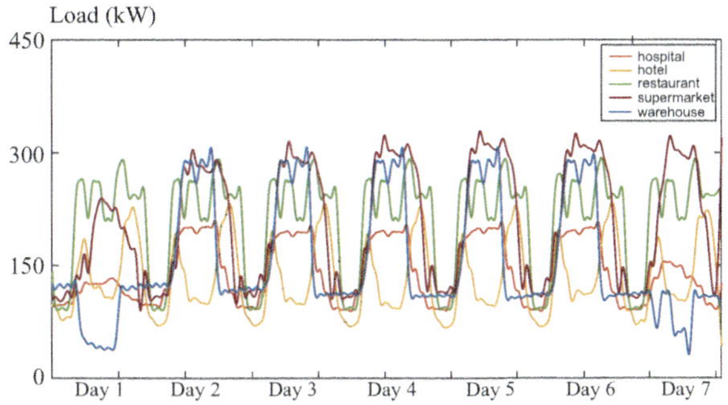

Fig. 6 One-week load profile

Table 1 Grid-connect devices setup

Device	Bus no.	PV area/m2
PV	10, 15	800
	22, 24, 27	1200
	30	1400
Device	Bus no.	Load type
Loads	7,25	Hospital
	8, 28	Htel
	12, 29	Restaurant
	17	Supermarket
	19	Warehouse

5.1 Training Setup and Results

The proposed DPe-TD3 algorithm contains two-step neural network training, i.e., the DNN projection training and the TD3 training.

For the DNN projection training, a three-layer DNN is used to map the states to the feasible actions, where the hidden layer contains 128 neurons. As the voltage of each bus changes as the active/reactive power varies, in the case study, 15000 data points of active/reactive power are selected in the DNN training dataset. For

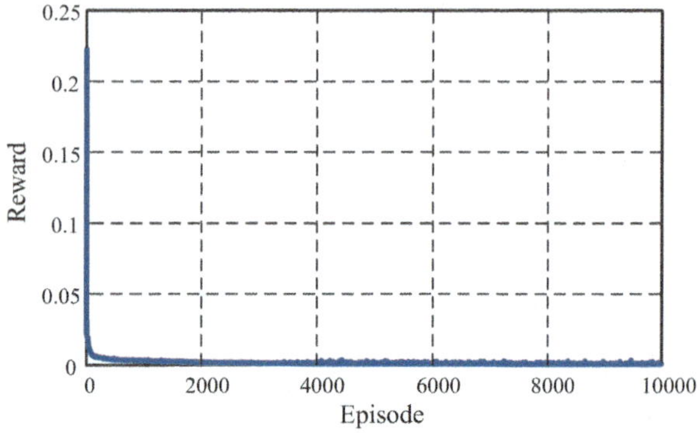

Fig. 7 Convergence of the reward of projection DNN training

Table 2 Hyperparameters of the training

Item	Value
Buffer size	2*105
Batch size	128
Discount factor	0.9
Max episodes	104
Actor network learning rate	1.00E-03
Critic network learning rate	2.00E-03
Target policy smoothing coefficient	0.1

the dataset preparation, the workflow can be descripted as: first, the actions are generated using a uniform distribution of feasible action space, and the states are sampled randomly from the state space, then the generated actions are fed into the environment calculation to generate the states (i.e., active power, reactive power and voltage), if the calculated voltage is within the voltage constraints, the generated actions are put into the training dataset, if the voltage does not within the constraints, the Algorithm 1 is adopted to generate the feasible actions, which are also put into the training dataset. With this strategy, the training dataset is established, where the states of the system (i.e., active/reactive power and voltage) are selected as the input and the projected actions are selected output.

After the dataset is established, the dataset is fed into the DNN for network training. The PyTorch framework is adopted for deep neural network training. The Microsoft Azure cloud platform with 4 AMD EPYC 7V12 CPUs is used for the training. The training process of the projection DNN is visualized as shown in Fig. 7. After almost 2200 episodes, the training converges. The trained DNN can be used to replace Algorithm 1 in the training of the proposed DPe-TD3 algorithm.

Then, the one-week active power data of inverters and load data are fed into the TD3 training. The hyperparameters in the training are presented as shown in Table 2 . The algorithm is coded using the PyTorch framework and trained with the same Microsoft Azure cloud platform. The convergence curve of the reward in each

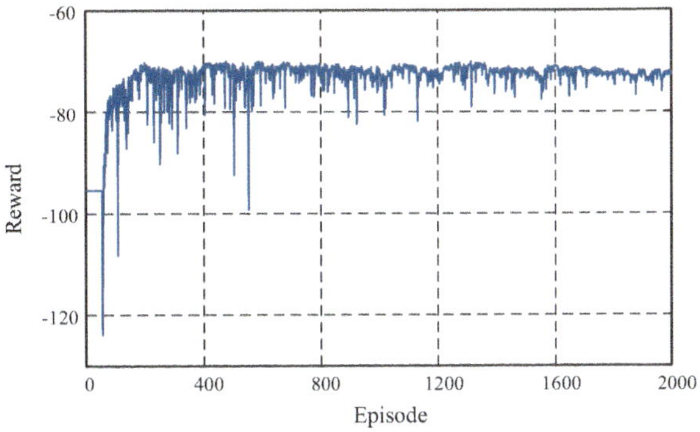

Fig. 8 Convergence of the reward of DNN-embedded TD3 training

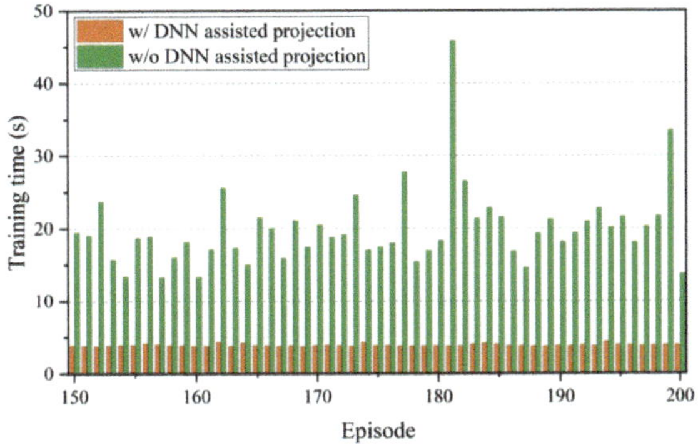

Fig. 9 Training time comparison between the DRL training with and without DNN-assisted projection layer

episode is shown in Fig. 8. From this curve, the agent does not start training during the first 55 episodes, as the training starts, the reward is reduced rapidly, and the reward converges after 400 episodes, which shows the developed algorithm has a fast rate of convergence. As in each time step, the agent selects actions with random noise, the reward curve has small fluctuations. To show the advantages of using DNN to improve computation efficiency, the training time comparison between scenarios with a projection DNN layer and scenarios without a projection DNN layer for each episode of training is shown in Fig. 9. As depicted in the comparison diagram, the trained DNN can significantly improve the training efficiency of reinforcement learning, reducing the training time by more than 300%.

After the training, the actor network is obtained. To show the training result, the trained network is tested with the 2 days (48 hours) data with 30 minutes resolution. The generated actions (i.e., reactive power setpoints) of six smart inverters are shown in Fig. 10.

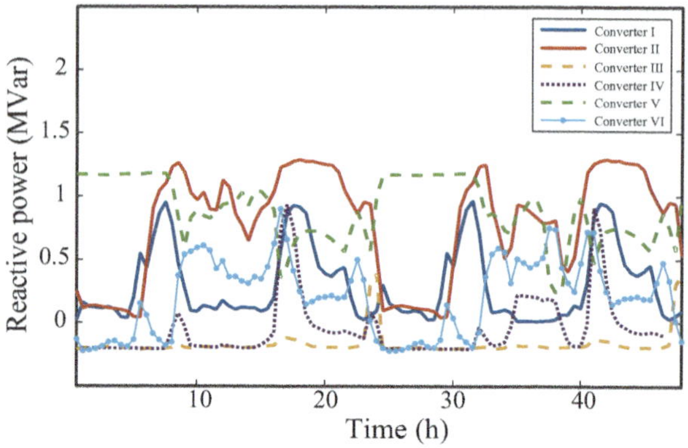

Fig. 10 48 hours reactive power setpoints of smart inverters

5.2 *Voltage Control Performance*

To show the voltage control performance of the proposed DPe-TD3 method, a comparison experiment of no extra voltage control with the same IEEE 33 bus distribution system and 2 days of data is carried out. The voltage profile of 33 buses without voltage is shown in Fig. 11. In the IEEE 33 bus system, the system rating voltage is 12.66 kV, the feasible voltage range is $[V_{lower} = 0.95 p.u., V_{upp} = 1.05 p.u.]$. From Fig. 12, if the voltage of the system is not under control, there are a lot of voltage violations during the system operation. When the system is controlled with the proposed DPe-TD3 method, the reactive power of PV inverters is controlled with the setpoint in Fig. 10, and the voltage profile of the system can be visualized as shown in Fig. 12, where the system does not have voltage violation. The voltage profile comparison shows that the proposed method can effectively restrain the voltage under the uncertain loads and solar irradiance profile.

Meanwhile, the transmission power loss of the distribution comparison between the proposed method and the no extra voltage control scenario is shown in Fig. 13.

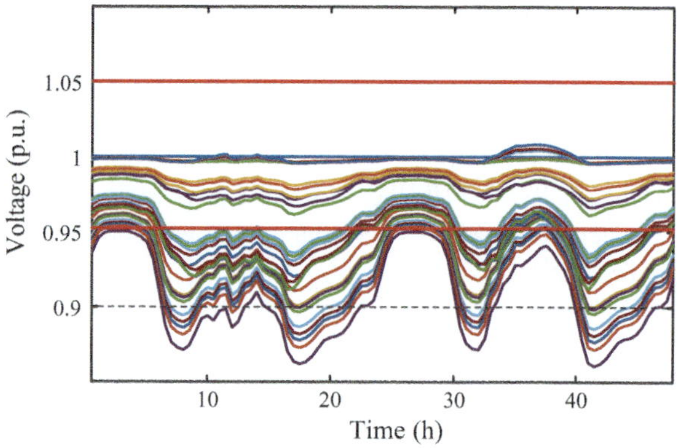

Fig. 11 Voltage profile of 33 bus without extra voltage control

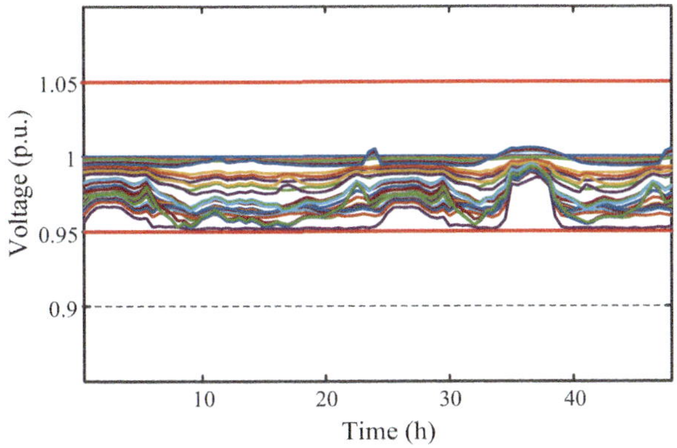

Fig. 12 Voltage profile of 33 bus system controlled by the proposed DPe-TD3 method

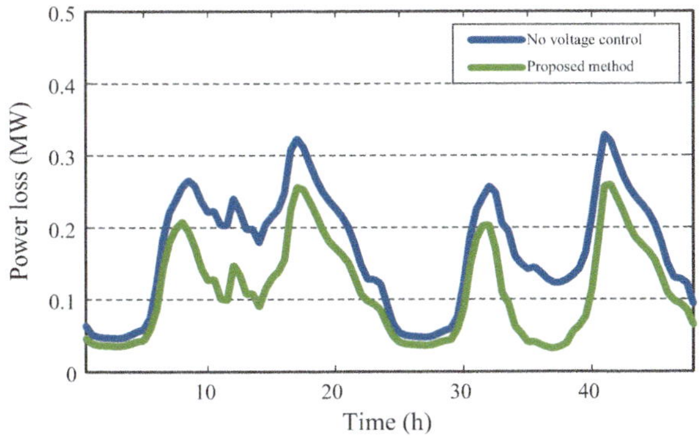

Fig. 13 Power loss comparison between the proposed DPe-TD3 method and no extra voltage control scenario

From the comparison, the power loss of the system controlled by the proposed method is lower every hour than the no extra voltage control scenario, which shows the proposed method has good capability to reduce the transmission loss of the system.

Hence, the proposed method can enable safe operation as well as minimize the transmission loss of the whole system.

5.3 Comparison with Existing Methods

To show the superiority of the proposed DPe-TD3 method against existing DRL-based voltage control methods, two comparison experiments are conducted, where the same training data and testing data are used. First, the comparison is made with the one group of DRL-based voltage control methods, e.g., ref. [21, 28], where the core idea is to minimize the voltage deviation V. The rewards of N-bus system can be written as:

$$r = -\sum_{i=1}^{N} |\Delta V_i| \tag{17}$$

The training hyperparameters are set the same as the TD3 training. The convergence is shown in Fig. 14, where the training converges after 500 episodes. The voltage profile of the method is shown in Fig. 15, and the power loss comparison between this method and the proposed method is shown in Fig. 16. From the voltage and power loss diagram, the voltage deviation reduction method has good capability to regulate the voltage of each bus, but introduces a larger transmission loss, even worse than the no voltage control scenario. Compared to this method, the proposed DPe-TD3 method can guarantee the safe operation while having better performance in reducing the transmission loss.

Next, the comparison is made with the group of existing safe DRL methods, e.g., Refs. [20, 23]. As mentioned in Sect. 1, this group of methods transfers the hard-constraints problem to the soft-constraints method using the additional penalty function in the reward function to regulate the voltage of the system. A Lagrange

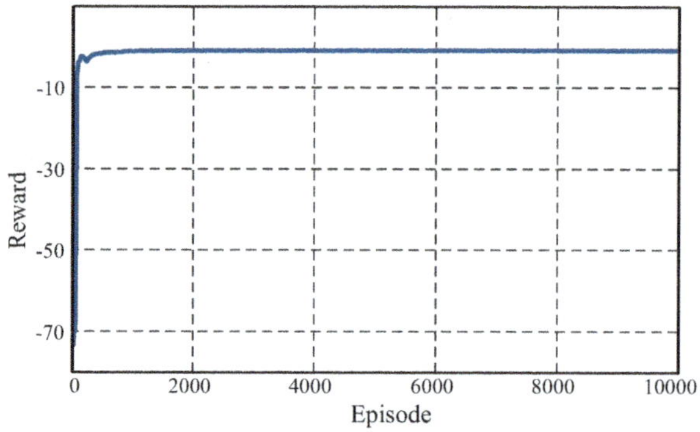

Fig. 14 Convergence of the reward of an existing DRL method in [28]

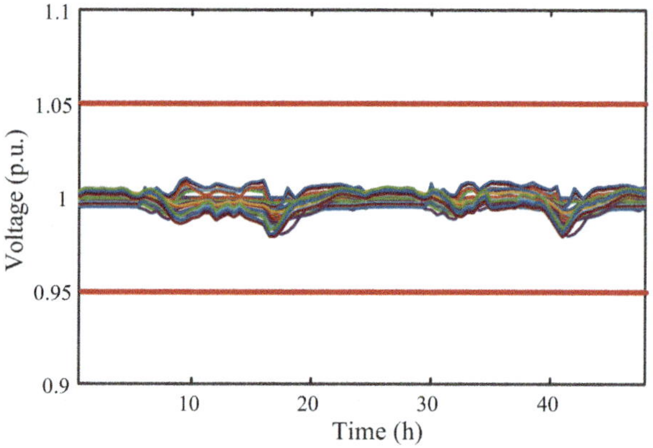

Fig. 15 Voltage profile of 33 bus system controlled by an existing DRL method in [28]

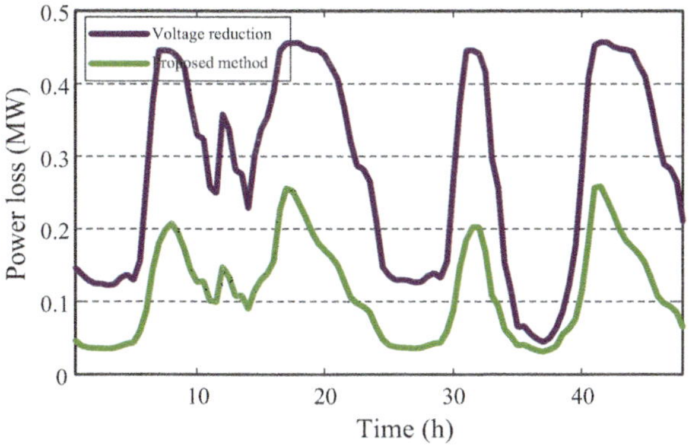

Fig. 16 Power loss comparison between the proposed DPe-TD3 method and an existing DRL method in [28]

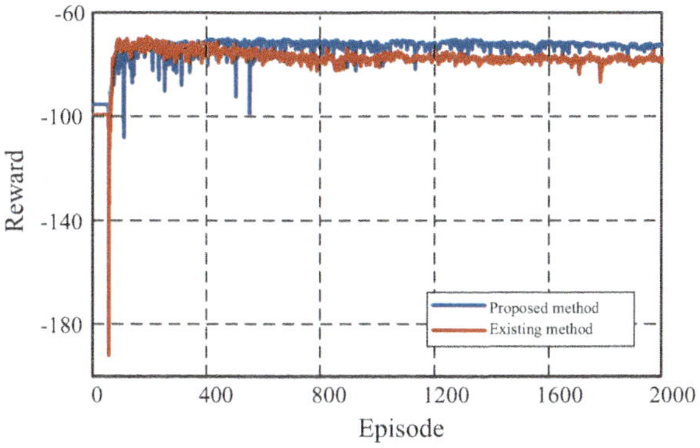

Fig. 17 Convergence of the reward of an existing DRL method in [20]

Multipliers-based soft-constraint DRL method is conducted in the distribution system [20]. To compare the training performance, the power loss reward is separated from the reward function of the soft-constraint method. The comparison of convergence of the training is visualized as shown in Fig. 17. From the convergence curves, the existing safe DRL method takes 800 episodes to converge, which is slower than the proposed DPe-TD3 method. Meanwhile, the trained reward is lower than the proposed method. The comparison indicates that the proposed DPe-TD3 method has better performance, as the training reward function of the existing safe DRL method has two sub-rewards to train, which may have some conflicts in some scenarios. After the training, the voltage profile of the soft-constraint DRL method is visualized in Fig. 18. From the diagram, this method has the capability to reduce the voltage violations of the system, but there are still several voltage violations during the two-day operation. The power loss comparison is shown in Fig. 19. It can be observed that the existing safe DRL method has a similar power loss reduction performance to the proposed method. From Figs. 16 and 19, compared to the existing safe DRL method, the proposed DPe-TD3 method has better performance to guarantee the safe operation while having similar performance in reducing the transmission loss.

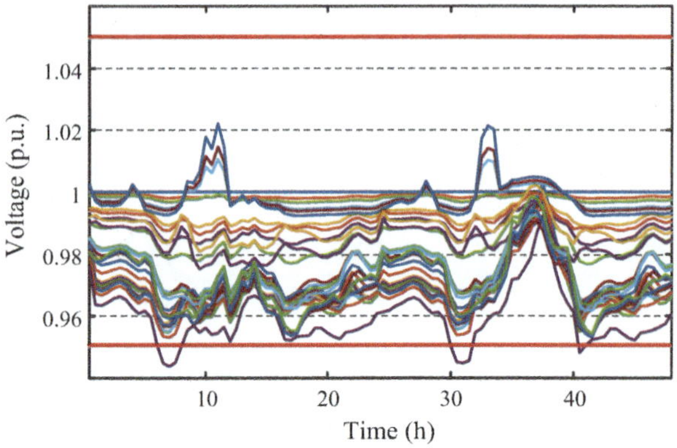

Fig. 18 Voltage profile of 33 bus system controlled by an existing DRL method in [20]

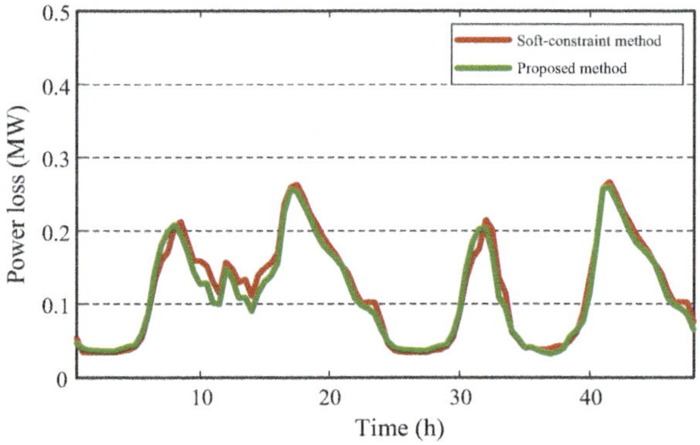

Fig. 19 Power loss comparison between the proposed DPe-TD3 method and an existing DRL method in [20]

Hence, the proposed DPe-TD3 method has superiority over existing DRL-based voltage control methods in achieving guaranteed safety and optimal solutions.

5.4 Applied in Different Load Scenarios

To further show the effectiveness of the proposed method to address scenarios with different loads, we conduct two more scenarios. The first scenario is that Bus 7, 8, 12, 17, 19, 25, 28, and 29 are all connected to hospital loads. After the training, we test the trained actor networks with the 2 days data. The voltage profile of the system without extra voltage control is shown in Fig. 20, while the system controlled with the proposed method, a voltage profile shown in Fig. 21. The comparison of the voltage profile shows that the proposed method can effectively regulate the voltage of each bus. The power loss comparison between the proposed DPe-TD3 method and the no voltage support scenario is shown in Fig. 22, which shows the proposed method can effectively reduce the power loss of the system operation.

Fig. 20 Voltage profile of IEEE 33 bus system connected to hospital loads without extra voltage control

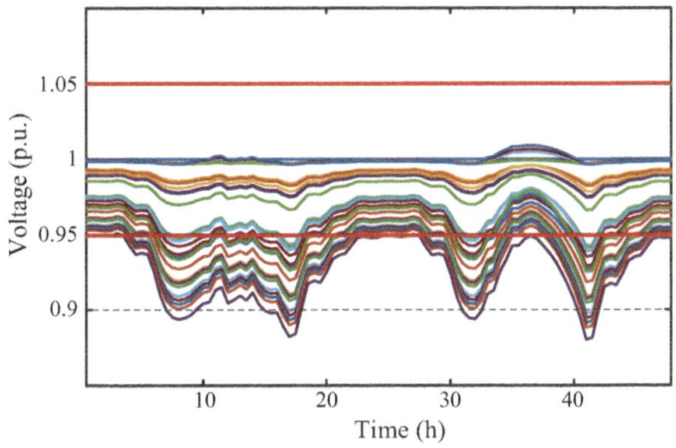

Fig. 21 Voltage profile of IEEE 33 bus system connected to hospital loads controlled by the proposed DPe-TD3 method

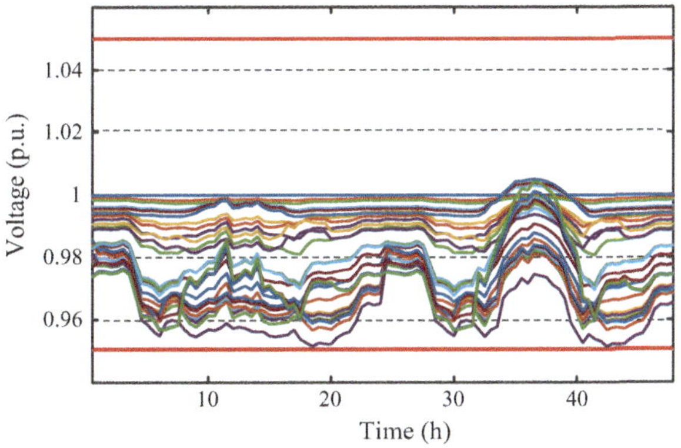

Fig. 22 Power loss of the system connected to hospital loads comparison between the proposed DPe-TD3 method and no extra voltage control scenario

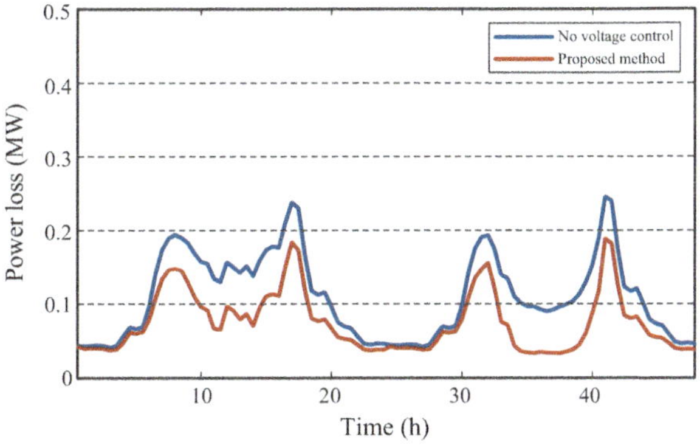

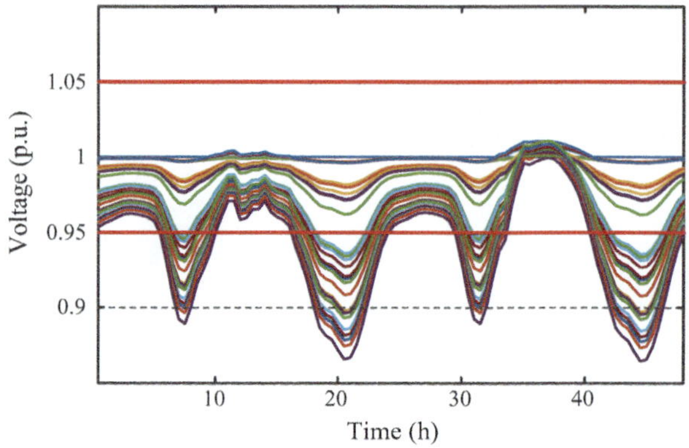

Fig. 23 Voltage profile of IEEE 33 bus system connected to restaurant loads without extra voltage control

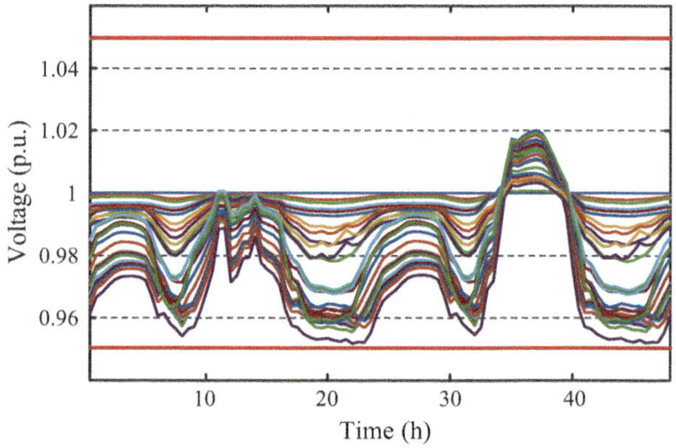

Fig. 24 Voltage profile of IEEE 33 bus system connected to restaurant loads controlled by the proposed DPe-TD3 method

The second scenario is that Bus 7, 8, 12, 17, 19, 25, 28, and 29 are all connected to restaurant loads. The voltage profile of the system without extra voltage control is shown in Fig. 23, while the system controlled with the proposed method, a voltage profile shown in Fig. 24. The comparison of the voltage profile shows that the proposed method can effectively regulate the voltage of each bus. The power loss comparison between the proposed DPe-TD3 method and the no voltage support scenario is shown in Fig. 25, which shows the proposed method can effectively reduce the power loss of the system operation.

From the experiment results, with different load profiles, the proposed DPe-TD3 algorithm can achieve optimal operation of distribution grids with 100% guaranteed safety.